ONCOGENES

Frontiers in Molecular Biology

Series editors

B.D.Hames
Department of Biochemistry, University of Leeds, Leeds LS2 9JT, UK

D.M.Glover
Cancer Research Campaign, Eukaryotic Molecular Genetics
Research Group, Department of Biochemistry, Imperial College
of Science and Technology, London SW7 2AZ, UK

ONCOGENES

Edited by

D.M.Glover

Cancer Research Campaign, Eukaryotic Molecular Genetics
Research Group, Department of Biochemistry, Imperial College
of Science and Technology, London SW7 2AZ, UK

and

B.D.Hames

Department of Biochemistry, University of Leeds, Leeds LS2 9JT, UK

OXFORD UNIVERSITY PRESS
Oxford New York Tokyo

Oxford University Press
Walton Street, Oxford OX2 6DP

Oxford is a trade mark of Oxford University Press

Published in the United States
by Oxford University Press, New York

©Oxford University Press 1989

First published 1989
Reprinted 1990

British Library Cataloguing in Publication Data

Oncogenes.—(Frontiers in molecular biology).
1. Immunochemistry
I. Glover, D.M. and Hames, B.D. II. Series
574.2'9   QR183.6

Library of Congress Cataloging in Publication Data
(Data available)

ISBN 0 19 963034 8
ISBN 0 19 963035 6 (Pbk)

Previously announced as:

ISBN 1 85221 081 8
ISBN 1 85221 077 X (Pbk)

Typeset and printed by Information Press Ltd, Oxford, England.

Oncogenes

This decade has seen an explosion in our knowledge of cellular genes which function abnormally during oncogenic transformation. Our intention in organizing this book was to review the major themes of research within this wide field, and to place the most recent developments into this context. The variety of cellular processes that can be affected by oncogenes are first introduced by Chris Marshall and subsequent chapters of the book focus on major facets of these processes. Throughout, the emphasis is on cellular oncogenes, and the involvement of oncogenic viruses is discussed with respect to their interactions with the cell. In Gordon Peters' chapter, for example, oncogenes found at viral integration sites are described. Retroviral insertions are just one example of how the rearrangement of chromosomal sequences can result in oncogene activation. Other chromosomal abnormalities, including translocations, inversions, and localized gene amplification, are also associated with oncogenesis. Such abnormalities, many of which are found in human tumors, are discussed in the Rabbitts' chapter. The gene products of the oncogenes have a wide range of biochemical functions, which are extensively discussed by Gordon Foulkes and his colleagues. The transforming proteins specified by the DNA tumor viruses have been under study for a long period of time. Recent interest has been directed to these proteins and their interactions with host cell proteins including the retinoblastoma gene product, discussed in the final chapter by David Lane. Our thanks go out to the authors, for their contributions and for their patience with each other and with the editors.

David Glover
David Hames

...ng

...will be

...stration

...pled how the

...lt in oncogene

...ding translocations

...lso associated with

...are found in human

...he gene products of the

...f mutations, which are

...and his colleagues. The

...tumor viruses have been

...ahway has been directed

...host cell proteins including

...in the final chapter by David

...r their contributions and for

...the editors.

David Glover

Contributors

J.Gordon Foulkes
Oncogene Science, Inc., 350 Community Drive, Manhasset, NY
11030, USA

Kenneth K.Iwata
Oncogene Science, Inc., 350 Community Drive, Manhasset, NY
11030, USA

David P.Lane
Imperial Cancer Research Fund, Clare Hall Laboratories,
South Mimms, Potters Bar, Herts EN6 3LD, UK

Christopher J.Marshall
Institute of Cancer Research, Royal Cancer Hospital, Chester Beatty
Laboratories, Fulham Road, London SW3 6JB, UK

Richard W.Michitsch
Oncogene Science, Inc., 350 Community Drive, Manhasset, NY
11030, USA

Gordon Peters
Imperial Cancer Research Fund, St Bartholomew's Hospital,
Dominion House, Bartholomew Close, London EC1A 7BE, UK

Pamela H.Rabbitts
MRC Laboratory of Molecular Biology, Hills Road, Cambridge
CB2 2QH, UK

Terry H.Rabbitts
MRC Laboratory of Molecular Biology, Hills Road, Cambridge
CB2 2QH, UK

Sydonia I.Rayter
Oncogene Science, Inc., 350 Community Drive, Manhasset, NY
11030, USA

John M.Sorvillo
Oncogene Science, Inc., 350 Community Drive, Manhasset, NY
11030, USA

David M. Valenzuela
Oncogene Science, Inc., 350 Community Drive, Manhasset, NY 11030, USA

Contents

4. Biochemical functions of oncogenes   113
S.I.Rayter, K.K.Iwata, R.W.Michitsch, J.M.Sorvillo,
D.M.Valenzuela and J.G.Foulkes

Abbreviations

AEV	avian erythroblastosis virus
AIDS	acquired immunodeficiency syndrome
ALL	acute lymphocytic leukemia
ALV	avian leukosis virus
AMP	adenosine monophosphate
A-T	ataxia telangiectasia
ATL	adult T-cell leukemia
ATLV	adult T-cell leukemia virus
BL	Burkitt's lymphoma
BLV	bovine leukemia virus
CAP	cAMP catabolite gene activator protein
CLL	chronic lymphocytic leukemia
CML	chronic myeloid leukemia
CSF	colony stimulating factor
CSV	chicken syncytial virus
DAG	diacyl glycerol
DM	double minute
DMBA	dimethyl-benzanthracene
EBV	Epstein – Barr virus
EF	elongaton factor
EGF	epidermal growth factor
FAP	familial adenomatous polyposis
FBJ	Finkel Biskis Jenkins
FBR	Finkel Biskis Reilly
FeLV	feline leukemia virus
FGF	fibroblast growth factor
F-MuLV	Friend murine leukemia virus
GRP	gastrin-releasing peptide
GTP	glycerol triphosphate
HEK	human embryonic kidney
hER	human estrogen receptor
hGR	human glucocorticoid receptor
HMSV	Harvey murine sarcoma virus
HOS	human osteosarcoma
HSR	homogeneously staining region

HTLV	human T-cell leukemia virus
IAP	intracisternal A-particle
IGF	insulin-like growth factor
IL	interleukin
IP3	inositol trisphosphate
KMSV	Kirsten murine sarcoma virus
LTR	long terminal repeat
M_r	relative molecular mass
MCF	mink cell focus (forming)
MMTV	mouse mammary tumor virus
MuLV	murine leukemia virus
NMU	*N*-nitroso-*N*-methylurea
PDGF	platelet-derived growth factor
PHA	phytohemaglutinin
PI	phosphoinositide
PIP	phosphatidyl inositol 4-phosphate
PIP2	phosphatidyl inositol 4,5-bisphosphate
PKA	protein kinase (cAMP-dependent)
PKC	protein kinase (Ca^{2+}/phospholipid-dependent)
Ptd.Cho	phosphatidylcholine
Ptd.Etn	phosphatidylethanolamine
Ptd.Ser	phosphatidylserine
RB	retinoblastoma
REV	reticuloendotheliosis virus
RFLP	restriction fragment length polymorphism
RPV	ring-neck pheasant virus
RSV	Rous sarcoma virus
SCLC	small-cell lung carcinoma
SDS-PAGE	sodium dodecyl sulphate-polyacrylamide gel electrophoresis
SFFV	spleen focus forming virus
snRNP	small nuclear ribonucleoprotein
SRE	serum response element
SRF	serum response factor
SSV	simian sarcoma virus
T3	triiodothyronine
TGF	transforming growth factor
TPA	12-*O*-tetradecanoyl phorbol 13-acetate

Oncogenes and cell proliferation: an overview
Christopher J.Marshall

1. Introduction

A general definition of an oncogene is a gene whose abnormal expression or altered gene product directly determines the production of the malignant phenotype. Cellular oncogenes are derived from normal cellular genes, the proto-oncogenes. The viral oncogenes carried by acutely transforming retroviruses are derived from cellular proto-oncogenes, whereas those of the DNA tumor viruses have no obvious cellular progenitor. The mechanisms which convert proto-oncogenes to oncogenes, a process known as oncogene activation, are thought to be the critical genetic events in neoplastic transformation. Oncogene research covers the following areas:

(i)   the identification of oncogenes,
(ii)   the elucidation of the differences between proto-oncogenes and their transforming counterparts,
(iii) the contribution of oncogenes to different stages of malignant transformation,
(iv)   the biochemical roles of proto-oncogene and oncogene products in controlling cell proliferation.

The use of oncogenes and their products has great clinical potential for identifying and classifying malignancies and as targets for therapy.

This chapter introduces the topic of oncogenes and summarizes the current areas of oncogene research. I will discuss our understanding of how proto-oncogene products may be involved in the control of cell proliferation, placing the emphasis on cellular oncogenes and on the viral oncogenes of acutely transforming retroviruses. The oncogenes of DNA tumor viruses (see Chapter 5) are not closely related to cellular genes, though their gene products clearly must be able to mimic and subvert the normal cellular mechanisms for the control of proliferation.

2. Identification of proto-oncogenes and their activation to oncogenes

Five main routes have led to the identification of cellular gene sequences which could function as oncogenes:

(i) the analysis of transduced cellular gene sequences in acutely transforming retroviruses,

(ii) identification of preferential integration sites of retroviruses (see Chapter 2),

(iii) characterization of chromosomal translocations (see Chapter 3),

(iv) characterization of amplified DNA sequences (see Chapter 3),

(v) direct gene transfer experiments to assay for the biological activity of cellular transforming genes (see Chapter 4).

Using these approaches at least 40 proto-oncogenes have been identified in the human genome (see *Table 1*). However, so far the vast majority of these genes have not been implicated in human malignancy and an involvement in human disease has only been clearly demonstrated in eight cases (c-Ha-*ras*-1, c-Ki-*ras*-2, N-*ras*, c-*erb*B1, c-*erb*B2, c-*myc*, N-*myc*, L-*myc* and c-*abl*).

With the exceptions of c-*abl* in chronic myeloid leukemia and of c-*myc* in Burkitt's lymphoma, particular oncogenes appear to be involved in only a fraction of cases of a given type of malignancy. Such observations raise important questions about the role of the oncogene. First, are there alternative pathways involving unknown genes leading to the same malignant phenotype? Second, does the involvement of a particular oncogene reflect subdivisions of the biology of a tumor that cannot be made at the pathological level? In addition, we know little about the chronology of genetic changes in malignant transformation. In some tumors a particular proto-oncogene may play a role in the early or initiating stages of disease, while in others the same gene could be involved in later stages leading to the clonal evolution of a more aggressive disease.

The characterization of oncogenes has shown that a few are derived from genes encoding proteins of known function. The v-*sis* oncogene is derived from the gene encoding the B chain of platelet-derived growth factor (1), the v-*erb*B gene from the gene for epidermal growth factor receptor (2), the v-*fms* gene from the gene for the receptor for Colony Stimulating Factor-1 (CSF-1) (3), and the v-*erb*A gene from the thyroid hormone receptor gene (4). Description of the relationships between these retroviral oncogenes and their normal cellular counterparts has given us the strongest evidence that oncogenes are derived from normal genes involved in growth control. However, the evidence for the involvement of the gene products of other proto-oncogenes in growth control is only circumstantial.

2.1 Acutely transforming retroviruses

More than 20 different proto-oncogene sequences have been identified by analysis of the cellular gene sequences transduced by acutely transforming retroviruses and this route is still leading to the identification of novel genes (5) (see *Table 1*). Acutely transforming retroviruses are able to transform cells in culture within a few days and induce tumors in suitable hosts with short latency periods (2 – 8 weeks). The viruses are in a sense laboratory constructs, since powerful selective forces have been exerted in their isolation. However, the feline leukemia viruses exemplify viruses that do occur in the wild which have transduced a cellular proto-oncogene. No single unifying mechanism is responsible for the transforming properties of the transduced retroviral oncogene compared with its non-transforming cellular proto-oncogene counterpart. In many of the cases which have been analyzed, gross structural changes resulting from truncation as well as individual point mutation distinguish the retroviral oncogene from its cellular homolog. Indeed the prevailing view of a few years ago that the role of retroviral transduction was to provide abnormal levels and sites of expression of a normal gene product has now been superseded by one involving structural changes for most of the retroviral oncogenes. The structural changes have been shown in some cases to lead to altered biochemical activity of the oncogene product (see Chapter 4), for example the constitutive tyrosine kinase activities of the oncogenes v-*erb*B and v-*fms* derived from growth factor receptors. However in some cases, for example v-*myc*, it seems that the role of retroviral transduction is to provide an environment for the abnormal expression of an essentially normal protein product.

2.2 Retrovirus insertions

A second route for the activation of cellular genes by retroviruses involves integration of the virus adjacent to the cellular gene leading to an alteration in that gene's expression. Characteristically the tumors which arise following this mode of action develop with long latent periods. Examination of the integration sites of a number of different viruses and tumor systems has identified several preferred integration sites adjacent to cellular genes. In some cases these integrations are adjacent to known proto-oncogenes, but in others previously unidentified genes are involved. These may be genes regarded as putative proto-oncogenes until more direct approaches are undertaken to show their role in malignant transformation. An extensive discussion of oncogenes at viral integration sites may be found in Chapter 2.

2.3 Chromosomal breakpoints in translocations

Cytogenetic analysis has identified a number of situations in which translocations with the same chromosomal breakpoints occur in a particular type of malignancy. In some situations, such as Burkitt's lymphoma or chronic myeloid leukemia (CML), the specific translocation is present in all or the vast majority of the cases analyzed. Molecular

Table 1. Proto-oncogenes

Proto-oncogene	Mode of initial identification[a]	Human chromosomal location[b]	Biochemical properties[c]
c-src-1	txrv (chicken RSV)	20q12-q13	cytoplasmic tyr k
c-src-2		1p36-p34	cytoplasmic tyr k
c-abl	txrv (mouse Ab MuLV)	9q34	cytoplasmic tyr k
c-fgr	txrv (cat GR-FeSV)	1p36.1-36.2	cytoplasmic tyr k
c-yes	txrv (chicken Y73,Esh)	1821.3	cytoplasmic tyr k
c-fes	txrv (cat ST-FeSV GaFeSV H21-FeSV)	15q25-q26	cytoplasmic tyr k
c-sea	txrv (chicken AEV-s13)		cytoplasmic tyr k
tck	rv insertion		cytoplasmic tyr k
trk	NIH transfection		cytoplasmic tyr k
c-fms	txrv (cat SM-FeSV H25-FeSV)	5q34	CSF-1 receptor tyr k
c-erb1	txrv (chicken AEV-H)	7p13-q11.2	EGF receptor tyr k
c-erb2	NIH transfection	17	receptor-like tyr k
c-kit	txrv (cat H24-FeSV)	4q11-q21	receptor-like tyr k
c-ros	txrv (chicken UR2)	6q22	receptor-like tyr k
c-met	NIH transfection	7p11-4	receptor-like tyr k
met	NIH transfection		receptor-like tyr k
c-raf1	txrv (chicken MH2/mouse 3611-MSV)	3p25	cytoplasmic ser/thr k
c-raf2			
c-mos	txrv (mouse MMSV)	8q22	cytoplasmic ser/thr k
pim-1	rv insertion		cytoplasmic ser/thr k
c-sis	txrv (monkey SSV)	22q12-q13	PDGF β chain
int-2	txrv insertion	17	growth factor/like
hst	NIH transfection		growth factor/like

N-*ras*	NIH transfection	1cen-1p21	membrane bound GTPase
c-Ha-*ras*1	txrv (rat Ha-MSV)	11p15	membrane bound GTPase
c-Ki-*ras*2	txrv (rat Ki-MSV)	12p12-pter	membrane bound GTPase
c-*myc*	txrv (chicken MC29 MH2)	8q24	nuclear proteins
N-*myc*	X-hybridization	2p23-pter	nuclear proteins
L-*myc*	X-hybridization	1p32	nuclear proteins
c-*ski*	txrv (chicken SKV-rASV)	1q22-qter	nuclear proteins
c-*ets*-1	txrv (chicken E26)	11q23-q24	nuclear proteins
c-*ets*-2		21q22.3	nuclear proteins
c-*myb*	txrv (chicken AMV)	6q22-q24	nuclear proteins
c-*fos*	txrv (mouse MMSV)	14q21-q31	nuclear proteins
c-*rel*	txrv (turkey REV-T)	2	nuclear proteins
p53	association with SV40 T antigen		nuclear proteins
c-*jun*	txrv (chicken ASV-17)		transcription factor
c-*erb*A1	txrv (chicken AEV-H)	17	thyroid hormone receptor
gli	gene amplification	12q13-q14.3	unknown
bcl-1	translocation breakpoint	11q13	unknown
bcl-2	translocation breakpoint	18q21	unknown
tcl-1	translocation breakpoint	14q32	unknown
fis-1	rv insertion		unknown
fim-1	rv insertion		unknown
dbl	NIH transfection		unknown

[a] txrv, acutely transforming retrovirus; NIH transfection, calcium phosphate-mediated gene transfer into NIH-3T3 cells; rv insertion, preferred integration site recognized by retrovirus insertion; X-hybridization, cross-hybridization of amplified sequences to oncogene probe.

[b] in some cases there is more than one human gene homologous to a retrovirus oncogene.

[c] tyr k, tyrosine kinase.

analysis of the Burkitt's lymphoma translocations and of the Philadelphia chromosome in CML has shown that they involve the proto-oncogenes, c-*myc* and c-*abl*, first recognized by their homology to retroviral oncogenes. Analysis of these translocations has therefore provided some of the strongest evidence for a role of proto-oncogenes in human malignancy. There is as yet no additional evidence to suggest that genes found at chromosomal breakpoints in other tumors, such as *bcl*-1, *bcl*-2 and *tcl*-1, play a role in oncogenesis. Alternative approaches will be required to show whether the translocations affecting these genes are essential to pathogenesis. Detailed discussion of specific chromosome translocations associated with malignancy and how these events alter the functions of the involved genes can be found in Chapter 3.

2.4 Gene amplification and over-expression

Both cytogenetic and molecular evidence show that some tumors contain amplified copies of genes. In some instances this amplification has been shown to involve known proto-oncogenes and has led to the identification of new proto-oncogenes, L-*myc* and N-*myc*, distantly related to c-*myc* (see Chapter 3).

The first example of the amplification of a proto-oncogene came from studies on the human HL60 promyelocytic leukemia cell-line in which the c-*myc* gene is amplified up to thirty times over normal (6) leading to elevated levels of gene expression. Subsequent investigations have shown that c-*myc* amplification is a relatively uncommon phenomenon in human leukemias and solid tumors except for breast carcinomas where approximately 30 per cent of the tumors contain amplified c-*myc* genes (7) and small-cell carcinoma of the lung where cells of a variant type contain amplified c-*myc*, N-*myc* or L-*myc* (8). Frequent rather than sporadic amplification of proto-oncogenes occurs in two other situations. The N-*myc* gene is amplified and expressed at a high level in the more aggressive forms of neuroblastoma (9,10) and amplification of the c-*erb*B2 gene and overexpression of its gene product occurs in about 30 per cent of breast carcinomas (11).

Although over-expression of proto-oncogene products can result from gene amplification, other mechanisms such as increased rates of transcription or mRNA stability are also possible. Increased levels of growth factor receptors have been observed on tumors where there is no apparent gene amplification, e.g. c-*erb*1, in squamous cell carcinoma (12). For many situations where gene amplification has been demonstrated, it has not been formally shown that the amplified gene encodes a perfectly normal product. Indeed there are examples where amplified normal genes co-exist with mutant genes in the same tumor (13). Therefore it is possible in some cases that amplification of the normal gene may provide a greater probability of generating mutant variants. Finally, although much of the work on gene amplification in tumors has concentrated on the known proto-oncogenes, the ability to selectively detect and isolate amplified gene sequences using gel renaturation techniques provides a means of identifying novel proto-oncogenes.

2.5 Gene transfer

The first unequivocal evidence for oncogenes in human malignancies came from the demonstration that DNA from a human bladder carcinoma cell-line contained a gene capable of morphologically transforming NIH-3T3 cells in gene transfer experiments (15). Since then extensive studies have shown that in the vast majority of cases, transforming genes detected by gene transfer are members of the *ras* gene family. Other genes have been detected by gene transfer sporadically. In some of these instances the activated oncogene was generated during the transfection procedure either by rearrangements (16 – 18) or by removing the gene from normal cellular influences (19). Even though such studies do not tell us anything about the genetic mechanism of transformation in the original tumor, they have served to identify genes affecting normal cellular growth control.

Molecular analysis of the human *ras* gene family shows that it consists of three closely related members, two of which, c-Ha-*ras*-1 and c-Ki-*ras*-2, are the human orthologs of the rat proto-oncogenes from which the viral oncogenes in Harvey murine sarcoma virus (HMSV) and Kirsten murine sarcoma virus (KMSV) were derived. The third gene, N-*ras*, was identified by gene transfer studies and has not been found in a retrovirus. The genes encode proteins of 188 or 189 amino acids which are highly homologous for all but a C-terminal sequence of 20 amino acids. The proteins are localized on the inner surface of the plasma membrane by virtue of alkyl and acyl residues covalently bound to cysteine residues at the C-terminus (20). They are able to bind and hydrolyze GTP (21). The GTPase activity is an intrinsic property of the p21^{ras} proteins, but *in vivo* the GTPase activity may be elevated by association with other factors (22). Related to the three *ras* genes are at least three other genes, *rho* (23), *ral* (24) and R-*ras* (25) encoding proteins of 189, 206 and 218 amino acids respectively which are predicted to be GTP-binding proteins and membrane-localized. However, none of these genes has been detected in an activated form by gene transfer experiments.

Analysis of the sequences of *ras* oncogenes detected by gene transfer shows that in almost all cases so far examined, activation of the *ras* proto-oncogenes arises by single base changes leading to an amino acid substitution at position 12, 13, or 61 (26). Mutations at codons 12 and 61 have been shown to lead to a reduction in the intrinsic capacity of the protein to hydrolyze GTP and this is presumed to account for the transforming capacity of the protein. Mutagenesis experiments *in vitro* have identified mutations at positions 59, 63, 116, and 119 which also lead to transforming activity (27,28). The mutations at codons 59 and 116 have been shown to affect the rate of guanine nucleotide exchange rather than GTP hydrolytic activity (28,29), showing that there are at least two independent mechanisms by which mutations can lead to *ras* proteins having the ability to transform cells.

Although experiments with gene transfer into NIH-3T3 cells allowed the discovery of *ras* proto-oncogenes, this assay has now been superseded by procedures which allow easier and more rapid detection of mutations. The two main techniques that have been employed are analysis by mutant-specific oligonucleotides (30), a technique whose sensitivity is much increased following polymerase chain amplification of the target sequences (31,32), and analysis by RNase H sensitivity (33). The use of these procedures together with gene transfer experiments has shown that some malignancies such as colon cancer and acute myeloid leukemia are characterized by high frequencies (25 – 40 per cent) of *ras* gene mutations, whereas tumors such as breast cancer show a very low incidence of *ras* gene mutation (reviewed in 34). Furthermore, the individual *ras* gene activated is strongly dependent on tissue type; c-Ki-*ras*-2 is activated in colon cancer (31,33), N-*ras* in acute myeloid leukemia (30,32) and c-H-*ras*-1 in bladder cancer. The reason for this relationship between type of malignancy and type of *ras* gene activated is unclear but interestingly parallels situations in rodent models where tumors are induced by a single specific carcinogen (35).

3. Multistage transformation and oncogenes

Much epidemiological and experimental evidence argues that malignant change is a process which results from multiple genetic alterations. It is therefore necessary to understand which of these steps involve oncogenes. Several lines of evidence suggest that changes to more than one proto-oncogene are involved in tumors. For example, some rodent tumors caused by viral integration contain integrations at two different putative oncogenes (see Chapter 2) and in a few human tumors in which there are changes to more than one proto-oncogene (36). Experiments with transgenic mice containing an activated oncogene show that the activity of a single oncogene is insufficient to cause a tumor since malignancies in such mice are clonal. This argues that additional events must occur to lead to tumor development (37). Perhaps the most persuasive evidence for the involvement of multiple oncogenes comes from studies in which the malignant transformation of non-immortal cell strains is facilitated by the introduction of two different oncogenes (38). The mechanism of such 'oncogene co-operation' is not clear. However, it appears that there is a group of oncogenes which are able to lead to the immortalization of non-immortal cell strains but which do not appear to be capable of producing altered growth control in established cell lines, while a second group of oncogenes causes altered growth control. Generally those genes capable of immortalization, such as c-*myc*, N-*myc* and p53, encode proteins that are localized in the nucleus, while the genes capable of producing altered growth control (e.g. v-*src*, *ras* genes) are cytoplasmic. This categorization is to some extent an over-simplification, since the viral *fos* gene in the Finkel Biskis Reilly (FBR) strain of murine osteosarcoma virus seems to encode a nuclear protein which is able both to immortalize and to produce altered growth control (40).

If tumors arise as a result of multistep events, then an important question is at which stage of malignant transformation is a particular oncogene involved. Since it is impossible to recreate at will the development of human tumors, the answers to such questions are necessarily indirect. From studies on the experimental activation of *ras* proto-oncogenes in carcinogen-treated animals, it has been argued that *ras* gene activation is an early and probably initiating event in carcinogenesis (35). This idea is supported in human disease by the analysis of certain pre-malignant stages. For example, adenomas in the colon and myelodysplastic syndromes have been found to contain mutant *ras* genes (31,41). Against these observations must be set findings of multiple *ras* gene activation within a tumor and of the absence of *ras* gene mutations in relapsed acute myeloid leukemia which originally contained a mutation (32,33). Such findings are difficult to reconcile with the idea of *ras* gene mutation always being an early or initiating event, but are more in keeping with a role of *ras* gene mutation in later stages of neoplastic progression.

Similarly there is no indication of the time at which the chromosome translocation involving c-*myc* and the immunoglobulin loci occur in Burkitt's lymphoma. Persuasive arguments can be made for the translocation occurring both before and after Epstein – Barr virus (EBV) infection (42,43). There is some evidence that the translocation of c-*abl* and the generation of the Philadelphia chromosome is not the initial step in CML. For example, B lymphoid cell lines derived from the patient's peripheral blood share the same glucose 6-phosphate dehydrogenase isozyme, a marker of clonality, as the leukemia but do not bear the Philadelphia chromosome (44). It has been argued that the amplification and over-expression of N-*myc* in neuroblastoma plays a role in tumor progression, since it is only the aggressive stages III and IV forms of disease which show amplification (9,10). Similarly, amplification of c-*myc* appears to be confined to the variant more aggressive forms of small-cell carcinoma of the lung (8). In summary, it seems that with the exception of the *ras* genes there is little indication of when proto-oncogenes are activated. In part this may be because only the *ras* genes can be reproducibly activated experimentally and their role in malignancy investigated.

4. Role of proto-oncogene and oncogene products in controlling cell division

The control of cell proliferation is exerted by growth stimulators called growth factors, and by growth inhibitors. Growth factors are analogous to hormones in that they are produced by one cell type to act on another through a cell surface receptor.

4.1 Signal transduction

The activation of growth factor receptors by binding of the ligand generates second messengers which lead to signals which eventually result in DNA synthesis. It is unlikely that there is one common second messenger system for all stimulators of DNA synthesis. The alternative mechanisms are summarized in *Figure 1*. Some growth factor receptors such as that for epidermal growth factor (EGF^R), insulin ($Insulin^R$), insulin-like growth factor ($IGF-1^R$), platelet derived growth factor ($PDGF^R$), and colony stimulating factor ($CSF-1^R$) have intrinsic tyrosine kinase activity, while others such as interleukin 2 receptor ($IL2^R$), granulocyte-macrophage colony stimulating factor receptor ($GMCSF^R$), interleukin 3 receptor ($IL3^R$), do not. Some growth factor receptors (e.g. Bombesin receptor) appear to work through phospholipase C-mediated breakdown of the minor membrane lipid phosphatidyl inositol 4,5 bisphosphate (PIP2) to generate inositol trisphosphate (IP3) and diacylglycerol (DAG). The water-soluble IP3 causes the release of Ca^{2+} from intracellular stores, and DAG causes the activation of protein kinase C. Still other growth factor receptors appear to be able to activate protein kinase C without causing the breakdown of PIP2, for example the receptors for embryonal carcinoma-derived growth factor (45) and IL3 (46). The relationship between tyrosine kinase activity and the breakdown of inositol phospholipids is complex. Stimulation of the $PDGF^R$ appears to cause both increased tyrosine phosphorylation and PIP2 breakdown (47). Whether breakdown of PIP2 is an immediate consequence of stimulating $PDGF^R$ is unclear; it remains possible that the primary effect is tyrosine kinase activity and that PIP2 breakdown is a secondary effect.

In 'classical' hormone responses, such as the activation of adenylate cyclase to yield cyclic AMP as a second messenger, the interaction between the receptor and the second messenger system is mediated by a guanine nucleotide-binding G protein (85). Receptor stimulation causes the bound GDP on the G protein to be exchanged for GTP, which results in the G protein-mediated activation of adenylate cyclase. The activation is terminated by the intrinsic GTPase activity of the G protein, which converts bound GTP to GDP. There is biochemical evidence that breakdown of PIP2 mediated by growth factor receptors is also mediated by G proteins, and G-like proteins may also couple receptors directly to ion channels (85). However, definitive evidence for G-proteins in growth factor responses requires purification of the components followed by reconstitution experiments.

DNA synthesis is a response which, unlike the responses to conventional hormones, occurs hours after the beginning of receptor stimulation. Furthermore, it appears that continued exposure to growth factor for several hours is needed to stimulate DNA synthesis. Studies in tissue culture show that the most effective way to stimulate DNA synthesis is to use two growth factors instead of one (48). This synergy between pairs of growth factors is most pronounced when one of them is insulin or IGF1, which on their own are generally only poorly mitogenic for cells in culture (48).

The identification of some oncogenes as being derived from genes encoding either growth factors or growth factor receptors has shown that at least some proto-oncogenes are involved in the control of normal cell proliferation. Thus the v-*sis* gene of Simian Sarcoma Virus (SSV) is derived from the gene for the B chain of PDGF (1); v-*erb*B is probably derived from the gene for the EGF receptor (2); v-*erb*A is derived from the gene for thyroid hormone receptor (4); and v-*fms* is derived from the CSF-1 receptor gene (3). Furthermore, two other proto-oncogenes, c-*erb*B2 and *met*, encode proteins which have the hallmarks of growth factor receptors.

Several events have been found to follow stimulation of cells by growth factor. First are rapid ionic changes, then changes in gene transcription and protein synthesis, and finally DNA synthesis itself.

4.1.1 Ionic changes

Within a few seconds of growth factor stimulation, intracellular pH rises through the activation of the Na^+/H^+ antiporter and the intracellular Ca^{2+} concentration increases. These events are common to all growth factors and are therefore likely to occur through multiple mechanisms. For example, stimulation with growth factors which lead to the breakdown of PIP2 activates protein kinase C, which may mediate the rise in intracellular pH. Other growth factor receptors, such as EGF^R, do not appear to be involved in kinase-C activation (49) and so must cause the ionic changes by different routes (50). It remains a controversial question as to how essential these ionic changes are to the initiation of DNA synthesis. Studies with inhibitors of cytoplasmic alkalinization have shown that this step is not essential (51), whereas experiments on cell lines which contain a mutant-defective antiporter argue that cytoplasmic alkalinization is necessary for DNA synthesis (52).

4.1.2 Changes in gene expression

While the time-scale of the ionic changes is seconds, changes in gene expression are detected some minutes after growth factor stimulation.

The earliest characterized change in gene expression is an elevation in the transcription rate of the proto-oncogene c-*fos* which starts at around 10–15 min after stimulation, peaks at 30 min, and then decreases to baseline once more by 1–1½ h (53). Following the rise in c-*fos* expression, a rise in c-*myc* expression occurs which peaks at around 1–3 h (53). Many other changes in gene expression have been found following growth factor stimulation (54). Most of these changes in gene expression have the important characteristic that they do not require new protein synthesis to take place.

The mechanisms underlying the changes in expression of genes regulated by growth factors have received much attention. For c-*fos*, the major regulatory event following growth factor stimulation has been shown by nuclear run-on assays to be an increase in gene transcription (53). Because the expression of c-*fos* is so transitory following growth factor stimulation, an important element in the mechanism of control is the short half-life of c-*fos* mRNA. Treisman has shown that there is a sequence at the 3′ end of c-*fos* mRNA which confers this instability on the c-*fos* mRNA (55). Interestingly this sequence is not present in the v-*fos* oncogenes of FBR and Finkel Biskis Jenkins (FBJ) strains of murine osteosarcoma viruses. Similar sequences have been identified in other genes regulated by growth factors (56). For c-*myc*, the regulation is complex including both transcriptional (53), and post-transcriptional control at the level of message stability (57).

Much recent work has focused on the regulatory elements controlling c-*fos* expression. Mutational analysis has defined a sequence at -297 to -317 upstream from the c-*fos* cap site that is the minimal element required for expression following serum stimulation; the serum response element (SRE) (58,59). A protein termed serum response factor (SRF) has been identified which binds to this site and presumably mediates the serum induction (60). However, no change in DNA binding affinity of this protein can be detected in stimulated cells. While the SRE is a minimal element required for the response to serum, there are also other levels of control. Downstream of the SRE are sequence elements required for basal transcription. Upstream of the SRE is a sequence essential for c-*fos* induction by medium conditioned by cells transformed by v-*sis* and it is likely that this is a PDGF-responsive element (61).

Expression of a number of genes is elevated following activation of protein kinase C by growth factors or phorbol esters such as 12-*O*-tetradecanoyl phorbol-13-acetate (TPA). An important element in these events is the transcription factor AP1 (62). Increased binding activity of this factor to DNA can be measured in cells treated with TPA (63). Recently it has become clear that the gene for AP1 is probably the progenitor of the v-*jun* oncogene of ASV17 (64,84). Presumably the differences between the v-*jun* product and AP1 result in alterations in gene expression that lead to transformation. Now that the first clear example of a transcription factor functioning as an oncogene has been found, considerable emphasis will no doubt be placed on searching for a role of other transcription factors in oncogenesis.

Since changes in gene expression following growth factor stimulation still occur in the presence of inhibitors of protein synthesis, the activity of transcription factors must be regulated at the post-translational level. Some transcription factors such as API and APII are substrates for protein kinase C (62,63). Others may be substrates for growth factor receptor tyrosine kinases, cyclic AMP-dependent kinases (63), or calcium-activated kinases. Post-translational modification might affect the ability of the transcription factor to bind to DNA, as appears to be the case for API in phorbol ester-treated cells (63), or may cause some modification which affects interactions with other components of the transcription machinery, as appears to be the case for the SRF of the c-*fos* gene.

Figure 1 provides in simplified schematic form a summary of some of the events following growth factor stimulation. The regulatory events following the early ionic changes and alterations in gene expression are still poorly analyzed. There is therefore a sizeable gap in our knowledge as to what is happening in growth factor-stimulated cells during the period of several hours between the early changes in gene expression and the induction of DNA synthesis.

4.2 Oncogene alterations and growth control

In this section I will discuss the ways in which specific components of the growth regulation system are changed so that they lead to oncogenesis.

The transduction of the PDGF B gene in SSV and the production of PDGF A and B chains in several types of human tumor show how the ectopic production of a growth factor gene could lead to cell proliferation. This autocrine mode of stimulating proliferation may also operate in malignant cells producing transforming growth factor-α (TGF-α), bombesin-like peptides, and hemopoietic growth factors. However, the production of growth factors by tumor cells does not necessarily imply an autocrine stimulation of growth. Some tumors produce a growth factor to which the tumor cells themselves are non-responsive. For example, breast carcinomas have been found to produce PDGF A or B chains but the malignant cells do not bear PDGF receptors (65). In such situations, the role of the growth factor production by the malignant cells may be paracrine, for example to promote the development of tumor vasculature by stimulating angiogenesis.

Two viral oncogenes, v-*erb*B and v-*fms*, have been shown to be derived from growth factor receptor genes for EGF and CSF-1 respectively. For each the structural difference of the viral oncogene product is thought to lead to an activated receptor tyrosine kinase that is active in the absence of ligand. However, different structural changes are involved in activating the two receptors. The v-*erb*B gene product of the AEV-H strain of avian

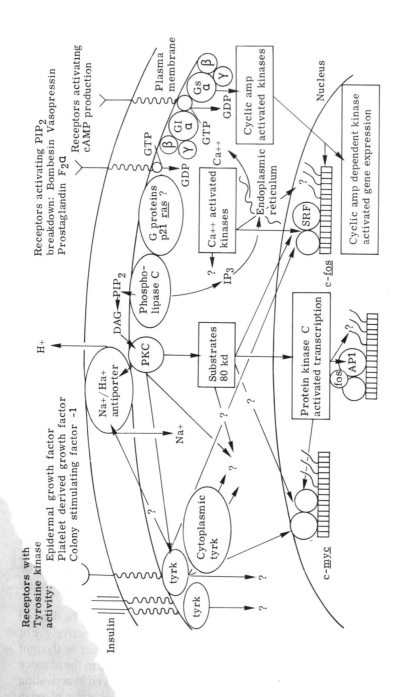

Figure 1. A schematic diagram of some of the signaling pathways and interactions following growth factor stimulation. DAG, diacylglycerol; IP₃, inositol 1,4,5 trisphosphate; PKC, protein kinase C; tyr k, tryosine kinase; SRF, serum response factor; AP1, transcription factor AP1 (c-*jun*).

erythroblastosis virus is truncated at the amino-terminus, with the consequent loss of the EGF-binding domain (66). A further truncation of the carboxy-terminus appears to confer fibroblast transforming activity to AEV-H but this is not found in other erythroblastosis virus isolates incapable of transforming fibroblasts (67). The v-*fms* protein retains the capacity to bind CSF-1, but the tyrosine kinase activity of the molecule is elevated in the absence of ligand (68). Both the c-*erb*B-2 and the *met* proto-oncogene products bear strong resemblances to tyrosine kinase growth factor receptors. The activated forms of onc-*neu* in rat neuroblastomas or onc-*met* in the MNNG-HOS human osteosarcoma cell line may be constitutively activated kinases. Constitutively-activated growth factor receptors have not yet been detected in human malignancies, though, as discussed earlier, over-expression of EGF^R or c-*erb*B2 through gene amplification or other mechanisms does occur in a wide range of human malignancies. How such over-expressed receptors contribute to abnormal growth control is not yet resolved. Over-expression may sensitize cells to low levels of growth factor (69), or alternatively, if there is background activity of receptors in the absence of ligand, elevation of receptor numbers to a sufficient level may produce a signal for proliferation.

While there are clear examples of growth factor or growth factor receptor genes becoming oncogenes, the relationship of oncogenes to components of second-messenger generating systems is less well defined. A number of viral oncogene products (e.g. v-*fes*, v-*src*, v-*abl*) have been shown to be derived from genes encoding cytoplasmic tyrosine kinases. The role of these enzymes in the control of cell proliferation remains obscure but it is generally believed that they are likely to be part of intracellular signalling pathways. For example, one of the substrates of c-*src* appears to be an inositol phospholipid kinase (70,71). Whatever the role of the proto-oncogene forms of the cytoplasmic tyrosine kinases, oncogenic alterations seem to result in an elevated enzymic activity. For example, both the v-*abl* protein and the bcr-*abl* protein found in CML appear to have greater tyrosine kinase activity than the c-*abl* protein (72). Even more obscure than the role of the tyrosine kinases is the function of the two proto-oncogene serine kinases, c-*raf* and c-*mos*. It remains to be determined how the alterations to these genes in v-*mos* and v-*raf/mil* result in transformation.

The properties of guanine nucleotide binding and GTP hydrolysis of the p21 *ras* proteins suggest, by analogy with classical G proteins, that the *ras* proteins are involved in signal transduction. The most direct evidence that p21 *ras* proteins mediate signals from growth factor receptors is provided by experiments in which serum-stimulated DNA synthesis is blocked by microinjection of antibodies against p21 *ras* (73). The time courses of these experiments suggest that p21 *ras* function is required in both early responses such as c-*fos* expression and in later stages of pathways to DNA synthesis (74). Thus there is a suggestive association between the prolonged exposure to growth factors required for initiation of DNA synthesis and the length of time p21 *ras* function is required. The demonstration that the mutations which convert the p21 *ras* proto-oncogene to an oncogene cause either a reduction in GTPase activity (26) or increase the exchange rate of guanine nucleotides (28,29) on the proteins also argues that the proteins function as regulatory G proteins. Either type of alteration is predicted to lead to an elevated level of the GTP-form of p21 *ras*. Since it is the GTP-form that is active (22), such mutant proteins are expected to be constitutively activated.

Several investigators have proposed that p21 *ras* proteins may function as a G protein to couple growth factor receptors to the breakdown of PIP2. The strongest evidence for this hypothesis comes from the T15 cell line in which a normal N-*ras* gene can be over-expressed under the control of the glucocorticoid-inducible promoter in the MMTV LTR. Elevating the level of expression of the normal N-*ras* gene in these cells causes an amplified response of inositol phospholipid breakdown following bombesin stimulation (75). This response appears to be a classical PIP2 breakdown because following stimulation, elevated levels of $1,4,5IP_3$ can be demonstrated and there is a rapid release of intracellular calcium (S. Davies and A. Lloyd, personal communication). Injection of transforming *ras* proteins into *Xenopus* oocytes results in a rapid generation of inositol phosphates and diacylglycerol, consistent with the idea that *ras* proteins are involved in the regulation of second messengers derived from phospholipids (76). Studies on cells containing transforming *ras* proteins show that they have elevated rates of inositol phospholipid breakdown (77; J. Hancock and M. Wakelam, personal communication). However, there is some discussion as to whether the quantitatively more important effect is the production of diacyglycerol rather than inositol phosphates, perhaps from the breakdown of other phospholipids such as phosphatidylcholine and phosphatidylethanolamine (78,79).

Whatever the mechanism of producing diacylglycerol, there is additional evidence that protein kinase C is required for the functioning of p21 *ras*. In cells which have had protein kinase C down-regulated by prolonged exposure to phorbol ester tumor promoters, the microinjection of p21 *ras* does not lead to the stimulation of DNA synthesis (80). More indirectly, some of the genes whose expression is elevated following *ras* transformation are also inducible by tumor promoters (81). Furthermore, while some enhancers function poorly in differentiated cells, this property can be overcome by co-transfection with a *ras*-containing plasmid or by treatment with tumor promoters (82).

Of the many changes in gene expression following growth factor stimulation, some have been shown to involve proto-oncogenes. It is likely therefore that the abnormal expression of c-*fos*, c-*myc* or p53 following viral oncogenesis or other genetic alterations would lead to altered growth control. The products of the c-*fos*, c-*myc* and p53 genes may be involved in controlling gene expression and DNA synthesis. Indeed, since the discovery that the v-*jun* oncogene is derived from the cellular gene (c-*jun*) encoding AP1 transcription factor (64), there is a distinct possibility that other gene products which play a role in controlling gene expression could function as oncogenes. However, at present we have little idea how alterations to the functioning of these genes and their products affects cell proliferation. Since the functioning of these gene products is downstream of the primary receptor-mediated events regulating cell proliferation, alterations to these genes might be expected to have fewer pleiotropic effects than, say, constitutive activation of a growth factor receptor. For example, 3T3 fibroblasts expressing a *myc* oncogene appear to have a much less disturbed pattern of growth regulation than the same cells expressing a *ras* oncogene (83).

5. Conclusion

In this introductory chapter, I have attempted to review some of the background and the ideas that are being pursued in oncogene research. Many of the areas are discussed in more detail elsewhere in this book. My aims have been to describe how oncogenes have been discovered and to show how oncogene research has become focused on the mechanisms that control cell proliferation, from the binding of growth factors to their receptors, through signal transduction to cell division. Overall, it is clear that we still have only a limited understanding of the control of cell proliferation. There are potentially many more functions in which genetic changes could lead to the generation of oncogenes that are yet to be discovered.

6. References

1. Waterfield,M.D., Scrace,G.T., Whittle,N., Stroobant,P., Johnson,A., Wasteson,A., Westermark,B., Heldin,C.H., Huang,J.S. and Deuel,T.F. (1983) Platelet-derived growth factor is structurally related to the putative transforming protein p28[sis] of simian sarcoma virus. *Nature*, **304**, 35–39.
2. Downward,J., Yarden,Y., Mayes,E., Scrace,G., Totty,N., Stockwell,P., Ullrich,A., Schlessinger,J. and Waterfield,M.D. (1984) Close similarity of epidermal growth factor receptor and v-erbB oncogene protein sequence. *Nature*, **307**, 521–527.
3. Sherr,C.J., Rettenmier,C.W., Saccia,R., Roussel,M.F., Look,A.T. and Stanley,E.R. (1985) The c-*fms* proto-oncogene product is related to the receptor for the mononuclear phagocyte growth factor, CSF-1. *Cell*, **41**, 665–676.
4. Sap,J., Munoz,A., Damm,K., Goldberg,Y., Ghysdael,J., Leutz,A., Beug,H. and Vennstrom,B. (1986) The c-erbA protein is a high affinity receptor for thyroid hormone. *Nature*, **324**, 635.
5. Maki,Y., Bos,T.J., Davis,C., Starbuck,M. and Vogt,P.K. (1987) Avian sarcoma virus 17 carries the *jun* oncogene. *Proc. Natl. Acad. Sci. USA*, **84**, 2848–2852.
6. Dalla-Favera,R., Wong-Staal,F. and Gallo,R.C. (1982) Onc gene amplification in promyelocytic leukaemia cell line HL-60 and primary leukaemic cells of the same patient. *Nature*, **299**, 61–63.

7. Escot,C., Thellet,C., Lidereau,R., Spyratos,F., Champeme,M.-H., Gest,J. and Callahan,R. (1986) Genetic alteration of the c-*myc* proto-oncogene (myc) in human primary breast carcinomas. *Proc. Natl. Acad. Sci. USA,* **83**, 4834–4838.
8. Little,C.D., Nau,M.N., Carney,D.N., Gazdar,A.F. and Minna,J.D. (1983) Amplification and expression of the c-*myc* oncogene in human lung cancer cell lines. *Nature,* **306**, 194–196.
9. Brodeur,G.M., Seeger,R.C., Schwab,M., Varmus,H.E. and Bishop,J.M. (1984) Amplification of N-*myc* in untreated human neuroblastomas correlates with advanced disease stage. *Science,* **224**, 1121.
10. Seeger,R.C., Brodeur,G.M., Sather,M., Dalton,A., Siegel,S.E., Wong,K.Y. and Hammond,D. (1985) Association of multiple copies of the N-*myc* oncogene with rapid progression of neuroblastomas. *New Engl. J. Med.,* **313**, 1111.
11. Venter,D.J., Tuzi,N.L., Kumar,S. and Gullick,W.J. (1987) Overexpression of the c-erbB-2 oncoprotein in human breast carcinomas: Immunohistological assessment correlates with gene amplification. *Lancet,* **ii**, 69.
12. Gullick,W.J., Marsden,J.J., Whittle,N., Wardl,B., Bobrow,L. and Waterfield,M.D. (1986) Expression of epidermal growth factor receptors on human cervical, ovarian and vulval carcinomas. *Cancer Res.,* **46**, 285–292.
13. Bos,J.L., Verlaan-de Vries,M., Marshall,C.J., Veeneman,G.H., van Boom,J.H. and van der Eb,A. (1986) A human gastric carcinoma contains a single mutated and an amplified normal allele of the K-*ras* oncogene. *Nucl. Acids Res.,* **14**, 1209–1217.
14. Kinzler,K.W., Bigner,S.H., Bigner,D.D., Trent,J.M., Law,M.L., O'Brien,S.J., Wong,A.J. and Vogelstein,B. (1987) Identification of an amplified, highly expressed gene in a human glioma. *Science,* **236**, 70–73.
15. Shih,C., Padhy,L.C., Murray,M. and Weinberg,R.A. (1981) Transforming genes of carcinomas and neuroblastomas introduced into mouse fibroblasts. *Nature,* **290**, 261–264.
16. Takahashi,M., Ritz,M. and Cooper,G.M. (1985) Activation of a novel human transforming gene, *ret*, by DNA rearrangement. *Cell,* **42**, 581–588.
17. Young,D., Waitches,G., Birchmeir,C., Fasano,O. and Wigler,M. (1986) Isolation and characterisation of a new cellular oncogene encoding a protein with multiple potential transmembrane domains. *Cell,* **45**, 711–719.
18. Eva,A., Vecchio,G., Diamond,M., Tronick,S.R., Dina,R., Cooper,G.M. and Aaronson,S.A. (1987) Independently activated *dbl* oncogenes exhibit similar yet distinct structural alterations. *Oncogene,* **1**, 355–360.
19. Taira,M., Yoshida,T., Miyagawa,K., Sakamoto,H., Terada,M. and Sugimura,T. (1987) cDNA sequence of human transforming gene *hst* and identification of the coding sequence required for transforming activity. *Proc. Natl. Acad. Sci. USA,* **84**, 2980–2984.
20. Hancock,J.F., Magee,A., Childs,J. and Marshall,C.J. (1989) All p21[ras] proteins are isoprenylated but only some are palmitoylated. **Cell,** in press.
21. McGrath,J.P., Capon,D.J., Smith,D.H., Chen,E.Y., Seeburg,P.H., Goeddel,D.V. and Levinson,A.D. (1983) Structure and organization of the human Ki-*ras* proto-oncogene and a related processed pseudogene. *Nature,* **304**, 501–504.
22. Trahey,M. and McCormick,F. (1987) A cytoplasmic protein stimulates normal N-*ras* p21 GTPase but does not affect oncogenic mutants. *Science,* **238**, 542–545.
23. Madaule,P. and Axel,R. (1985) A novel *ras*-related gene family. *Cell,* **41**, 31–40.
24. Chardin,P. and Tavitian,A. (1986) The *ral* gene: a new *ras* related gene isolated by the use of a synthetic probe. *EMBO J.,* **5**, 2203–2208.
25. Lowe,D.G., Capon,D.J., Delwart,E., Sakaguchi,A.Y., Naylor,S.L. and Goeddel,D.V. (1987) Structure of the human and murine R-ras genes novel genes closely related to *ras* proto-oncogenes. *Cell,* **48**, 137–146.
26. Marshall,C.J. (1986) The *ras* gene family. In *Oncogenes and Growth Factors,* Kahn,P. and Graf,T. (eds.), Springer-Verlag, Berlin, pp. 192–199.
27. Fasano,O., Aldrich,T., Tamanoi,F., Taparowsky,E., Furth,M. and Wigler,M. (1984) Analysis of the transforming potential of human H-ras by random mutagenesis. *Proc. Natl. Acad. Sci. USA,* **81**, 4008–4012.
28. Walter,M., Clark,S.G. and Levinson,A.D. (1986) The oncogenic activation of human p21 *ras* by a novel mechanism. *Science,* **233**, 649–652.

29. Lacal,J.C. and Aaronson,S.A. (1986) Activation of *ras* p21 transforming properties associated with an increase in the release rate of bound guanine nucleotide. *Mol. Cell Biol.*, **6**, 4214–4220.
30. Bos,J.L., Toksoz,D., Marshall,C.J., Verlaan-de Vries,M., Veeneman,G.H., van der Eb,A., van Boom,J.H., Janssen,J.W.G. and Steenvoorden,A.C.M. (1985) Amino-acid substitution at codon 13 of the N-*ras* oncogene in human acute myeloid leukaemia. *Nature*, **315**, 726–730.
31. Bos,J.L., Fearon,E.R., Hamilton,S.R., Verlaan-de Vries,M., van Boom,J.H., van der Eb,A.J. and Vogelstein,B. (1987) Prevalence of *ras* gene mutations in human colorectal cancers. *Nature*, **327**, 293–297.
32. Farr,C., Saiki,R., Ehrlich,H., McCormick,F. and Marshall,C.J. (1988) Analysis of ras Gene Mutations in Acute Myeloid Leukemia using the Polymerase Chain Reaction and Oligonucleotide Probes. *Proc. Natl. Acad. Sci. USA*, **85**, 1629–1632.
33. Forrester,K., Almoguera,C., Han,K., Grizzle,W.E. and Perucho,M. (1987) Detection of high incidence of K-ras oncogenes during human colon tumorigenesis. *Nature*, **327**, 298–303.
34. Bos,J.L. (1988) The ras-gene family and human carcinogenesis. *Mutation Research*, **195**, 255–271.
35. Barbacid,M. (1987) *ras* Genes. *Annual Review of Biochemistry*, **56**, 779–827.
36. Murray,M.J., Cunningham,J.M., Parada,L.F., Dautry,F., Leibowitz,P. and Weinberg,R.A. (1983) The HL60 transforming sequence: A ras oncogene co-existing with altered *myc* genes in hematopoietic tumors. *Cell*, **33**, 749–757.
37. Sinn,E., Muller,W., Pattengale,P. and Leder,P. (1987) Coexpression of MMTV.v-Ha-ras and MMTC/c-myc genes in transgenic mice: Synergistic action of oncogenes *in vivo*. *Cell*, **49**, 465–475.
38. Land,H., Parada,L.F. and Weinberg,R.A. (1983) Cellular oncogenes and multistep carcinogenesis. *Science*, **222**, 771–778.
39. Ruley,H.E. (1983) Adenovirus early region 1A enables viral and cellular transforming genes to transform primary cells in culture. *Nature*, **304**, 602.
40. Jenuwein,T. and Muller,R. (1987) Structure-function analysis of fos protein: a single amino acid change activates the immortalizing potential of v-*fos*. *Cell*, **48**, 647–657.
41. Hirai,H., Kobayashi,Y., Mano,H., Hajiwara,K., Maru,Y., Omine,M., Mizoguchi,H., Nishida,J. and Takatu,F. (1987) A point mutation at codon 13 of the N-*ras* oncogene in myelodysplastic syndrome. *Nature*, **327**, 430–432.
42. Lenoir,G.M. and Bornkamm,G.W. (1987) Burkitt's Lymphoma, a human cancer model for the study of the multistep development of cancer: Proposal for a new scenario. *Adv. Vir. Oncol.*, **7**, 173–206.
43. Klein,G. (1987) In defence of the 'old' Burkitt Lymphoma scenario. *Adv. Vir. Oncol.*, **7**, 207–211.
44. Fialkow,P.J., Martin,P.J., Nayfeld,V., Penfold,G.K., Jacobsen,R.J. and Hansen,J.A. (1981) Evidence for a multistep pathogenesis of chronic myelogenous leukaemia. *Blood*, **58**, 158–163.
45. Mahdevdan,L.C., Aitken,A., Heath,J. and Foulkes,J.G. (1987) Embryonal carcinoma-derived growth factor activates protein kinase C *in vivo* and *in vitro*. *EMBO J.*, **6**, 921–926.
46. Whetton,A.D., Monk,P.N., Consalvey,S.D., Huang,S.J., Dexter,T.M. and Downes,C.P. (1988) Interleukin 3 stimulates proliferation via protein kinase C activation without increasing inositol lipid turnover. *Proc. Natl. Acad. Sci. USA*, **85**, 3284–3288.
47. Berridge,M.J., Heslop,J.P., Irvine,R.F. and Brown,K.D. (1984) Inositol trisphosphate formation and calcium mobilization in response to platelet-derived growth factor. *Biochem. J.*, **222**, 195–201.
48. Rozengurt,E. and Mendoza,S.A. (1985) Synergistic signals in mitogenesis: role of ion fluxes, cyclic nucleotides and protein kinase C in Swiss 3T3 cells. *J. Cell Sci. [Suppl.]*, **3**, 229–242.
49. Rozengurt,E. (1986) Early signals in the mitogenic response. *Science*, **234**, 161–166.
50. Vara,F. and Rozengurt,E. (1985) Stimulation of Na^+/H^+ antiport activity by epidermal growth factor and insulin occurs without activation of protein kinase C. *Biochem. Biophys. Res. Commun.*, **130**, 646.
51. Besterman,J.M., Tyrey,S.J., Cragoe,E.J. and Cuatrecasas,P. (1984) Inhibition of

epidermal growth factor-induced mitogenesis by amiloride and an analog: Evidence against a requirement for Na$^+$/H$^+$ exchange. *Proc. Natl. Acad. Sci. USA*, **81**, 6762–6766.

52. Pouyssegur,J., Chambard,J.C., Franchi,A., Paris,S. and Van Obberghen-Schilling,E. (1985) Growth factor activation of the sodium$^+$ -H$^+$ anti-porter controls growth of fibroblast by regulating intra-cellular pH. *Cancer Cells*, **3**, 409–415.

53. Greenberg,M.E. and Ziff,E.B. (1984) Stimulation of 3T3 cells induces transcription of the c-*fos* proto-oncogene. *Nature*, **311**, 433–438.

54. Linzer,D.I.H. and Nathans,D. (1983) Growth-related changes in specific mRNAs of cultured mouse cells. *Proc. Natl. Acad. Sci. USA*, **80**, 4271–4275.

55. Treisman,R. (1985) Transient accumulation of c-*fos* RNA following serum stimulation requires a conserved 5′ element and c-fos 3′ sequences. *Cell*, **42**, 889–902.

56. Shaw,G. and Kamen,R. (1986) A conserved AU sequence from the 3′ untranslated region of GM-CSF mRNA mediates selective mRNA degradation. *Cell*, **46**, 659–667.

57. Blanchard,J.-M., Piechaczyk,M., Dani,C., Chambard,J.-C., Franchi,L., Pouyssegur,J. and Jeanteur,P. (1985) c-*myc* gene is transcribed at high rate in Go-arrested fibroblasts and is post-transcriptionally regulated in response to growth factors. *Nature*, **317**, 443–445.

58. Verma,I.M. and Sassone-Corsi,P. (1987) Proto-oncogene *fos*: Complex but versatile regulation. *Cell*, **51**, 513–514.

59. Mohun,T., Garrett,N. and Treisman,R. (1987) Xenopus cytoskeletal actin and human c-*fos* gene promoters share a conserved protein-binding site. *EMBO J.*, **6**, 667–673.

60. Treisman,R. (1987) Identification and purification of a polypeptide that binds to the c-*fos* serum response element. *EMBO J.*, **6**, 2711–2717.

61. Hayes,T.E., Kitchen,A.M. and Cochran,B.H. (1987) Inducible binding of a factor to the c-*fos* regulatory region. *Proc. Natl. Acad. Sci. USA*, **84**, 1272–1276.

62. Lee,W., Mitchell,P. and Tjian,R. (1987) Purified transcription factor AP-1 interacts with TPA-inducible elements. *Cell*, **49**, 741–752.

63. Angel,P., Imagawa,U., Chin,R., Stein,B., Imbra,R.J., Rahmsdorf,H.J., Jonot,C., Herrlich,P. and Karin,M. (1987) Phorbol ester-inducible genes contain a common *cis* element recognized by a TPA-modulated trans-acting factor. *Cell*, **49**, 729–739.

64. Bohman,D., Bos,T.J., Admont,A., Nishimura,T., Vogt,P.K. and Tjian,R. (1987) Human proto-oncogene c-*jun* encodes a DNA binding protein with structural and functional properties of transcription factor AP-1. *Science*, **238**, 1386–1392.

65. Bronzert,D.A., Pantazis,P., Antoniades,H.N., Kasid,A., Davidson,N., Dickson,R.B. and Lippman,M.E. (1987) Synthesis and secretion of platelet-derived growth factor by human breast cancer cell lines. *Proc. Natl. Acad. Sci. USA*, **84**, 5763–5767.

66. Ullrich,A., Coussens,L., Hayflick,J.S., Dull,T.J., Gray,A., Tam,A.W., Lee,J., Liberman,T.A., Schlessinger,J., Downward,J., Mayes,E.L., Whittle,N., Waterfield,M.D. and Seeburg,P.H. (1984) Human epidermal growth factor receptor cDNA sequence and aberrant expression of the amplified gene in A431 epidermoid carcinoma cells. *Nature*, **309**, 418–425.

67. Garrett,D.C., Tracy,S.E. and Robinson,H.L. (1986) Differences in sequences encoding the carboxyl-terminal domain of the epidermal growth factor receptor correlate with differences in the disease potential of viral erbB genes. *Proc. Natl. Acad. Sci. USA*, **83**, 6053–6057.

68. Wheeler,E.F., Rettenmier,C.W., Look,A.T. and Sherr,C.J. (1986) The v-fms oncogene induces factor independence and tumorigenicity in a CSF-1 dependent macrophage cell line. *Nature*, **324**, 377–380.

69. Velu,T.J., Beguinot,L., Vass,W.C., Willingham,M.C., Merlino,G.T., Pastan,I. and Lowy,D.R. (1987) Epidermal growth factor dependent transformation by a human EGF receptor proto-oncogene. *Science*, **238**, 1408–1410.

70. Courtneidge,S.A. and Heber,A. (1987) An 81 kd protein complexed with middle T antigen and pp60$^{c\text{-}src}$: A possible phosphatidylinositol kinase. *Cell*, **50**, 1031–1037.

71. Kaplan,D.R., Whitman,M., Schaffhausen,B., Pallas,D.C., White,M., Cantley,L. and Roberts,T.M. (1987) Common elements in growth factor stimulation and oncogenic transformation: 85 kd phosphoprotein and phosphatidylinositol kinase activity. *Cell*, **50**, 1021–1029.

72. Konopka,J.B., Watanabe,S.M. and Witte,O.N. (1984) An alteration of the human c-

abl protein in K562 leukemia cells unmasks associated tyrosine kinase activity. *Cell,* **37**, 1035–1042.

73. Mulcahy,L.S., Smith,M.R. and Stacey,D.W. (198 .) Requirement for *ras* proto-oncogene function during serum-stimulated growth of NIH-3T3 cells. *Nature,* **313**, 241–243.

74. Stacey,D.W., Watson,T., Kung,H.-F. and Curran,T. (1987) Microinjection of transforming *ras* protein induces c-*fos* expression. *Mol. Cell. Biol.,* **7**, 523–527.

75. Wakelam,M.J.O., Davies,S.A., Houslay,M.D., McKay,I.A., Marshall,C.J. and Hall,A. (1986) Normal p21^{N-ras} couples bombesin and other growth factor receptor to inositol phosphate production. *Nature,* **323**, 173–176.

76. Lacal,J.C., de la Pena,P., Moscat,J., Garcia-Barreno,P., Anderson,P.S. and Aaronson,S.A. Rapid stimulation of diacylglycerol production in Xenopus Oocytes by microinjection of H-*ras* p21. *Science,* **238**, 533–536.

77. Fleischman,L.F., Chawala,S.B. and Cantley,L. (1986) Ras-transformed cells: altered levels of phosphatidylinositol-4,5-bisphosphate and catabolites. *Science,* **231**, 407–410.

78. Wolfman,A. and Macara,I.G. (1987) Elevated levels of diacylglycerol and decreased phorbol ester sensitivity in ras-transformed fibroblasts. *Nature,* **325**, 359–361.

79. Lacal,J.-C., Moscat,J. and Aaronson,S.A. (1987) Novel source of 1,2 diacylglycerol elevated in cells transformed by Ha-*ras* oncogene. *Nature,* **330**, 269–272.

80. Lacal,J.-C., Fleming,T.P., Warren,B.S., Blumberg,P.M. and Aaronson,S.A. (1987) Involvement of functional protein kinase C in the mitogenic response to the H-ras oncogene product. *Mol. Cell. Biol.,* **7**, 4146–4149.

81. Doherty,P.J., Hua,L., Liau,G., Gal,S., Graham,D.E., Sobel,M. and Gottesman,M.M. (1985) Malignant transformation and tumor promoter treatment increase levels of a transcript for a secreted glycoprotein. *Mol. Cell. Biol.,* **5**, 466–473.

82. Wasylyk,C., Imler,J.L., Perez-Mutul,S. and Wasylyk,B. (1987) The c-Ha-*ras* oncogene and a tumor promoter activate the polyoma virus enhancer. *Cell,* **48**, 525–534.

83. Falcone,G., Summerhayes,I.C., Paterson,H., Marshall,C.J and Hall,A. (1987) Partial transformation of mouse fibroblastic and epithelial cell lines with the v-*myc* oncogene. *Exp. Cell Res.,* **168**, 273–284.

84. Bohmann,D., Bos,T.J., Admon,A., Nishimura,T., Vogt,P.K. and Tjian,R. (1988) Human proto-oncogene c-*jun* encodes a DNA binding protein with structural and functional properties of transcription factor AP-1. *Science,* **239**, 1386–1392.

85. Gilman,A.G. (1987) G proteins: Transducers of receptor-generated signals. *Ann. Rev. Biochem.,* **56**, 615–649.

2

Oncogenes at viral integration sites

Gordon Peters

1. Introduction

Cancer is believed to result from genetic alterations in a restricted set of cellular genes, and agents that damage DNA or introduce novel DNA into cells feature prominently among proven carcinogens (1,2). The purpose of this chapter is to review one family of agents that have the potential to do both, the *retroviruses*, and to discuss the ways in which they activate the latent oncogenic properties of particular cellular genes, the *proto-oncogenes*. As the intention is to be illustrative rather than encylopedic, initial emphasis will be placed on basic concepts and theoretical considerations. However, in the latter sections, more detailed information will be presented for specific genes and retroviral systems that have been the most informative sources of experimental evidence.

2. Cellular oncogenes

Retroviruses can cause cancer in their natural hosts with an efficiency and rapidity that ranks them among the most potent carcinogens known. However, the oncogenic properties of most retroviruses are mediated by genetic information that is of cellular rather than viral origin, and it was this realization that first pointed to the existence of cellular proto-oncogenes in the DNA of all somatic cells (reviewed in 3 and 4). Whether of viral or cellular origin, the exact definition of an oncogene is at best fluid, and at times controversial (5). For the purposes of this chapter, the term *oncogene* will be taken to mean any piece of DNA whose capacity to encode protein is responsible for changing the behavior of a cell such that it has an increased probability of becoming malignant.

By definition, the genetic information contained in a cellular oncogene pre-exists in all normal cells, so that some stimulus or alteration must be required to unmask its oncogenic properties. Simplistically, such changes are of two types: qualitative modifications that alter the function or specificity of the resultant protein, and quantitative modifications in the level of expression of the protein. While there are now several recognized routes by which such changes can be brought about, some of which operate in the absence of any viral input (2, and see Chapter 3), the overwhelming majority of cellular oncogenes identified thus far have come to light through the analysis of retrovirus-induced tumors in animal models. Why then are retroviruses such uniquely potent yet experimentally accessible carcinogens? The crux lies in their replicative cycle which requires the formation of a DNA intermediate that is inserted as a *provirus* into the chromosomal DNA of the host cell (*Figure 1*). Once established, the provirus becomes perpetuated as part of the genetic complement of the cell and is inherited by all subsequent daughter cells. Retroviruses are therefore obligatory mutagens that cause sustained disruptions of cellular DNA. If such events create novel, oncogenic combinations of genetic information, or activate the latent properties of a proto-oncogene, then it is clear that a cell in which this occurs, and all its progeny, may be destined to become neoplastic.

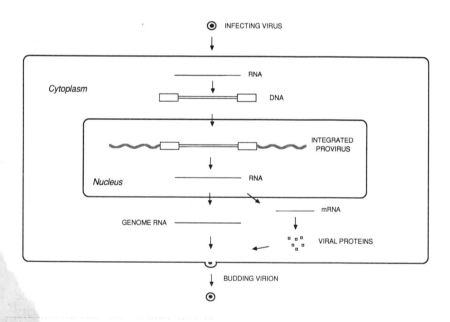

Figure 1.   The figure outlines the major steps and flow of genetic information in the normal replicative cycle of a standard retrovirus. Following establishment of the DNA provirus within the cellular genome, multiple copies of viral RNA are generated and packaged into progeny virus particles. Virions bud continuously from the infected cell without causing lysis.

3. The retrovirus life cycle

Before discussing how oncogene activation may be brought about, it is first necessary to outline the essential features of retrovirus replication that bear on the process. For more comprehensive coverage of this topic, the reader is referred to recent review articles (6,7). As their name implies, retroviruses undertake the converse of the normal flow of genetic information in that the sequence of their single-stranded RNA genome is first copied into DNA (*Figure 1*). This is accomplished by the viral RNA-dependent DNA polymerase (reverse transcriptase) through a complex series of priming, RNA hydrolysis, and template switching events. The net result is a double-stranded DNA intermediate that is slightly longer than the viral genome, due to duplication of specific sequences that occur at the 5′ and 3′ ends of the viral RNA. These duplicated sequences make up identical structures, referred to as the long terminal repeats (LTRs), that are present at each end of the DNA intermediate (*Figures 1* and *2*). Retroviral DNA therefore displays striking similarities to the transposable genetic elements of prokaryotes and lower eukaryotes, a feature that is further underscored by the presence of short, inverted repeats at the boundaries of each LTR, and the duplication of four to six base pairs of DNA at the sites of integration (8, and *Figure 2*).

The sequences assembled in the LTRs are involved in two crucial aspects of the viral life cycle, the integration of proviral DNA into the host chromosome and the control of viral gene expression. The former is catalyzed by a virally encoded endonuclease, while the latter relies almost exclusively on the transcriptional machinery of the host cell (6,7,9). The exact details of integration and gene expression need not concern us here, but there are several key features that do. The first is that although integration of the proviral DNA is very precise with regard to viral nucleotide sequences, it appears to be random with respect to cellular DNA sequences (8–10). Thus, while integration invariably creates a provirus that is co-linear with the viral genome, and bounded exactly by two LTRs, there is no discernible control over where in the cellular genome the provirus ends up. This does not exclude the possibility that particular regions of chromatin might adopt configurations that are more or less accessible to proviral integration, perhaps in a tissue-specific or developmentally regulated pattern, but no general principle has yet become apparent.

The other relevant features are the elements within the LTRs that orchestrate viral gene expression. All retroviral LTRs contain recognizable variants of the DNA sequence motifs commonly associated with initiation of eukaryotic mRNA synthesis, and the signals for processing and polyadenylation (6,7 and *Figure 2*). They also contain transcriptional enhancer elements that can act as potent modulators of the activity of *cis*-linked promoters (11), and there is growing evidence that it is the

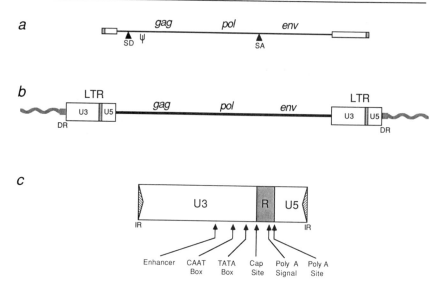

Figure 2.   Structure of a typical retroviral genome, provirus and LTR. (a) The RNA genome of a typical retrovirus is depicted showing the linear order of the three major structural and replicative genes *gag*, *pol*, and *env*. The boxed sections identify sequences that are duplicated in the formation of the long terminal repeat (LTR) in the provirus. The shaded boxes represent short direct repeats (designated R) present at the 5′ and 3′ termini of the genome. SD and SA mark the approximate locations of the splice donor and acceptor sites, respectively, that are involved in generating the subgenomic mRNA for the *env* gene products. ψ represents the major signal required for the packaging of genome RNA into viral particles. (b) The corresponding DNA provirus is shown, integrated within cellular DNA (wavy line) and bounded by two LTRs (boxes). U3 and U5 refer to sequences derived from the 3′ and 5′ end of the viral genome and reflect the boxed segments in (a). The DR motifs identify short (4 – 6 bp) direct repeats in the cellular DNA that are generated during proviral integration. (c) A prototype retroviral LTR is depicted in which the various signals for initiation and termination of viral RNA synthesis are located relative to U3, R and U5. IR identifies short inverted repeats that are present at the 5′ and 3′ boundaries of each LTR.

enhancer that confers the inherent tissue specificity displayed by many retroviruses (see references 12 – 17, for example).

Although under the control of the LTR, the provirus in all other respects behaves as if it were a normal cellular gene and is transcribed by the host RNA polymerase II (6,7). The resulting RNA is capped and poly-adenylated, but serves two distinct functions: as the genome RNA for progeny virions and as a messenger RNA for some of the viral proteins. The sequences that encode the viral proteins are organized into discrete domains or genes that have been designated *gag*, standing for group-specific antigens but encompassing all of the internal structural proteins of the viral core, *pol*, for RNA-dependent DNA polymerase, and *env*, for the viral envelope proteins (6,7,18). The definition of these domains as genes and the boundaries between them are somewhat arbitrary since the individual viral proteins are initially expressed as large precursors, some of which span more than one reading frame. Moreover, two additional functions have to be accommodated in this scheme, namely the viral protease that is encoded beteween *gag* and *pol*, and the endonuclease or 'integrase' that is at the 3′ end of *pol* (9,18). It has also become apparent that some retroviruses, typified by the human T-cell leukemia viruses (HTLV-1 and II) and bovine leukemia virus (BLV) have the capacity to encode additional, non-structural proteins that are expressed from complex spliced RNAs and play a role in regulating viral gene expression (19 – 22; reviewed in 23).

The major spliced RNA common to all retrovirus-infected cells is the sub-genomic mRNA for the *env* proteins (6,7). Expression of *env* in this way ensures an adequate level of these products to populate the unit membrane that surrounds the viral particle (18). However, splicing also ensures that this sub-genomic RNA does not become packaged into virions since the major determinant for packaging lies downstream of the splice donor site within the intron of the *env* mRNA (6,7 and *Figure 2*). Genomic RNAs that do carry the packaging signal become assembled into viral cores that bud through the plasma membrane of infected cells. Although some retroviruses can be cytopathic, the vast majority exit the cell without concomitant lysis. At the cellular level, therefore, retrovirus infection is essentially innocuous and indeed there are many situations in which animals carrying severe virus loads show no outward manifestations of disease or malaise (24).

4. Modes of retroviral oncogenesis

How can this generally benign life cycle be reconciled with the proven oncogenicity of retroviruses? In the first place, it is essential to recognize that there are distinct categories of retrovirus-induced neoplasms, the most obvious distinction being the speed at which they develop (24). Thus, some retrovirus infections result in lethal tumors within a matter of days or weeks, and the agents involved are commonly referred to as *acutely oncogenic*. This contrasts with the second category of tumors that arise in viremic animals long after the initial infection, perhaps several months or even years later, a situation that has been variously described as *non-acute, chronic, long-latency,* or even *weakly oncogenic*. The latter is something of a misnomer, since in many of these situations the tumor incidence among infected individuals can approach 100 per cent. Moreover, the overwhelming majority of tumors associated with naturally occurring retroviruses fall into the long-latency category. Although it is on these latter systems, and the underlying molecular events, that the bulk of this chapter will focus, it is instructive to review all of the mechanistic options identified thus far in retrovirus-induced malignancies, and their etiological implications.

4.1 Transduction of cellular sequences

Historically, the early thinking on retrovirus-induced tumorigenesis was understandably influenced by analogies with the DNA tumor viruses, such as polyoma and SV40, where it had been shown that virally encoded proteins were directly responsible for cell transformation (25 and Chapter 5). Indeed, many of the analogies appeared to be justified in the pioneering studies on Rous sarcoma virus (RSV) when it was shown that viral mutants could be isolated that were impaired in their ability to transform cells (26). A viral gene, given the name *src*, was therefore directly implicated in tumorigenesis. The dramatic change in thinking came when it was realized that *src* was not a strictly viral gene, but was in fact derived from a counterpart present and expressed in normal cells (27,28). These studies heralded the discovery of cellular proto-oncogenes and were followed by the identification of many further examples of viral oncogenes, the v-*oncs*, that have been created by the capture or transduction of genetic information from their cellular progenitors, the c-*oncs* (3,4).

Details of all of these genes and their functions are beyond the remit of this chapter, but there are a number of general features that warrant discussion.

(i)   With the exception of RSV, the acquisition of v-*onc* sequences has occurred at the expense of part of the viral genome (*Figure 3*), so that the recombinant retrovirus that is formed is unable to replicate autonomously. Its continued propagation is therefore dependent on viral structural proteins and enzymes supplied in *trans* by an associated, non-defective helper virus.

(ii) The v-*oncs* invariably represent modified forms of their cellular progenitors that have been truncated, mutated, or fused to viral coding sequences (3,4,5). In some instances, pieces of genetic information from two quite separate cellular genes have been combined in the formation of the v-*onc*.

(iii) The transduced v-*onc* sequences are expressed at relatively high levels, as part of the provirus, irrespective of its location within the host cell genome (3,4). Thus, every cell that contains a transcriptionally active provirus is exposed to the oncogenic properties of the v-*onc* product and liable to display the phenotypic effects.

Viruses that contain transduced oncogenes are therefore capable of transforming individual cells in tissue culture and are associated with very rapid tumorigenesis in animals (24). Significantly, the tumors that arise are polyclonal since they can comprise multiple independently infected cells. However, although they have been highly informative in unearthing cellular oncogenes and their functions, these acutely oncogenic agents are exceedingly rare relative to the total numbers of infectious retroviruses in natural populations. Tumorigenesis has provided a very powerful

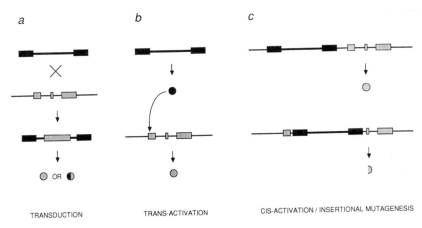

a   b   c

TRANSDUCTION   TRANS-ACTIVATION   CIS-ACTIVATION / INSERTIONAL MUTAGENESIS

Figure 3.   Modes of oncogene activation by integrated retroviral genomes. Three distinct scenarios are presented for the potential interactions between a retroviral provirus, represented in black, and a typical cellular gene, the exons of which are depicted as shaded boxes. In (a) the virus has recombined with the cellular gene such that some of these exon sequences become incorporated into a replication defective, acutely transforming viral genome. In such a situation, the cellular information (now a v-*onc*) is expressed from viral promoters irrespective of the location of the provirus in the chromosomal DNA. Two possibilities are envisaged for the resultant oncogene product (represented as a circle), either a fusion of viral and cellular information or a protein comprising exclusively cellular information. This latter situation does not preclude truncation or some structural alteration relative to the normal product of the gene. In (b) the interaction between provirus and gene occurs in *trans* and should therefore operate irrespective of their relative positions in the genome. Here a strictly viral gene product (black circle) is envisaged as having some regulatory influence on the expression of the cellular product (shaded circle). In (c), two alternative situations are presented in which the interaction between virus and gene occurs in *cis*. In the first example, the provirus is located in chromosomal DNA adjacent to the cellular gene and is influencing the expression of the normal gene product from its normal promoter. The second example places the integrated provirus within the body of the gene resulting in the expression of a truncated or structurally altered protein.

selection for their detection, but in practice there are only of the order of 100 known isolates, mostly observed in laboratory settings or in domesticated animal stocks. Indeed it has been argued that such agents would normally be subject to extinction unless intentionally maintained in laboratories (5).

4.2 Viral gene products that are oncogenic

Although the majority of the retroviral v-*onc* sequences are known to be derived from genetic information of cellular origin, this is not an invariant rule. For example, at least some of the tumorigenic properties of spleen focus forming virus (SFFV) and other rapidly oncogenic members of the murine leukemia virus family appear to be attributable to strictly viral sequences (24,29,30). Here recombination or transduction has generated a mosaic *env* gene, comprising elements acquired from proviral DNA that is endogenous to the mouse genome. How such recombinant glycoproteins influence the growth properties of infected cells is open to conjecture, but it seems unlikely to involve direct interaction with a cellular oncogene (but see section 7.2.5).

With HTLV and BLV, on the other hand, cellular oncogenes are again invoked. In the absence of alternative explanations, such as those discussed below, it has been suggested that leukemogenesis by these agents may involve the virally encoded regulatory proteins (the so-called *trans*-activating functions) acting on specific cellular genes (31–39 and *Figure 3*). However, supportive evidence for such a mechanism remains tentative, and it is not intuitively consistent with the disease pattern associated with HTLV and BLV. Since *trans*-activation should operate in every cell, irrespective of the relative positions of the provirus and target gene (they could for instance be on separate chromosomes) the expected outcome would be the rapid development of polyclonal tumors. In practice, HTLV – and BLV – associated tumors appear after prolonged latency and are generally monoclonal (40–42).

4.3 *Cis*-activation and insertional mutagenesis

Most of the tumors associated with natural retrovirus infections develop after long periods of viremia, and any infectious particles released from the tumor cells have the same oncogenic potential as the initial virus stock (24). They evidently do not contain an acutely oncogenic agent, have not acquired a transduced oncogene, and do not induce morphological transformation of cultured cells. Consequently, such systems have proved inherently less amenable to study since many of the standard techniques of experimental virology, such as the isolation of mutants, are not applicable. In more recent years, however, the advent of restriction mapping, blot hybridization, and gene cloning have provided alternative avenues for research.

What has emerged is a common theme in which cellular oncogenes again figure, this time activated by a *cis*-acting retrovirus (4,24,43,44). As depicted simplistically in *Figure 3*, the essential feature of this theme is that the integrated provirus is physically linked to the target gene, but has not undergone the recombinatorial events required for transduction. Thus, in contrast to the acutely oncogenic viruses, the position of the provirus in the cellular genome is all important. There are many corollaries, which will be discussed in the context of specific examples, but the critical point is that the juxtaposition of provirus and oncogene, in whatever configuration, is a purely stochastic event. Tumorigenesis will therefore be very rare relative to the total number of virally infected cells and randomly integrated proviruses. This can, in part, account for the long latency of tumor development, since it is first necessary to establish a large pool of infected cells. It also makes a key prediction, namely that the resultant tumor will be dominated by a clone or clones of cells in which a provirus has integrated in a specific location (43,44).

5. Insertional mutagenesis: theory and practice

The notion that retroviruses can activate cellular oncogenes by what has been termed *insertional mutagenesis* rests on a series of tenets concerning the randomness of proviral integration, the dominance of resulting mutations and the clonality of the ensuing tumors (43). Before reviewing the details and intricacies of actual examples, it will be instructive to expand on these general features.

The first is the accepted, though not formally proven, premise that the sites of proviral integration within the host genome can be entirely random (7 – 10). If correct, then every cellular gene can be regarded as a potential target for mutagenesis by proviral insertion. Assuming very crude figures for the size of a cellular gene (say 3 kb) relative to the total mammalian genome (3×10^6 kb), the probability that a provirus might integrate within or near a particular gene would be of the order of 1 in 10^6. As it would not be at all uncommon for a viremic animal to carry a burden of 10^6 infected cells, at least at some time during its lifespan, the probability is quite high that a cell carrying a specifically mutated gene might arise in the normal course of retroviral infection. Latency may therefore reflect the time required for this probability to reach unity, although it is unlikely that this is the whole story.

The second consideration is that for a provirally induced mutation to be detectable, it would have to have a tangible effect on the phenotype of the cell. The most likely outcome of proviral intrusion into an active cellular gene would be disruption and cessation of its normal function. In diploid cells, such negative effects are likely to be compensated for by the remaining normal allele on the undisturbed chromosome (43). The only exceptions would arise if the afflicted cell happened to be functionally hemizygous for that gene (45,46) or in the unlikely event that proviruses should simultaneously knock out both alleles (47). Although highly improbable in somatic cells, this latter situation does arise when a provirus becomes established in cells of the germ line and hence inheritable in a Mendelian fashion. Inbreeding can render such endogenous proviruses homozygous, so that both alleles become affected, and there are well-documented examples of genetic traits traceable to just such insertional mutagenesis events (48 – 51).

Nevertheless, the overwhelming majority of proviral insertions would be expected to be phenotypically silent, which is exactly what is observed. Overtly viremic animals can enjoy perfect health, and their productively infected cells behave normally. Thus, the exceptional cases that give rise to tumors must reflect dominant mutational events that give the cell a selective advantage over its neighbors. Such dominant effects obviously echo the quantitative and qualitative alterations that accompany oncogene transduction.and, as will be seen below, can operate on some of the same cellular genes that are familiar as the progenitors of viral oncogenes.

But perhaps the most significant feature of insertional mutagenesis is that the resultant tumor should be dominated by the progeny of the single cell that sustained the initial mutation. A simple impression of clonality can be gleaned by examining the acquired proviruses in tumor DNA by restriction enzyme digestion and Southern blotting (52–59). As exemplified in *Figure 4*, the introduction of proviral DNA into the cellular genome will create novel restriction fragments, the most informative of which will be those that span the junctions between viral and cellular DNA. In a mixed population of infected cells, proviruses will be integrated at many different sites and create novel junction fragments of many different sizes, each of which will therefore be much less abundant than fragments from the single copy genes common to all cells. In a clonal population, on the other hand, the novel junction fragments will be the same in every cell, and present at an abundance equivalent to a haploid gene.

As evidenced by the specific examples discussed in sections 6 and 7, most of the naturally occurring tumors that result from infection by a non-acutely oncogenic retrovirus comply with these predictions, or approximately so. In practice, the quantitation can be distorted by the inclusion of some normal cells in the tumor biopsy, by superinfection of the original clone during tumor progression, or, less commonly, in situations where the tumor may be oligoclonal rather than strictly monoclonal. Nevertheless, it is clear that tumors invariably contain at least one provirus, and frequently several, that have been acquired by infection. On Southern blots, these new proviral bands are usually superimposed on the pattern of endogenous proviral sequences that are characteristic of the particular animal and hence present in all somatic cells. The implicit assumption, however, is that it is one of the newly acquired proviruses that must be responsible for the dominant mutation that led to tumorigenesis. Furthermore, tumors of similar pathology and viral etiology might be expected to have sustained mutations in the same gene.

Such notions prompted researchers to examine the sites in tumor DNA at which proviruses had integrated. Minimally, this permits the identification of loci that are common targets for insertional mutagenesis, whether they be previously anonymous regions of cellular DNA or already recognizable genes. The seminal discovery in this regard was the demonstration by Hayward and his colleagues that in bursal lymphomas induced by avian leukosis virus (ALV), proviral DNA can be detected within or adjacent to the c-*myc* gene (60). In the ensuing years, and as discussed in sections 6 and 7, numerous additional examples have come to light in which oncogenes have been the targets for insertional mutations. More importantly, however, the same experimental approaches have uncovered new genes that have not been encountered in viral form, and have securely underpinned the premise that *cellular* proto-oncogenes are directly involved in neoplasia (2,3,4,44).

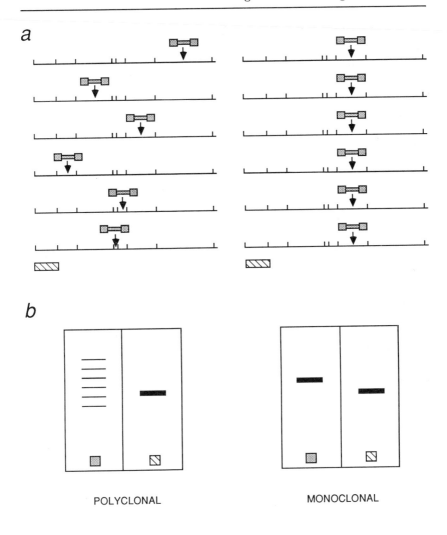

Figure 4. Provirus integration and tumor clonality. Panel (a) is a diagrammatic representation of two contrasting populations of virus-infected cells. The first comprises a mixture of randomly infected cells in which the acquired proviruses (shaded boxes) have integrated at many different sites within the cellular DNA (horizontal line). Each provirus might therefore be expected to reside within a different restriction fragment, as exemplified by cleavage at the sites marked by vertical bars. On Southern blots of total DNA from such a population, represented graphically in panel (b) a viral probe (shaded box) would detect fragments of many different sizes. Each of these would be present at concentrations far below that of a restriction fragment common to all cells, for example one identified by a unique cellular DNA probe (hatched box). In contrast, in a monoclonal population, each cell would contain a provirus within the same restriction fragment, reflecting the situation in the original parental cell. Viral DNA probes would therefore detect a unique restriction fragment on Southern blots, at a concentration comparable to that of a cellular gene. In practice, acquired proviruses reside on only one of any chromosome pair and thus appear at only half the intensity of a diploid gene.

6. Proviral activation of known oncogenes

Given the catholic definition of an oncogene at the beginning of this chapter, the distinction between oncogenes that are *known* and those that are therefore *unknown* requires justification. The intention of this sub-division is to distinguish between the experimental logic that led to the discovery of these genes. Known genes refer to those that, although now recognized as frequent targets of insertional mutagenesis, were originally identified by other means. Some have no previous connections with retrovirology, but most have previously featured as v-*onc* sequences in acutely transforming viruses. In all likelihood the list of examples provided in *Table 1* will continue to grow as it now seems certain that insertional mutagenesis in its broadest sense is an essential prelude to transduction (3,4). In contrast, section 7 will spotlight genes whose identification and characterization were a direct consequence of proviral insertion and activation in virally induced tumors.

6.1 The c-*myc* gene

The c-*myc* gene has proved to be one of the most illuminating examples of a cellular proto-oncogene: it can be activated in a number of different ways, both viral and non-viral, and it has been implicated in tumors of a variety of different species, including chickens, mice, rats, cats and humans (3,4 and Chapter 3). In all of these cases, the general topology of the *myc* gene is the same, comprising two coding exons that are highly conserved among species, preceded by a non-coding exon that is presumed to serve some regulatory function in c-*myc* expression, and is less well conserved (61–69). For example, although the first exons of mouse and human *myc* are reasonably homologous, that of chicken *myc* appears quite different (70,71).

6.1.1 Myc activation by ALV

Like many cellular oncogenes, the initial identification of *myc* was as a v-*onc* component, in this case in a series of acutely transforming avian retroviruses, typified by MC29 (72). Transduction of *myc* coding sequences from exons II and III rendered these viruses defective, so that their replication depends on the presence of competent helper viruses (63,72). The natural helpers for these acutely oncogenic agents all belong to the same virus family, the avian leukosis viruses (ALV), that contain only the standard viral genes but are themselves oncogenic after prolonged latency (24,73). It is these latter viruses, rather than their defective derivatives, that are major etiological agents of leukemia in domestic fowl. Depending on the strain, some 20 to 80 per cent of infected chickens will succumb to B-cell lymphomas, in the bursa of Fabricius, at between 4 and 12 months of age (24, 73). There is, however, a comforting circularity

Table 1.   Targets of proviral insertion[a]

	Gene/locus	Activating virus	Associated tumors
Group a	*myc*	ALV, MuLV, REV, FeLV	B-lymphoid, T-lymphoid
	myb	MuLV, ALV	Myeloid, B-lymphoid
	erbB	ALV	Erythroblastic
	mos	IAP	B-lymphoid
	ras^H	ALV	Nephroblastic
	ras^K	F-MuLV	Erythroid
	fms/fim-2	F-MuLV	Myeloid
	IL-2	GALV	T-lymphoid
	IL-3	IAP	Myeloid
	CSF-1	MuLV(BALB/c)	Myeloid
	p53	SFFV, F-MuLV	Erythroleukemia
Group b	*int*-1	MMTV	Mammary
	int-2	MMTV	Mammary
	int-3	MMTV	Mammary
	int-4	MMTV	Mammary
	pim-1	MuLV	T-lymphoid
	lck	MuLV	T-lymphoid
	spi-1	SFFV	Erythroid
	evi-1	MuLV(AKXD)	Myeloid
Group c	*pvt/mis*-1	M-MuLV	B-lymphoid, T-lymphoid
	ahi-1	MuLV	B-lymphoid
	dsi-1	M-MuLV	T-lymphoid
	M*lvi*-1,2,3	M-MuLV	T-lymphoid
	gin-1	G-MuLV	T-lymphoid
	fis-1	F-MuLV	Myeloid, lymphoid
	fim-1	F-MuLV	Myeloid
	fim-3	F-MuLV	Myeloid
	evi-2	MuLV(BXH-2)	Myeloid

[a]The table lists cellular loci that are reported sites of retroviral integration in primary tumors and tumor cell lines, and distinguishes three separate categories. Those included in *a* represent 'known' genes that were originally identified as components of acutely transforming retroviruses, or by other criteria. Group *b* comprises genes whose initial identification depended on their being activated by proviral insertion, while group *c* represents common proviral integration loci for which there is as yet no evidence for an expressed cellular gene. The abbreviations for the various activating viruses are as follows: ALV, avian leukosis virus; FeLV, feline leukemia virus; F-MuLV, Friend murine leukemia virus; GALV, gibbon ape leukemia virus; G-MuLV, Gross murine leukemia virus; IAP, intracisternal A-particle; M-MuLV, Moloney murine leukemia virus; MMTV, mouse mammary tumor virus; REV, reticuloendotheliosis virus; SFFV, spleen focus forming virus. Abbreviations in parenthesis refer to strains of inbred or back cross mice.

in the system since in virtually all of these lymphomas, ALV proviral sequences are located within or adjacent to the c-*myc* gene (60,74–81). It was this convincing endorsement of insertional mutagenesis that first lent it credence and underlined the relevance of cellular as opposed to

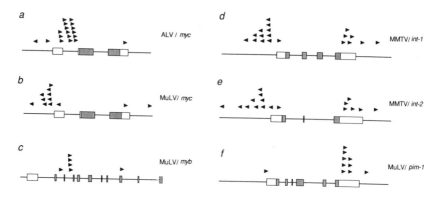

Figure 5.   Distribution of activating proviruses in specific cellular oncogenes. Six specific examples are shown to illustrate the reported locations and orientations of proviruses (depicted by the arrowheads) relative to the coding (shaded boxes) and non-coding exon (open boxes) of particular genes. Thus, each arrowhead represents a separate tumor in which proviral DNAs from the indicated viruses have been mapped relative to the respective genes. Note, however, that the genes are not necessarily drawn to scale and the positions of the various proviruses should not be taken as representative of real tumor cases. The dotted line surrounding the most 3' c-*myb* exon indicates remaining uncertainty over the exact number and location of *myb* exons.

viral oncogenes in naturally occurring malignancies.

The same system continues to provide important pointers to the molecular mechanisms involved. In the first place, the vast majority of ALV proviral insertions in chicken c-*myc* occur either in the intron between exons I and II or within exon I (70,71,77,80,81 and *Figure 5a*). This effectively dissociates the body of the *myc* gene from its normal transcriptional promoters, and may have other deregulatory consequences. Secondly, almost all of these proviruses are in the same transcriptional orientation as the c-*myc* gene (*Figure 5a*), such that expression of *myc* RNA is initiated from the viral promoter (60,75–81). This scenario is commonly referred to as *promoter insertion* and was, in fact, instrumental in the initial discovery since it was argued and later shown that RNA transcripts initiating within the 3' LTR of the provirus may read through into flanking cellular DNA (60,75,78,82). These ALV tumors therefore contain chimeric transcripts comprising both LTR and *myc* sequences (*Figures 6a* and *6b*). Thirdly, all ALV proviruses detected at the c-*myc* locus have sustained major deletions (74,76,78–81, and *Figure 6b*). The reasons for this remain uncertain since the defects do not shown a consistent pattern, ranging from small deletions in the *gag* domain to excisions that leave only a single, residual LTR (79–81).

Detailed surveys of a large number of ALV-induced lymphomas have recently confirmed that the most common deletions remove sequences next to but not including the 5' LTR (80,81). This region of the provirus encompasses a cluster of sequence elements involved in initiation of

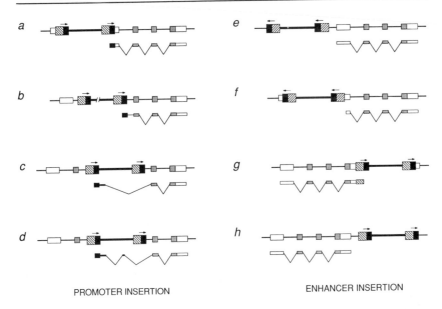

PROMOTER INSERTION   ENHANCER INSERTION

Figure 6. Different modes of *cis*-activation by integrated proviruses. The figure shows various conformations in which an integrated provirus might activate an adjacent cellular oncogene. Although depicted in a generalized format, each reflects a conformation observed in different retrovirus-induced tumors, as discussed in the text. Proviral sequences are represented by either hatched (U3) or solid (U5) boxes, and heavy lines. The prototype oncogene is shown as comprising four exons in which the protein-coding sequences are shaded. The corresponding transcripts are represented by smaller boxes linked by RNA splicing events. The direction of viral transcription is indicated by an arrow above each LTR. In situations (a) through (d), the expression of the target gene is directed by the retroviral promoter, whereas in (e) through (h), expression from the normal cellular promoter is influenced by the presence of the provirus in adjacent DNA.

reverse transcription, packaging of virion RNA, splicing of sub-genomic RNA, and possibly in regulating gene expression and RNA stability (6,7,83,84). One apparent consequence of these deletions is that the function of the 5' LTR may be impaired (81). Hybrid virus-*myc* RNAs therefore initiate within the 3' LTR, and it has been suggested that loss of function of one LTR is selected for since it abrogates promoter occlusion or interference effects (74,75,85). A second consequence is that the provirus cannot sustain expression of the *gag*, *pol*, and *env* gene products. As well as ruling out any direct effects of viral gene products on c-*myc* expression (82), this latter observation has been interpreted as a strategy by which the tumor cells can escape immune surveillance (74,75,81,82). Alternatively, it has been argued that cells that do not express *env* remain susceptible to superinfection since the surface receptors will not be blocked (81). As cancer is viewed as a multistep process, of which activation of c-*myc* may be only one step, then additional insertional mutagenesis events by subsequent viral infection may also contribute to tumor progression (section 8). Finally, one additional consequence of these deletions, though probably fortuitous rather than selected for, is that the typical LTR-*myc* hybrid RNAs cannot be packaged into virus particles. This would account for the low frequency of *myc* transduction events noted in naturally occurring ALV-induced lymphomas (80). Nevertheless, the fact that it can happen is evidence that other types of deletions, involving 3′ rather than 5′ sequences, may also occur (80, 81). The one common denominator is the presence of an ALV LTR within or close to c-*myc*.

Whatever the reasons behind the proviral deletions, the immediate effect of promoter insertion is a substantial elevation, routinely 20- to 100-fold, in the levels of *myc* RNA (60,75,78–81,86). Although these transcripts may be structurally different from the normal c-*myc* mRNAs, there is no evidence for loss or deletion of *myc* sequences, and in no instance analyzed to date has a provirus perturbed the protein coding domains in the second and third exons (*Figure 5a*). It is therefore generally accepted that the oncogenic effects of promoter insertion are attributable to quantitative rather than qualitative changes in the c-*myc* product, though it remains possible that some enhancement of oncogenic function may also accrue through point mutations within the protein itself (79).

However, promoter insertion is not the only way in which elevated levels of *myc* RNA can be achieved in ALV-induced lymphomas, although it does account for over 90 per cent of the examples analyzed in detail. In most of the remaining cases, an ALV provirus can be detected 5′ to the *myc* coding sequences but in the opposite transcriptional orientation (71,78–81 and *Figure 5a*). In such a configuration, it is impossible for c-*myc* transcripts to initiate at the normal viral promoter. Moreover, there are also rare cases in which the ALV provirus is located 3′ of the *myc* coding domains, so that indirect effects on *myc* expression must be invoked (71,78 and *Figure 5a*). The most tenable explanation for these scenarios is that the ALV LTR is influencing the normal c-*myc* promoters via *cis*-acting enhancer effects (44,78, and *Figures 6e – 6h*). Since enhancers can operate over variable distances (11), such a theory would be compatible with less stringent positioning of proviruses relative to the regulatory elements of the gene. However, some of the inverted proviruses are located downstream of exon I, presumably entailing enhancement of transcription from cryptic promoters within the first intron (71). Conversely, as enhancers operate irrespective of orientation, the same type of effects may also prevail when proviruses are in a promoter insertion mode. For example, in some cell lines developed from ALV-induced lymphomas, there is evidence that *myc* transcription initiated at the normal cellular promoter(s) may be enhanced by proviral sequences that are ideally placed for promoter insertion, within the first intron (86).

Other possible explanations must of course be countenanced, the most compelling being the suggestion that sequences within the first untranslated exon of c-*myc*, perhaps in conjunction with other sequence elements, play important roles in regulating *myc* expression (65–71).

Whether such effects operate at the levels of initiation, attenuation or RNA stability (65,68,69,87,88), the disruption of exon I or its divorce from the remainder of the c-*myc* gene could obviously influence the steady-state levels of *myc* mRNA, or indeed its translational efficiency (66,67). These theories were originally invoked in situations where the mammalian c-*myc* locus has been activated by chromosomal translocation rather than proviral insertion (see Chapter 3). However, in both contexts more global explanations have also been proffered, such as alterations in chromatin structure (89). Here the notion would be that insertion of a provirus or disruption of the locus by other means might alter the local nucleosome configurations and affect the accessibility of nearby genes to the transcription machinery. Paradoxically, similar notions have also been considered to explain the clustering of ALV proviruses in the vicinity of DNase I hypersensitive sites (80). In other words, provirus integration may be facilitated at regions of open chromatin, or may induce the formation of open chromatin.

6.1.2 Myc activation by other retroviruses

As well as its inherent relevance to the mechanism of bursal lymphomagenesis in chickens, the realization that c-*myc* could be activated by ALV had important implications, not least of which is the possibility that the cellular gene may be similarly activated by other means. Among these, the chromosomal translocations of human Burkitt's lymphomas and mouse plasmacytomas have understandably achieved most prominence, but for the present purposes, c-*myc* activation in other retrovirus-induced tumors is more relevant.

The initial description of promoter insertion by ALV was quickly followed by similar findings with another type of avian retrovirus, chicken syncytial virus (CSV, reference 90). Although CSV is a member of the reticuloendotheliosis virus (REV) family, which are completely unrelated to the ALVs, the mode of activation of chicken c-*myc* appears to follow the same principles; the predominant situation is promoter insertion by proviruses that have sustained deletions near the 5' LTR (91,92). However, in some of these tumors, it is apparent that the provirally activated *myc* allele has also undergone amplification, potentially adding to the enhancement of expression conferred by virus insertion (90).

With the mammalian c-*myc* gene, on the other hand, a different picture has emerged. Infection of mice or rats by murine leukemia viruses (MuLV), such as the Moloney and AKR strains, can result in T-cell lymphomas after variable latencies (24,29). In some but not all instances, the virus first undergoes recombination with *env* sequences from endogenous proviral elements to generate so-called MCF (mink cell focus forming) variants (29). The complexities of these systems are beyond the remit of this chapter, but what is important is that in many of these lymphomas the c-*myc* gene is activated by an adjacent MuLV provirus.

Reported frequencies of *myc* involvement range from a few per cent up to 65 per cent, presumably dependent on the strains of mouse and virus, and the stage of the disease (93–97). Nevertheless, the pattern of proviral insertion that prevails is in sharp contrast to that seen in chicken c-*myc*. The vast majority (90 per cent) of these proviruses are clustered in approximately 2 kb of DNA upstream of exon I of c-*myc* and most of them are in the opposite transcriptional orientation to that of the gene (*Figure 5b*). The predominant situation therefore appears to be enhancement of transcription rather than promoter insertion (*Figures 6e* and *6f*), resulting in relatively modest increases, commonly between 5- and 30-fold, in the levels of the normal c-*myc* transcripts (95).

As always, there are no hard and fast rules since around 10 per cent of the proviruses disrupt the first exon and hence lie downstream of the c-*myc* promoters (there are also rare examples of proviruses 3' of the gene) and a minority of the proviruses are in the same transcriptional orientation as the gene (93,95,97, and *Figure 5b*). Whether the latter situations permit downstream promotion is not yet clear but, unlike their ALV counterparts, few of the MuLV proviruses have sustained obvious deletions. Interpretation of these variations on a theme must remain speculative, but they might reflect differences in the potency of ALV and MuLV promoters and enhancers, differences in regulation of the chicken and mammalian c-*myc* genes by the upstream sequences, or differences in the optimal levels of c-*myc* required for tumorigenesis.

Variations of a different kind are observed in T-cell tumors induced by feline leukemia virus (FeLV) where again c-*myc* activation occurs in a small but significant percentage of cases. A common situation here, however, is one in which the *myc* gene is essentially transduced by the FeLV provirus, although not all of these transductions give rise to transmissable viruses (98–100). FeLV activation of *myc* therefore spans the grey area between promoter insertion and transduction and shows examples of both these extremes. One of the prerequisites for transduction is thought to be a proviral integration that permits the synthesis of hybrid RNAs initiated in the 5' LTR and retaining the packaging signals in the *gag* leader. Fusion to the coding exons of *myc* presumably requires the loss of 3' viral sequences but these can be regenerated during the subsequent reverse transcription of packaged, hybrid RNAs (3,4). The diagnostic features of transduction are therefore *myc* sequences, from which introns have been removed, flanked by viral sequences and two LTRs, but the net effect, as before, is an increase in the levels of *myc*-encoding RNA.

6.2 The c-*erbB* gene

Proviral insertions in c-*myc* have predominantly quantitative effects since the virus *per se* does not cause structural alterations in the gene product. That insertional mutagenesis can also have qualitative effects is best exemplified by the proviral activation of another known oncogene, c-*erb*B.

The *erb*B gene first came to light as a transduced v-*onc* in defective avian retroviruses that induce rapid erythroblastosis (72,101) but its fame stems from the realization that the cellular progenitor, c-*erb*B, encodes the cell surface receptor for epidermal growth factor (102 and Chapter 4). The transduced version, v-*erb*B, specifies a truncated form of the receptor that lacks both the external ligand-binding domain at the amino terminus and an important site of auto-phosphorylation at the carboxy terminus (103,104, reviewed in 105). Tumorigenesis is therefore presumed to result from constitutive receptor activity and growth stimulation irrespective of the presence of the growth factor.

The viruses in which transduced *erb*B was identified were again derivatives of standard, replication-competent ALVs (72). Although principally associated with bursal lymphomas, ALVs can also induce other, long latency neoplasms, including erythroblastosis, nephroblastomas, and fibrosarcomas (24,73,101). For reasons that are not yet clear, the spectrum and incidences of these malignancies are influenced by the strain of both ALV and chicken. Significantly, in one particular line (RPRL 15) and its derivatives, infection by the RAV-1 strain of ALV is associated with an incidence of erythroblastosis as high as 70 to 80 per cent (106). Moreover, analyses of the sites of proviral insertion in these malignancies showed that in virtually every case of erythroblastosis, an ALV provirus is present within the c-*erb*B locus, in either of two configurations (106 – 109).

Over half of the tumors contain an apparently intact provirus located upstream of the region of c-*erb*B that is represented in v-*erb*B (107,108). These insertions therefore disrupt the structure of the gene product as well as presumably influencing the levels of expression. Significantly, all of the proviruses are in the same transcriptional orientation as the gene, appropriate for promoter insertion, and elicit novel *erb*B-containing RNAs of the same size despite some variability in the sites of integration. The apparent explanation is that transcripts initiated in the 5′ LTR extend through the 3′ LTR into adjacent *erb*B sequences. RNA splicing then eliminates most of the intervening sequences, linking a splice donor in the viral sequences to the splice acceptor that marks the boundary of the first v-*erb*B-homologous exon (109 and *Figure 6c*). The situation therefore re-creates the amino-terminal truncation of c-*erb*B that is characteristic of the v-*onc*, but not the carboxy-terminal truncation. One curious twist is that the spliced RNAs incorporate part of the *env* region of ALV by joining the normal splice donor at the 5′ end of the genome to the splice acceptor for *env*, then exploiting a second cryptic donor to effect linkage to *erb*B (109 and *Figure 6d*).

The other common situation in these tumors is again the transduction of *erb*B, presumably preceded by proviral insertions as above but resulting from recombination between viral and cellular sequences rather than splicing (108,109a). Indeed, it seems likely that the initial stages of the disease may be elicited by the original proviral insertion, but that more aggressive, polyclonal outgrowths eventually develop as a consequence of transduction (108). Nevertheless, the net effect on c-*erbB* expression is both qualitative and quantitative, consistent with the prevailing theories on oncogenicity by modified growth factor receptors (105).

6.3   The c-*myb* gene

Although the product of the c-*myb* proto-oncogene is a nuclear protein rather than a cell surface receptor (25 and Chapter 4), there are nevertheless some obvious parallels between proviral activation of c-*myb* and c-*erbB*. Both genes contain multiple exons of which only some are transduced in the corresponding v-*oncs*, suggesting that the oncogenic properties of both genes can only be unmasked by truncation (105, 110 – 114). With *myb*, initial identification again relied on analyses of two acutely transforming avian retroviruses that induce tumors of the myeloid lineage (72, 101) and it has been from myeloid malignancies and cell lines, mostly derived from mice infected with the Abelson leukemia virus complex, that the majority of reports on insertional mutagenesis of the c-*myb* locus have emanated (115 – 122). MuLV proviral insertions have occurred predominantly in a region upstream of the exons that are represented in v-*myb*, and in the same transcriptional orientation as the gene (*Figure 5c*). The resultant read-through RNAs are processed so that a cryptic splice donor in *gag* is linked to the splice acceptor of an internal *myb* exon (118,120,121 and *Figure 6c*). Despite the obvious similarities to *erbB*, there have been no reports of *myb* transduction by MuLV. This may in part reflect the frequent deletion of some of the proviral sequences, but the numbers are as yet too small to derive a consistent picture. The one alternative scenario that has been observed is proviral insertion near the 3′ end of the gene, within the penultimate v-*myb*-homologous exon (118,119,121,122 and *Figure 5c*).

Taken together, the obvious parallels between the recombination junctions for the transduction of v-*myb* (110,111) and the alterations imposed by proviral insertion in c-*myb* (116 – 122), suggest that truncation of the *myb* protein at its amino- or carboxy-terminus may be the critical determinant of oncogenicity, rather than significantly elevated levels of expression. However, a recent report in which the chicken c-*myb* locus has been the target for insertional mutagenesis, by an artificially constructed recombinant of two ALVs, questions such a conclusion (123). Here, several of the integrations occur upstream of the presumed initiation codon. Thus, although aberrant *myb* transcripts may be generated, the *myb* protein may not be structurally affected. Curiously, the tumors induced by this recombinant virus arise very rapidly, more reminiscent of an acutely oncogenic agent, and involve lymphoid rather than myeloid lineages. The design of the recombinant in which the *env* gene of ring-neck pheasant virus (RPV) was substituted by that of a different ALV, tempts some speculative analogies to the rapidly oncogenic SFFVs (see sections 4.2 and 7.2.5).

6.4   Other known genes

These examples of known oncogenes are consistent targets of proviral activation and serve to illustrate many of the mechanisms involved. However, there have also been less frequent cases involving other recognizable genes, not all of which have previously earned the accolade of *oncogene*. For example, there is a single case in which the insertion of gibbon ape leukemia virus DNA within the 3' untranslated region of the interleukin-2 (IL2) gene has rendered IL2 expression constitutive, a situation that may well have contributed to oncogenicity in this leukemic T-cell line, through autocrine stimulation (124). Similarly, the mouse myelomonocytic leukemia line WEHI-3B constitutively produces IL3. Here, activation has been traced not to a retrovirus *per se*, but to the genome of an intracisternal A-particle (IAP, reference 125). These entities are close relatives of retroviral proviruses, and other transposable elements, since they are flanked by LTRs that display the expected inverted terminal repeats and contain regulators of gene expression (126,127). There are around 1000 copies of IAP DNA in the haploid genome of *Mus musculus*, a proportion of which is transcriptionally active (128). The resultant RNA becomes packaged into particulate structures, IAPs, that accumulate in the cisternae of the endoplasmic reticulum. Distinction between IAPs and retroviruses therefore hinges on the absence of an extracellular, infectious branch in their life cycle. Such structures may nevertheless have the capacity to transpose to different regions of the genome and indulge in insertional mutagenesis. In WEHI-3B cells an IAP genome has become inserted so that its 5' LTR is adjacent to the promoter of the IL3 gene (125).

An analogous situation occurs in two independent mouse plasmacytoma cell lines where IAP genomes have been shown to disrupt the c-*mos* proto-oncogene (129 – 134). The positions of these insertions within the amino terminus of the c-*mos* coding domain echo the truncation that is characteristic of the transduced v-*mos* gene in Moloney murine sarcoma virus. However, it is not clear whether truncation of the c-*mos* protein is a critical factor, since linking of the normal c-*mos* sequences to an MuLV LTR can also render them oncogenic (135). In most tissues, c-*mos* expression appears to be repressed by *cis*-linked regulatory elements (136), so that even modest levels of synthesis may be sufficient for oncogenesis. Interestingly, the two plasmacytoma cell lines differ in this regard; the line in which the IAP is in the same orientation as c-*mos* is a more active expressor than the line where the IAP is in the inverse orientation (131). In the case of *mos*, the notion that the IAP activation played a significant role in tumorigenesis seems fairly secure since the rearranged *mos* sequences are active in cell transformation assays (129). However, the amplification of specific IAP genes is a relatively common feature of murine plasmacytomas and may not always constitute a common cause (137,138). Thus, the occasional sightings of IAP elements in, for example, the c-*myc* and immunoglobulin genes serve to illustrate their potential for insertional mutagenesis rather than establishing an etiological significance (137,139,140).

Finally, in this section on known genes, there are two isolated reports of insertional activation of *ras* genes. The first of these occurred in an ALV-induced nephroblastoma and serves as a timely reminder of the pleiotropic influences of ALVs, both in terms of disease spectrum and target genes (141). Somewhat surprisingly, there are no indications that the ras^H gene in question was mutated. Since the oncogenic properties of *ras* genes are more commonly attributable to specific point mutations, they seem unlikely targets for insertional activation. Nevertheless, there is also a single example of enhanced ras^K expression resulting from Friend murine leukemia virus (F-MuLV) insertion in an infected bone marrow cell line (142). Such sporadic examples of insertional mutagenesis are likely to become a recurrent theme as probes for more cellular genes become available and the analyses are extended to an increasing catalog of cell lines and tumors (142a).

7. Genes and proviruses in common integration loci

One of the central tenets of insertional mutagenesis is the expectation that tumors which share the same pathology and viral etiology will result from proviral activation of the same oncogenes. In the examples presented thus far, the locations of the activating proviruses were inferred from their proximity to known cellular genes, for which nucleic acid probes were already available. In the examples to be discussed next, the locations of potential oncogenes are inferred from their proximity to integrated proviruses. In essence, however, the two situations are equivalent, since both hinge on the repeated finding of acquired proviruses within the same limited regions of cellular DNA. This section therefore describes how the physical linkage between proviral and cellular DNA can be exploited to retrieve probes specific for the integration region (and hence the gene), and to test for similarly located proviruses in other tumors. Such approaches have their origin in the *transposon tagging* methods used by *Drosophila* geneticists to isolate genes mutated by mobile genetic elements (143), and would undoubtedly have led to the discovery of *myc*, *myb*, and *erbB* had they not been preceded by studies on the acutely transforming retroviruses.

7.1 Mouse mammary tumor virus (MMTV) and the *int* genes

The generalized approach for identifying targets of insertional mutagenesis is perhaps best exemplified by the analyses of sites of proviral integration in tumors induced by MMTV. By the criteria established in section 3, MMTV is a fairly typical example of a non-acutely oncogenic retrovirus, causing adenocarcinomas of the mammary alveoli with an average latency of between 4 and 12 months (24,144). The tumors are also, typically, clonal as judged by the stoichiometry of the proviral–cellular DNA junctions (*Figure 4*) and contain at least one, though more usually several, acquired MMTV proviruses (54–56, reviewed in 145 and 146). Since the prediction is that one of these acquired proviruses should lie close to a critical cellular gene, the strategy is to exploit viral sequence probes to isolate recombinant DNA clones that comprise the junctions of viral and cellular DNA. After propagation and characterization of these cloned junction fragments, it is then possible to isolate pieces of non-repetitive DNA from the cellular sequences adjacent to the provirus (*Figure 7*). These so-called 'flanking sequence probes' are specific for a particular site of proviral integration. Each intact provirus should yield two such probes, from the 5' and 3' junctions repectively, that will identify the same integration site (*Figure 7*). If the same restriction enzyme is used throughout, Southern blots of normal DNA will display a unique restriction fragment that hybridizes to both these probes. With the corresponding tumor DNA, however, each probe should detect two fragments, one reflecting the normal, uninterrupted allele on the unaffected chromosome, the other the novel junction created by proviral insertion (*Figure 7*). In a truly clonal population of tumor cells, the stoichiometry of these two bands should be 1:1, and most MMTV-induced tumors approximate this situation (54–56, 147–149). However, in practice, most MMTV-induced tumors contain multiple proviruses, so that this approach alone will not necessarily identify the culpable provirus or integration site.

In this context, there are two critical questions that can be asked:
(i)   whether the same site is disrupted by proviral insertion in other MMTV-induced tumors, and
(ii) whether the site comprises part of a functional cellular gene.
Of these, the first is conceptually the simpler question since it can be addressed by straightforward Southern blotting of tumor DNA, using the flanking sequence probes. Moreover, by the judicious choice of restriction enzymes, or by 'chromosome walking' techniques, it is possible to survey substantial stretches of cellular DNA on either side of the prototype integration site. When applied to proviruses in MMTV-induced tumors, this approach revealed that there are indeed regions of DNA that are common targets of proviral integration in independent mammary tumors. However, in contrast to the monopoly of insertions in c-*myc* in ALV-induced lymphomas and c-*erb*B in erythroblastosis, the analyses revealed

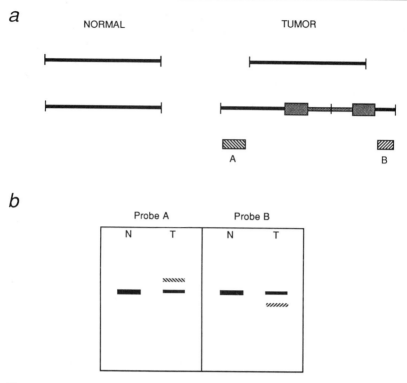

Figure 7. Junction fragment probes for viral integration sites. (a) A unique fragment of cellular DNA is represented (solid line), bounded by restriction enzyme cleavage sites (vertical bars). Two identical copies of this fragment are present in normal diploid cells. In the tumor DNA, however, one copy of this fragment has been disrupted by the insertion of a provirus (shaded boxes). Since the provirus also carries a site for the restriction endonuclease, cleavage of tumor cell DNA generates two novel fragments comprising the junctions between viral and cellular DNA. By using the viral sequences as probes, it is possible to isolate these junctions by recombinant DNA techniques. The cellular DNA sequences either 5' or 3' to the provirus can then be characterized and used as probes (cross-hatched boxes) that are specific for the site of integration. On Southern blots, depicted schematically in panel (b), such probes detect the same restriction fragment in normal DNA (N), in diploid amounts (bold line). Tumor DNA (T) on the other hand will show one normal copy of the integration site from the uninterrupted chromosome, plus the novel 5' or 3' junction fragments from which the probes were derived (hatched lines). In a truly clonal tumor, the stoichiometry of normal and rearranged alleles should be 1:1. Probes isolated in this way from one tumor can be used to screen other tumors for proviral disruption of the same cellular domain.

not one common integration locus but several designated *int*-1, *int*-2 etc (*Table 1* and references 146–151). The frequencies with which these regions are disrupted are variable, apparently depending on the strains of both mouse and virus, but the most consistently involved and extensively studied are the *int*-1 and *int*-2 loci (145–148, 152–155).

Despite the similarities in their names, reflecting the similar ways in which they were indentified, the *int*-1, *int*-2, *int*-3 and *int*-4 loci are completely different and reside on different mouse chromosomes (146,

156 – 158). Nevertheless, it is convenient to consider at least *int*-1 and *int*-2 together, in view of the remarkable parallels in the disposition of the activating proviruses. As presently defined, each of these domains extends for over 30 kb of cellular DNA and it is not yet clear what the maximum limits will be. However, by exploiting probes specific for parts of the viral and cellular sequences, it was possible to determine roughly where and in what orientation each MMTV provirus had integrated in independent tumors (156, 159). For both *int*-1 and *int*-2, the picture that emerged was quite striking in that the proviruses are distributed in two distinct clusters on either side of a stretch of cellular DNA that has remained undisturbed in all the tumors analyzed to date (*Figures 5d* and *5e*). Proviruses within each of the groupings are oriented the same way, but the two clusters are in opposite orientations such that viral transcription is directed outwards from the centre of the domain (*Figures 5d* and *5e*). Based on the analogies with c-*myc*, it therefore seemed likely that this central portion might represent a cellular gene. The proof came when it was shown that tumors in which a provirus had integrated within *int*-1 or *int*-2 contained low levels of RNA derived from the relevant segments of cellular DNA (147,156,159). Significantly, neither region is detectably transcribed in normal mammary tissues, leading to the conclusion that the MMTV provirus in adjacent DNA is activating the expression of a gene, *int*-1 or *int*-2, that is normally silent in mammary cells. The inappropriate expression of these genes may therefore confer some selective advantage to the cell and contribute to neoplastic growth.

Much has happened since the initial indentification of *int*-1 and *int*-2; both genes are now sequenced (160,161), the exon structures have been elucidated (160 – 163), homologs have been detected and cloned from other species (164 – 167), and inroads have been made into understanding the nature and functions of the respective protein products (168 – 170). However, while they represent interesting and novel examples of potential oncogenes, the most illuminating features for the present purposes are their modes of activation. With very few exceptions, the activating MMTV proviruses are intact, so that deletion is not a prerequisite; they are arranged on either side of the transcripion unit, so that initiation of RNA synthesis occurs at the normal cellular promoters; and they are oriented so as to allow the enhancer functions of the viral LTRs most ready access to the promoters of the *int* genes (*Figures 6e – 6h*, reviewed in 145 and 146). As optimum enhancer effects are experienced by proximal promoters, interposition of a strong MMTV promoter between the enhancer and the target gene could obviously deplete these effects.

It should be stressed, however, that this interpretation is merely a rationalization of the data and there is no direct evidence for enhancement. Furthermore, the notion that enhancers operate at a distance has to be taken to considerable extremes since proviruses as much as 10 to 15 kb from the start of the gene are capable of activating transcription.

At the other extreme, activating proviruses are frequently to be found within the untranslated trailer at the 3' end of the gene, such that RNA transcripts terminate at the polyadenylation signal in the viral LTR (155,156,160,161; *Figures 5d, 5e* and *6g*). Indeed, in some instances, a provirus resides only a few basepairs downstream of the termination codon for the predicted protein (160). Moreover, in both *int*-1 and *int*-2 there are rare cases of proviral insertion within the 5' end of the transcribed region in a promoter insertion mode (155,160,161). However, despite these variations on the general theme, there are no documented mammary tumors in which an MMTV provirus has disrupted the protein coding capacities of either *int*-1 or *int*-2. Thus, the signs are that it is the normal products that are involved in oncogenesis, as a result of expression at elevated levels or in an inappropriate setting.

For *int*-3 and *int*-4, the situation is much less clear, as to date there have been too few examples to draw firm conclusions (146,149). Nevertheless, both qualify as common, albeit infrequent, integration regions and both appear to include *bona fide* cellular genes as evidenced by RNA transcripts in the relevant tumors. It is therefore clear that the induction by MMTV of mammary tumors that are indistinguishable by histological criteria can arise through proviral activation of any of a number of discrete cellular genes, and that there may be more *int*-like genes awaiting discovery.

7.2 Common integration sites for MuLV

ALV-activation of c-*myc* set the precedent for insertional mutagenesis of a cellular proto-oncogene and the MMTV-activation of the *int* genes can be regarded as the precedent for transposon tagging approaches. However, the successful elucidation of both systems was aided by the high and consistent frequency with which the respective genes are implicated. Such simple solutions are not always forthcoming. The analysis of leukemias associated with infection by various members of the MuLV family is a case in point. Here the situation is inherently complex since different isolates of MuLV induce different spectra of diseases (24). Different strains of mice also display varied susceptibility to virus infection and tumorigenesis, attributable to a complex set of host genetic factors (24). Moreover, MuLV-induced tumors typically contain multiple acquired proviruses and may not always be strictly monoclonal (97,171). To make matters worse, most laboratory mice also contain large numbers of endogenous MuLV-related proviruses of varied tropism, with which the input virus can recombine (24,29). To overcome some of these complications, some groups have elected to examine tumorigenesis in the rat, which is susceptible to MuLV infection but presents a less complex background (172–176). However, this can create additional problems in equating events observed in the rat with those applicable in the natural host (174).

Large numbers of MuLV-induced tumors of various lineages have now been subjected to transposon tagging approaches. What has emerged is a steadily increasing series of common integration loci, some of which are clearly recognizable as cellular genes but many of which have not yet graduated to this status since evidence for RNA expression is lacking (172–184a,b,c and *Table 1*). The distances over which proviruses can activate transcription might explain some of the negative findings, but a further discomfiting feature is the generally low and sometimes inconsistent frequencies with which the common integration loci surface in different tumors.

The most extensively studied systems have been the T-cell lymphomas in both mice and rats, induced primarily by the Moloney strain of MuLV. As already discussed in section 6, between 10 and 65 per cent of these thymomas have been reported to show proviral activation of the c-*myc* locus (93–97). However, transposon tagging approaches have also identified a second candidate oncogene, namely *pim*-1, among the several common integration loci.

7.2.1 The pim-1 gene

The acronym *pim*-1, standing for *proviral integration site for MuLV*, was given to a common insertion region uncovered in mouse thymomas induced by the Moloney, AKR, and MCF strains of MuLV in several different mouse backgrounds (177). Disruption of *pim*-1 is most evident in lymphomas that develop with a latency of under 6 months, where the frequency is around 50 per cent, but is less prevalent in late-developing tumors. Proviral insertion within the locus results in increased expression of a *pim*-1-specific RNA, compared to the low levels detectable in normal thymus and spleen (184). More recent studies on the architecture of the gene have established that the majority of proviral insertions have occurred within the 3′ untranslated portion of the gene and in the same transcriptional orientation (184,185 and *Figure 5f*), reminiscent of the downstream insertions in *int*-1 and *int*-2. Although enhancement of the normal *pim*-1 promoter is therefore the most obvious interpretation (*Figures 6g* and *6h*), the concentration of proviruses in this limited segment of the transcription unit might also be indicative of effects on RNA stability. In this context, the levels of *pim*-1 RNA achieved when proviruses are in this configuration exceed those seen in cases where the proviruses lie upstream of the gene, in either orientation (184, 185). Continuing the analogy to the *int* genes, the protein coding domain of *pim*-1 is not disrupted in any of these tumors, arguing for a role for the normal product in tumorigenesis (185). However, the predicted product of the gene displays closer analogies to other well known oncogenes since the *pim*-1 sequences show extensive similarities to those of protein kinases (see Chapter 4) particularly those with specificity for serine residues (185).

7.2.2 The lck gene

It was this frequent association between oncogene products and protein kinase activity that initially led to the identification of the *lck* gene, rather than transposon tagging approaches. It had been observed that particular mouse T-cell lines, derived from MuLV-induced thymomas, contain elevated levels of phosphotyrosine when compared to typical hemopoietic cell lines (186). This was attributable to an abnormally high level of a cellular protein kinase for which it was eventually possible to isolate cDNA clones (187,188). However, the justification for including *lck* (or *tck* as it has also been known) in this section is that transcriptional activation of the gene in these cell lines resulted from promoter insertion by integrated MuLV (188). To date, there are only two reported examples of this phenomenon but *lck* has generated considerable interest as a lymphoid-specific gene (189,189a).

7.2.3 The mis-1/pvt locus

It has been my intention in this section to concentrate on provirally activated *genes*, rather than discussing each of the common integration *loci* that are now recognized, as listed in *Table 1*. While these loci hold promise for future understanding, they presently offer few insights into the modes of oncogene activation. However, several features of the *mis-1/pvt* locus warrant making an exception.

This locus was identified by two separate lines of research, the first being the analysis of sites of proviral insertion in MuLV-induced rat thymomas, hence the acronym *mis*-1, for *Moloney integration site* (173,175,190). The alternative acronym *pvt* derives from *plasmacytoma variant translocation*, and was coined to described a region on mouse chromosome 15 that was involved in the less frequent forms of chromosomal translocation characteristic of plasmacytomas (178,190,191). The more common translocations link the c-*myc* gene on chromosome 15 to the immunoglobulin heavy chain gene on chromosome 12, a direct parallel with the typcial t(8;14) translocations in human Burkitt's lymphoma (see Chapter 3). Similar parallels can now be drawn between the t(6;15) translocations in mice that involve *pvt* and the t(2;8) variants of Burkitt's lymphoma that involve the human homolog of *pvt* (192,193).

The fact that *mis*-1 and *pvt* are one and the same (190) indicates that the locus is conserved in mouse, rat, and human DNA and that it may play a central role in lymphomagenesis. However, as there is as yet no evidence for a *pvt*-specific transcript, it is considered likely that these rearrangements may be exerting their effects via activation of c-*myc* (178,190,191). The distance between *pvt* and *myc* is now estimated to be of the order of 100 to 500 kb, yet c-*myc* expression is evidently elevated in plasmacytomas that have alterations in *pvt* (191). The same holds true for insertions at *Mlvi*-1 which also resides on the same region of mouse chromosome 15 (195).

7.2.4 Common integration sites for MuLV in myeloid tumors

Among the other major categories of MuLV-induced tumors that have been analyzed in some detail are those associated with infection by the Friend strain, F-MuLV. This virus was initially recognized as the replication competent helper in the Friend virus complex. The complex induces rapid erythroleukemias in susceptible mice, attributable to the acutely oncogenic component, SFFV, but as with most helper viruses, F-MuLV was itself shown to be oncogenic after long latency (24,30). A wide spectrum of leukemias can result, depending on the strain of mouse, including T- and B-cell leukemias, erythroblastosis and in some cases a high percentage of myeloid leukemias (24). It is probably this latter feature that prompted most interest in insertional mutagenesis by F-MuLV, given the more frequent associations of the Moloney strain and some other MuLVs with lymphoid tumors.

Four common sites of proviral insertion have now been reported (see *Table 1*). The first of these, *fis*-1, is disrupted relatively infrequently, of the order of 5 – 10 per cent or less, in both myeloid and lymphoid neoplasms (179). Proviruses within this locus are clustered in about 1.5 kb of DNA in the same transcriptional orientation, but no *fis*-1-specific transcript has yet been detected. The same features are true for the locus designated *fim*-1, standing for *Friend integration myeloid*, the probe for which was isolated from an F-MuLV-transformed bone marrow cell line (180). However, the same cell line also contained proviruses in two other common integration domains, designated *fim*-2 and *fim*-3, which seem to be more frequently implicated (180,181). In the case of *fim*-2, one of the threads running through this chapter comes full circle. Although identified as an otherwise anonymous locus by transposon tagging, it has recently become apparent that *fim*-2 is not only expressed in tumors, but corresponds to the c-*fms* proto-oncogene. The product of this gene has in turn been identified as the cell surface receptor for the hemopoietic growth factor CSF-1 (196,197). Thus, proviral insertion is viewed as indirectly enhancing the expression of the CSF-1 receptor from its normal promoter and thereby influencing the growth properties of the affected cells.

A second common thread may also apply to *fim*-3 since it has recently been mapped to the same region of mouse chromosome 3 as the *evi*-1 gene (198,199). The latter was first identified as a common integration locus in myeloid tumors arising in AKXD mice (182) but was soon promoted to the status of a gene that encodes a zinc finger protein presumably involved in binding to DNA (200). Although the two loci are over 100 kb apart, it may well be that *evi*-1 is the elusive target gene for insertions at *fim*-3.

7.2.5 Friend virus-induced erythroleukemias

In section 4, and most standard texts on retroviruses, the replication defective component of the Friend virus complex, SFFV, is assigned to a separate category, since it does not conform to the stereotypes of either acutely or non-acutely oncogenic viruses. Thus, although it does not contain a transduced cellular oncogene, the SFFV genome, and in particular the recombinant *env* gene, is evidently responsible for the very rapid induction of erythroleukemia in infected animals (24,29,30). It is however possible to reconcile these apparent anomalies by distinguishing between different stages in the progression of the disease, an early proliferative phase, which is probably attributable to the mosaic *env* protein, followed by a clonal expansion of tumorigenic cells (30). In this latter phase, two separate types of event appear to be implicated. In the first instance, SFFV has been shown to behave as a typical insertional mutagen since around 95 per cent of erythroid tumors in a recent survey were found to have an SFFV provirus within a common 15 kb domain, given the acronym *Spi*-1, for *SFFV proviral integration* (183). All of the proviruses mapped thus far are in the same transcriptional orientation, and the vast majority appear structurally intact. More importantly, proviral insertions within *Spi*-1 are associated with expression of RNA from a presumed *Spi*-1 gene (183).

In contrast, the second common observation is disruption or complete inactivation of a gene, in this case the previously described nuclear oncogene p53 (47,201 – 203). Significantly, some of the perturbations of p53 are caused by the insertion of either SFFV or helper F-MuLV DNA within one or both alleles of the gene (47,203). One possible interpretation is that the normal p53 product is a suppressor of cell growth, now commonly referred to as an 'anti-oncogene', and that there is a positive selection for cells in which its activity is depleted, whether by proviral insertion or some other means (204 and Chapter 5).

7.2.6 Lack of common integration sites for HTLV and BLV

It is clear then that SFFV, and perhaps other rapidly oncogenic variants, such as the recombinant ALV described in section 6.3, can now be regarded as *bona fide* insertional mutagens. Nevertheless, the acutely and non-acutely oncogenic stereotypes still cannot accommodate the HTLV and BLV families (see section 4.2). While tumors associated with these viruses are operationally clonal, as judged by the patterns of integrated proviruses, there are as yet no indications of common integration sites (40 – 42). The only tenous connection to a mutagenic event concerns an HTLV-I positive T-cell leukemia in which an interstitial deletion has occurred adjacent to the *pvt* locus, but there is no formal proof that the deletion was in any way related to HTLV integration (193).

7.2.7 Insertional mutagenesis by hepatitis-B virus

Although this chapter is principally concerned with retroviruses, it is perhaps relevant to consider the prospects for insertional mutagenesis by other viruses. In contrast to the RNA tumor viruses, integration is not an essential step in the replication of DNA viruses. If and when it does occur, in tumors for example, it is seen as an aberrant event whose function is to establish a heritable source of some oncogenic viral gene product rather than perturbation of a cellular oncogene (see Chapter 5). Nevertheless, it is becoming increasingly evident that there are some remarkable parallels between the hepatitis-B family of DNA viruses and the retroviruses. The most striking are the requirement for reverse transcription and some specificity in the viral sequences involved in integration (205). It is therefore intriguing that there have now been several reports of viral insertions at specific loci in hepatocellular carcinomas, including genes such as c-*myc* and a novel retinoic acid receptor gene (206–210). It will be interesting to see how reproducible such observations prove given the enormous health problem posed by liver cancers related to hepatitis-B.

8. Multistep carcinogenesis and dual activation of oncogenes

Among the unexpected observations to emerge from studies on insertional mutagenesis were examples of tumors in which proviruses were located in more than one common integration locus. The identification of *fim*-1, *fim*-2, and *fim*-3 is a case in point since all three were originally detected within the same F-MuLV-transformed bone marrow cell line, although it remains to be seen how reproducible the phenomenon is in primary leukemias (180,181). In other systems it can occur with a remarkably high frequency, perhaps best exemplified by MMTV insertions within both *int*-1 and *int*-2 in up to 50 per cent of the mammary tumors in some mouse strains (155), and the activation of both c-*myc* and *pim*-1 in MuLV-induced thymomas (95,171).

Two possible explanations can be envisaged. The first is simply that these tumors may not be strictly clonal but may comprise mixtures of cells that have sustained different insertional mutations. In some of the MuLV-thymomas this appears to be the case as judged by the stoichiometry of acquired proviral junctions and the segregation of cells with different patterns of proviral insertion after transplantation of the tumor (95,171). However, the same criteria support the alternative explanation in other thymomas and in the MMTV-induced mammary tumors, namely that the tumor is dominated by cells that have sustained two critical insertion mutations (95,155).

On the basis of truly random integration, such double hits would have an extremely low probability of happening by chance, suggesting that there must be a positive selection for cells of this type. Thus, the chance activation of one oncogene may confer a selective advantage that is insufficient to cause frank malignancy, but will result in an expanded population susceptible to further mutation. Each mutational event can therefore be viewed as increasing the probability that neoplastic growth will ensue. Such a stochastic accumulation of mutations is entirely in keeping with accepted notions of multistep carcinogenesis, and the manipulation of retroviral systems with distinct stages of insertional mutagenesis holds promise for future understanding of such processes.

Obviously, not all of the multiple steps invoked in tumor progression need involve a viral agent. Indeed, it is generally accepted that the long latency characteristic of many retrovirally induced tumors must in part reflect the time required for additional events to occur. In examples discussed above, the frequent deletion of ALV proviruses in c-*myc* may represent one such second event. Thus, while proviral insertion may provide an initial growth advantage, this may be followed by clonal selection of a cell in which *myc* levels have become optimized or virus expression abrogated by the deletion of parts of the provirus. Similarly, the different stages of SFFV-induced disease envisage a hyperplastic phase attributable to virus expression, an insertional mutation of the *Spi*-1 gene, and possibly a third phase involving rearrangements of the p53 oncogene.

9. Summary and prospects

One of the major aspirations of this chapter has been to provide some unifying themes for the ways in which provirus integration can activate the latent properties of cellular oncogenes. As most of these themes rely on rationalizing the experimental observations there are bound to be flaws and gaps in understanding, and many of the experimental realities continue to defy interpretation. One of the most glaring is how to reconcile random integration and chance mutation with the manifestly non-random association between proviruses and their oncogene targets. Thus, genes such as c-*myc* can be activated in several different ways by several different viruses (ALV, REV, MuLV, and FeLV for example) and many retroviruses appear capable of activating different genes by different mechanisms (for example ALV and *myc*, *myb*, *erb*B, etc). Why then are there no cases of ALV activation of *pim*-1, *int*-2, or c-*fms*, and why does MMTV not act via *myc* when it can be shown that artificial linkage of MMTV and *myc* can lead to mammary tumorigenesis (211)?

Speculative answers can be sought at two different levels. In the first place, there may be constraints as to which viruses can influence expression of which genes. Not only are retroviruses restricted in the tissues they can infect, but it is quite feasible that in situations encountered by retroviruses during their natural life cycles, not all regions of chromatin may be equally accessible to integration. In more specific terms one could envisage matching of the viral enhancer to particular regulatory features found on only certain genes, for example through interactions with transcription factors or elements of the nuclear matrix.

A second level of constraint may be that only cells of a particular lineage or differentiation status can respond to a particular oncogene product in a way that promotes neoplasia. For example, there would be little benefit to a cell that is induced to express a growth factor for which it has no receptors. Phenotypic selection could be imposed and effectively sharpened at each step in multistage carcinogenesis leading to an apparently stringent association between retrovirus, oncogene, and tumor type.

A similar type of selection may also influence the mechanisms by which integrated retroviruses are seen to activate their target genes. There may, for instance, be an optimum level of expression that is encountered when proviruses lie in a particular configuration relative to the coding sequences. Too much of a gene product could prove deleterious and, unless the target is a suppressor or anti-oncogene, too little is unlikely to have a dominant phenotypic effect. To date, the inactivation of p53 is the only real example for such a situation, although it could be argued that truncation of a gene product is a close approximation. Thus, proviral insertions in c-erbB clearly block its normal function, but the net effect is to change the specificity of the biochemical consequences rather than prevent them altogether. In this regard, we remain too ignorant of the functions and properties of oncogene products to appreciate the nuances of structural alterations. For example, c-erbB and c-fms both encode growth factor receptors but are activated in completely different ways, truncation on the one hand and increased expression of the normal product on the other; similarly for the two nuclear proteins encoded by c-myb and c-myc.

Whatever the status of the target gene, the levels of expression attained after provirus insertion are almost certainly constrained by inherent features of the gene coupled to the integrity, location, and functional properties of the activating virus. Thus the various modes of activation of c-myc in chickens and mice are likely to reflect differences in the regulatory features in the two species, differences in the virus LTRs, and perhaps different requirements for transformation of B- and T-lymphocytes.

In truth, therefore, we still have a lot to learn about insertional mutagenesis and can only take comfort in knowing that tumorigenesis has selected and conveniently presented to us the features that are most important. All we have to do is understand what they are telling us. The information is certainly there for the taking, as are a rich seam of potential oncogenes that for various reasons have not or cannot be candidates for transduction by the transforming retroviruses.

10. References

1. Cairns,J. (1981) The origins of human cancers. *Nature*, **289**, 353–357.
2. Bishop,J.M. (1987) The molecular genetics of cancer. *Science*, **235**, 305–311.
3. Bishop,J.M. (1983) Cellular oncogenes and retroviruses. *Ann. Rev. Biochem.*, **52**, 301–354.
4. Varmus,H.E. (1984) The molecular genetics of cellular oncogenes. *Ann. Rev. Genet.*, **18**, 553–612.
5. Duesberg,P.H. (1985) Activated proto-onc genes: Sufficient or necessary for cancer? *Science*, **228**, 669–677.
6. Hughes,S.H. (1983) Synthesis, integration, and transcription of the retroviral provirus. *Curr. Top. Microbiol. Immunol.*, **103**, 23–49.
7. Varmus,H. and Swanstrom,R. (1984) Replication of retroviruses. In *RNA tumor viruses*. Weiss,R., Teich,N., Varmus,H. and Coffin,J. (eds.), Cold Spring Harbor Laboratory Press, New York, pp. 369–512, and pp. 75–134 (Supplement).
8. Varmus,H.E. (1983) Retroviruses. In *Mobile genetic elements*, Shapiro,J. (ed.), Academic Press, New York, pp. 411–503.
9. Panganiban,A.T. (1985) Retroviral DNA integration. *Cell*, **42**, 5–6.
10. Brown,P.O., Bowerman,B., Varmus,H.E. and Bishop,J.M. (1987) Correct integration of retroviral DNA in vitro. *Cell*, **49**, 347–356.
11. Khoury,G. and Gruss,P. (1983) Enhancer elements. *Cell*, **33**, 313–314
12. Chatis,P.A., Holland,C.A., Silver,J.E., Frederickson,T.N., Hopkins,N. and Hartley,J.W. (1984) A 3′ end fragment encompassing the transcriptional enhancers of nondefective Friend virus confers erythroleukemogenicity on Moloney leukemia virus. *J. Virol.*, **52**, 248–254.
13. Des Groseillers,L. and Jolicoeur,P. (1984) The tandem direct repeats within the long terminal repeat of murine leukemia viruses are the primary determinant of their leukemogenic potential. *J. Virol.*, **52**, 945–952.
14. Lenz,J., Celander,C., Crowther,R.L., Patarca,R., Perkins,D.W. and Haseltine,W.A. (1984) Determination of the leukaemogenicity of a murine retrovirus by sequences within the long terminal repeat. *Nature*, **308**, 467–470.
15. Ishimoto,A., Adachi,A., Sakai,K. and Matsuyama,M. (1985) Long terminal repeat of Friend-MCF virus contains the sequences responsible for erythroid leukemia. *Virology*, **141**, 30–42.
16. Yoshimura,F.K., Davison,B. and Chaffin,K. (1985) Murine leukemia virus long terminal repeat sequences can enhance gene activity in a cell-type-specific manner. *Mol. Cell. Biol.*, **5**, 2832–2835.
17. Short,M.K., Okenquist,S.A. and Lenz,J. (1987) Correlation of leukemogenic potential of murine retroviruses with transcriptional tissue preference of the viral long terminal repeats. *J. Virol.*, **61**, 1067–1072.
18. Dickson,C., Eisenmann,R., Fan,H., Hunter,E. and Teich,N. (1984) Protein biosynthesis and assembly. In *RNA tumor viruses*. Weiss,R., Teich,N., Varmus,H. and Coffin,J. (eds.), Cold Spring Harbor Laboratory Press, New York, pp. 513–648 and pp. 135–145 (Supplement).
19. Seiki,M., Hattori,S., Hirayama,Y. and Yoshida,M. (1983) Human adult T-cell leukemia virus: complete nucleotide sequence of the provirus genome integrated in leukemia cell DNA. *Proc. Natl. Acad. Sci. USA*, **80**, 3618–3622.
20. Haseltine,W.A., Sodroski,J., Patarca,R., Briggs,D., Perkins,D. and Wong-Staal,F. (1984) Structure of 3′ terminal region of type II human T-lymphotropic virus: evidence for new coding region. *Science*, **225**, 419–421.
21. Sagata,N., Yasunaga,T., Tsuzuku-Kawamura,J., Ohishi,K., Ogawa,Y. and Ikawa,Y. (1985) Complete nucleotide sequence of the genome of bovine leukemia virus: its evolutionary relationship to other retroviruses. *Proc. Natl. Acad. Sci. USA*, **82**, 677–681.
22. Shimotohno,K., Takahashi,Y., Shimizu,N., Gojobori,T., Golde,D.W., Chen,I.S.Y., Miwa,M. and Sugimura,T. (1985) Complete nucleotide sequence of an infectious clone of human T-cell leukemia virus type II: an open reading frame for the protease gene. *Proc. Natl. Acad. Sci. USA*, **82**, 3101–3105.
23. Wong-Staal,F. and Gallo,R.C. (1985) Human T-lymphotropic retroviruses. *Nature*, **317**, 395–403.

24. Teich,N., Wyke,J., Mak,T., Bernstein,A. and Hardy,W. (1984) Pathogenesis of retrovirus-induced disease. In *RNA tumor viruses,* Weiss,R., Teich,N., Varmus,H. and Coffin,J. (eds.), Cold Spring Harbor Laboratory Press, New York, pp. 785–998, and pp. 188–248 (Supplement).
25. Bishop,J.M. (1985) Viral oncogenes. *Cell,* **42**, 23–38.
26. Martin,G.S. (1970) Rous sarcoma virus: a function required for the maintenance of the transformed state. *Nature,* **227**, 1021–1023.
27. Stehelin,D., Varmus,H.E., Bishop,J.M. and Vogt,P.K. (1976) DNA related to the transforming gene(s) of avian sarcoma viruses is present in normal avian DNA. *Nature,* **260**, 170–173.
28. Spector,D.H., Smith,K., Padgett,T., McCombe,P., Roulland-Dussoix,D., Moscovici,C., Varmus,H.E. and Bishop,J.M. (1978) Uninfected avian cells contain RNA related to the transforming gene of avian sarcoma viruses. *Cell,* **13**, 371–379.
29. Famulari,N. (1983) Murine leukemia viruses with recombinant *env* genes: a discussion of their role in leukemogenesis. *Curr. Top. Microbiol. Immunol.,* **103**, 75–108.
30. Ruscetti,S. and Wolff,L. (1984) Spleen focus-forming virus: relationships of an altered envelope gene to the development of a rapid erythroleukemia. *Curr. Top. Microbiol. Immunol.,* **112**, 21–44.
31. Rosen,C.A., Sodroski,J.G., Kettman,R., Burny,A. and Haseltine,W.A. (1984) *Trans*-activation of the bovine leukemia virus long terminal repeat in BLV-infected cells. *Science,* **227**, 320–322.
32. Sodroski,J.G., Rosen,C.A. and Haseltine,W.A. (1984) *Trans*-acting transcriptional activation of the long terminal repeat of human T lymphotropic viruses in infected cells. *Science,* **225**, 381–385.
33. Chen,I.S.Y., Slamon,D.J., Rosenblatt,J.D., Shah,N.P., Quan,S.G. and Wachsman,W. (1985) The X gene is essential for HTLV replication. *Science,* **229**, 54–58.
34. Felber,B.K., Paskalis,H., Kleinman-Ewing,C. and Pavlakis,.G.N. (1985) The pX protein of HTLV-I is a transcriptional activator of its long terminal repeats. *Science,* **229**, 675–679.
35. Seiki,M., Inoue,J., Takeda,T. and Yoshida,M. (1986) Direct evidence that p40^x of human T-cell leukemia virus type I is a *trans*-acting transcription activator. *EMBO J.,* **5**, 561–565.
36. Greene,W.C., Leonard,W.J., Wano,Y., Svetlik,P.B., Peffer,N.J., Sodroski,J.G., Rosen,C.A., Goh,W.C. and Haseltine,W.A. (1986) *Trans*-activator gene of HTLV-II induces IL-2 receptor and IL-2 cellular gene expression. *Science,* **232**, 877–880.
37. Inoue,J., Seiki,M., Taniguchi,T., Tsuru,S. and Yoshida,M. (1986) Induction of interleukin 2 receptor gene expression by p40^x encoded by human T-cell leukemia virus type 1. *EMBO J.,* **5**, 2883–2888.
38. Maruyama,M., Shibuya,H., Harada,H., Hatakeyama,M., Seiki,M., Fujita,T., Inoue,J., Yoshida,M. and Taniguchi,T. (1987) Evidence for aberrant activation of the interleukin-2 autocrine loop by HTLV-1-encoded p40^x and T3/Ti complex triggering. *Cell,* **48**, 343–350.
39. Nerenberg,M., Hinrichs,S.H., Reynolds,R.K., Khoury,G. and Jay,G. (1987) The *tat* gene of human T-lymphotropic virus type 1 induces mesenchymal tumors in transgenic mice. *Science,* **237**, 1324–1329.
40. Kettmann,T., Deschamps,J., Conez,D., Claustriaux,J.J., Palm,R. and Burny,A. (1983) Chromosome integration domain for bovine leukemia provirus in tumors. *J. Virol.,* **47**, 146–150.
41. Yoshida,M., Seiki,M., Yamaguchi,K. and Takatsuki,K. (1984) Monoclonal integration of human T-cell leukemia provirus in all primary tumors of adult T-cell leukemia suggests causative role of human T-cell leukemia virus in the disease. *Proc. Natl. Acad. Sci. USA,* **81**, 2534–2537.
42. Seiki,M., Eddy,R., Shows,T.B. and Yoshida,M. (1984) Nonspecific integration of the HTLV provirus genome into adult T-cell leukaemic cells. *Nature,* **309**, 640–642.
43. Varmus,H.E. (1982) Recent evidence for oncogenesis by insertion mutagenesis and gene activation. *Cancer Surveys,* **1**, 309–320.
44. Nusse,R. (1986) The activation of cellular oncogenes by retroviral insertion. *Trends in Genetics,* **2**, 244–247.
45. Varmus,H.E., Quintrell,N. and Ortiz,S. (1981) Retroviruses as mutagens: Insertion

and excision of a nontransforming provirus alter expression of a resident transforming provirus. *Cell*, **25**, 23–36.

46. King,W., Patel,M.D., Lobel,L.I., Goff,S.P., Nguyen-Huu,M.C. (1985) Insertion mutagenesis of embryonal carcinoma cells by retroviruses. *Science*, **228**, 554–558.

47. Hicks,G.G. and Mowat,M. (1988) Integration of Friend murine leukemia virus into both alleles of the p53 oncogene in an erythroleukemic cell line. *J. Virol.*, **62**, 4752–4755.

48. Jenkins,N.A., Copeland,N.G., Taylor,B.A. and Lee,B.K. (1981) Dilute (d) coat colour mutation of DBA/2J is associated with the site of integration of an ecotropic MuLV genome. *Nature*, **293**, 370–374.

49. Copeland,N.G., Jenkins,N.A. and Lee,B.K. (1983) Association of the lethal yellow (A^y) coat colour mutation with an ecotropic murine leukemia virus genome. *Proc. Natl. Acad. Sci. USA*, **80**, 247–249.

50. Jaenisch,R., Harbers,K., Schnieke,A., Lohler,J., Chumakov,I., Jahner,D., Grotkopp,D. and Hoffmann,E. (1983) Germline integration of Moloney murine leukemia virus at the Mov13 locus leads to recessive lethal mutation and early embryonic death. *Cell*, **32**, 209–216.

51. Harbers,K., Kuehn,M., Delius,H. and Jaenisch,R. (1984) Insertion of retrovirus into the first intron of α1(I) collagen gene leads to embryonic lethal mutation in mice. *Proc. Natl. Acad. Sci. USA*, **81**, 1504–1508.

52. Steffen,D. and Weinberg,R.A (1978) The integrated genome of murine leukemia virus. *Cell*, **15**, 1003–1010.

53. Canaani,E. and Aaronson,S.A. (1979) Restriction enzyme analysis of mouse cellular type C viral DNA: emergence of new viral sequences in spontaneous AKR/J lymphomas. *Proc. Natl. Acad. Sci. USA*, **76**, 1677–1681.

54. Cohen,J.C., Shank,P.R., Morris,V.L., Cardiff,R. and Varmus,H.E. (1979) Integration of the DNA of mouse mammary tumor virus in virus-infected normal and neoplastic tissue of the mouse. *Cell*, **16**, 333–345.

55. Cohen,J.C. and Varmus,H.E. (1980) Proviruses of mouse mammary tumor virus in normal and neoplastic tissues from GR and C3Hf mouse strains. *J. Virol.*, **35**, 298–305.

56. Groner,B. and Hynes,N.E. (1980) Number and location of mouse mammary tumor virus proviral DNA in mouse DNA of normal tissue and of mammary tumors. *J. Virol.*, **33**, 1013–1025.

57. Jahner,D., Stuhlmann,H. and Jaenisch,R. (1980) Conformation of free and of integrated Moloney leukemia virus proviral DNA in preleukemic and leukemic BALB/Mo mice. *Virology*, **101**, 111–123.

58. Neiman,P., Payne,L.N. and Weiss,R.A. (1980) Viral DNA in bursal lymphomas induced by avian leukosis viruses. *J. Virol.*, **34**, 178–186.

59. Van der Putten,H., Terwindt,E., Berns,A. and Jaenisch,R. (1979) The integration sites of endogenous and exogenous Moloney murine leukemia virus. *Cell*, **18**, 109–116.

60. Hayward,W.S., Neel,B.G. and Astrin,S.M. (1981) Activation of a cellular *onc* gene by promoter insertion in ALV-induced lymphoid leukosis. *Nature*, **290**, 475–480.

61. Battey,J., Moulding,C., Taub,R., Murphy,W., Stewart,T., Potter,H., Lenoir,G. and Leder,P. (1983) The human c-*myc* oncogene: structural consequences of translocation into the IgH locus in Burkitt lymphoma. *Cell*, **34**, 779–787.

62. Bernard,O., Cory,S., Gerondakis,S., Webb,E. and Adams,J.M. (1983) Sequence of the murine and human cellular *myc* oncogenes and two modes of *myc* transcription resulting from chromosome translocation in B lymphoid tumors. *EMBO J.*, **2**, 2375–2383.

63. Watson,D.K., Reddy,E.P., Duesberg,P.H. and Papas,T.S. (1983) Nucleotide sequence analysis of the chicken c-*myc* gene reveals homologous and unique coding regions by comparison with the transforming gene of avian myelocytomatosis virus MC29, delta-gag-*myc*. *Proc. Natl. Acad. Sci. USA*, **80**, 2146–2149.

64. Watt,R., Stanton,L.W., Marcu,K.B., Gallo,R.C., Croce,C.M. and Rovera,G. (1983) Nucleotide sequence of cloned cDNA of human c-*myc* oncogene. *Nature*, **303**, 725–728.

65. Leder,P., Battey,J., Lenoir,G., Moulding,C., Murphy,W., Potter,H., Stewart,T. and Taub,R. (1983) Translocations among antibody genes in human cancer. *Science*, **222**, 765–771.

66. Saito,H., Hayday,A.C., Wiman,K., Hayward,W.S. and Tonegawa,S. (1983) Activation

of the c-*myc* gene by translocation: a model for translational control. *Proc. Natl. Acad. Sci. USA*, **80**, 7476–7480.

67. Darveau,A., Pelletier,J. and Sonenberg,N. (1985) Differential efficiencies of *in vitro* translation of mouse c-*myc* transcripts differing in the 5′ untranslated region. *Proc. Natl. Acad. Sci. USA*, **82**, 2315–2319.

68. Eick,D., Piechaczyk,M., Henglein,B., Blanchard,J.-M., Traub,B., Kofler,E., Wiest,S., Lenoir,G.M and Bornkamm,G.W. (1985) Aberrant c-*myc* RNAs of Burkitt's lymphoma cells have longer half-lives. *EMBO J.*, **4**, 3717–3725.

69. Rabbits,P.H., Forster,A., Stinson,M.A. and Rabbitts,T.H. (1985) Truncation of exon 1 from the c-*myc* gene results in prolonged c-*myc* stability. *EMBO J.*, **4**, 3727–3733.

70. Shih,C.-K., Linial,M., Goodenow,M.M. and Hayward,W.S. (1984) Nucleotide sequence 5′ of the chicken c-*myc* coding region: localization of a noncoding exon that is absent from *myc* transcripts in most avian leukosis virus-induced lymphomas. *Proc. Natl. Acad. Sci. USA*, **81**, 4697–4701.

71. Nottenburg,C. and Varmus,H.E. (1986) Features of the chicken c-*myc* gene that influence the structure of c-*myc* RNA in normal cells and bursal lymphomas. *Mol. Cell. Biol.*, **6**, 2800–2806.

72. Graf,T. and Stehelin,D. (1982) Avian leukemia viruses. Oncogenes and genome structure. *Biochem. Biophys. Acta*, **651**, 245–271.

73. Crittenden,L.B. (1980) New hypotheses for viral induction of lymphoid leukosis in chickens. In *Viruses in Naturally Occurring Cancers*. Essex,M., Todaro,G. and zur Hausen,H. (eds.), Cold Spring Harbor Laboratory Press, New York, pp. 529–541.

74. Fung,Y.-K.T., Fadly,A.M., Crittenden,L.B. and Kung,H.-J. (1981) On the mechanism of retrovirus-induced avian lymphoid leukosis: Deletion and integration of the provirus. *Proc. Natl. Acad. Sci. USA*, **78**, 3418–3422.

75. Neel,B.G. and Hayward,W.S. (1981) Avian leukosis virus-induced tumors have common proviral integration sites and synthesize discrete new RNAs: oncogenesis by promoter insertion. *Cell*, **23**, 323–334.

76. Fung,Y.-K., Crittenden,L.B., Kung,H.-J. (1982) Orientation and position of avian leukosis virus DNA relative to the cellular oncogene c-*myc* in B-lymphoma tumors of highly susceptible $151_5 \times 7_2$ chickens. *J. Virol.*, **44**, 742–746.

77. Neel,B.G., Gasic,G.P., Rogler,C.E., Skalka,A.M., Fu,G., Hishinuma,F., Papas,T., Astrin,S.M. and Hayward,W.S. (1982) Molecular analysis of the c-*myc* locus in normal tissue and in avian leukosis virus-induced lymphomas. *J. Virol.*, **44**, 158–166.

78. Payne,G.S., Bishop,J.M. and Varmus,H.E. (1982) Multiple arrangements of viral DNA and an activated host oncogene in bursal lymphomas. *Nature*, **295**, 209–214.

79. Westaway,D., Payne,G. and Varmus,H.E. (1984) Proviral deletions and oncogene base-substitutions in insertionally mutagenized c-*myc* alleles may contribute to the progression of avian bursal tumors. *Proc. Natl. Acad. Sci. USA*, **81**, 843–84.

80. Robinson,H.L. and Gagnon,G.C. (1986) Patterns of proviral insertion and deletion in avian leukosis virus-induced lymphomas, *J. Virol.*, **57**, 28–36.

81. Goodenow,M.M. and Hayward,W.S. (1987) 5′ long terminal repeats of *myc*-associated proviruses appear structurally intact but are functionally impaired in tumors induced by avian leukosis viruses. *J. Virol.*, **61**, 2489–2498.

82. Payne,G.S., Courtneidge,S.A., Crittenden,L.B., Fadly,A.M., Bishop,J.M. and Varmus,H.E. (1981) Analysis of avian leukosis virus DNA and RNA in bursal tumors: viral gene expression is not required for maintenance of the tumor state. *Cell*, **23**, 311–322.

83. Broome,S. and Gilbert,W. (1985) Rous sarcoma virus encodes a transcriptional activator. *Cell*, **40**, 537–546.

84. Arrigo,S., Yun,M. and Beemon,K. (1987) *cis*-Acting elements within the *gag* genes of avian retroviruses. *Mol. Cell. Biol.*, **7**, 388–397.

85. Cullen,B.R., Lomedico,P.T. and Ju,G. (1984) Transcriptional interference in avian retroviruses - implications for the promoter insertion model of leukaemogenesis. *Nature*, **307**, 241–245.

86. Linial,M. and Groudine,M. (1985) Transcription of three c-*myc* exons is enhanced in chicken bursal lymphoma cell lines. *Proc. Natl. Acad. Sci. USA*, **82**, 53–57.

87. Nepveu,A. and Marcu,K.B. (1986) Intragenic pausing and anti-sense transcription within the murine c-*myc* locus. *EMBO J.*, **5**, 2859–2865.

88. Bentley,D.L. and Groudine,M. (1986) A block to elongation is largely responsible for decreased transcription of c-*myc* in differentiated HL60 cells. *Nature,* **321**, 702–706.
89. Schubach,W. and Groudine,M. (1984) Alteration of c-*myc* chromatin structure by avian leukosis virus integration. *Nature,* **307**, 702–708.
90. Noori-Daloii,M.R., Swift,R.A. and Kung,H.-J. (1981) Specific integration of REV proviruses in avian bursal lymphomas. *Nature,* **294**, 574–576.
91. Swift,R.A., Shaller,E., Witter,R.L., Kung,H.-J., (1985) Insertional activation of c-*myc* by reticuloendotheliosis virus in chicken B lymphoma: Nonrandom distribution and orientation of the proviruses. *J. Virol.,* **54**, 869–872.
92. Swift,R.A., Boerkoel,C., Ridgway,A., Fujita,D.J., Dodgson,J.B. and Kung-H.-J. (1987) B-lymphoma induction by reticuloendotheliosis virus: Characterization of a mutated chicken syncytial virus provirus involved in c-*myc* activation. *J. Virol.,* **61**, 2084–2090.
93. Corcoran,L.M., Adams,J.M., Dunn,A.R. and Cory,S. (1984) Murine T lymphomas in which the cellular *myc* oncogene has been activated by retroviral insertion. *Cell,* **37**, 113–122.
94. Li,Y., Holland,C.A., Hartley,J.W. and Hopkins,N. (1984) Viral integration near c-*myc* in 10–20% of MCF 247-induced AKR lymphomas. *Proc. Natl. Acad. Sci. USA,* **81**, 6808–6811.
95. Selten,G., Cuypers,H.T., Zijlstra,M., Melief,C. and Berns,A. (1984) Involvement of c-*myc* in MuLV-induced T cell lymphomas in mice: frequency and mechanism of activation. *EMBO J.,* **3**, 3215–3222.
96. Steffen,D. (1984) Proviruses are adjacent to c-*myc* in some murine leukemia virus-induced lymphomas. *Proc. Natl. Acad. Sci. USA,* **81**, 2097–2101.
97. O'Donnell,P.V., Fleissner,E., Lonial,H., Koehne,C.F. and Reicin,A. (1985) Early clonality and a high-frequency proviral integration into the c-*myc* locus in AKR leukemias. *J. Virol.,* **55**, 500–503.
98. Levy,L.S., Gardner,M.B. and Casey,J.W. (1984) Isolation of a feline leukaemia provirus containing the oncogene *myc* from a feline lymphosarcoma. *Nature,* **308**, 853–856.
99. Mullins,J.I., Brody,D.S., Binari,R.C. and Cotter,S.M. (1984) Viral transduction of c-*myc* gene in naturally occurring feline leukaemias. *Nature,* **308**, 856–858.
100. Neil,J.C., Hughes,D., McFarlane,R., Wilkie,N.M., Onions,D.E., Lees,G. and Jarrett,O. (1984) Transduction and rearrangement of the *myc* gene by feline leukaemia virus in naturally occurring T-cell leukaemias. *Nature,* **308**, 814–820.
101. Graf,T. and Beug,H. (1978) Avian leukemia viruses. Interaction with their target cells in vivo and in vitro. *Biochim. Biophys. Acta,* **516**, 269–299.
102. Downward,J., Yarden,Y., Mayes,E., Scrace,G., Totty,N., Stockwell,P., Ullrich,A., Schlessinger,J. and Waterfield,M.D. (1984) Close similarity of epidermal growth factor receptor and v-*erb-B* oncogene protein sequences. *Nature,* **307**, 521–527.
103. Ullrich,A., Coussens,L., Hayflick,J.S., Dull,T.J., Gray,A., Tam,A.W., Lee,J., Yarden,Y., Libermann,T.A., Schlessinger,J., Downward,J., Mayes,E.L., Whittle,N., Waterfield,M.D. and Seeburg,P.H. (1984) Human epidermal growth factor receptor cDNA sequence and aberrant expression of the amplified gene in A431 epidermoid carcinoma cells. *Nature,* **309**, 418–425.
104. Downward,J., Parker,P. and Waterfield,M.D. (1984) Autophosphorylation sites on the epidermal growth factor receptor. *Nature,* **311**, 483–485.
105. Martin,G.S. (1986) The *erb* B gene and the EGF receptor. *Cancer Surveys,* **5**, 199–219.
106. Fung,Y.-K.T., Lewis,W.G., Crittenden,L.B. and Kung,H.-J. (1983) Activation of the cellular oncogene c-*erb*B by LTR insertion: molecular basis for induction of erythroblastosis by avian leukosis virus. *Cell,* **33**, 357–368.
107. Raines,M.A., Lewis,W.G., Crittenden,L.B. and Kung,H.-J., (1985) c-*erb*B activation in avian leukosis virus-induced erythroblastosis: clustered integration sites and the arrangement of provirus in the c-*erb*B alleles. *Proc. Natl. Acad. Sci. USA,* **82**, 2287–2291.
108. Miles,B.D. and Robinson,H.L. (1985) High-frequency transduction of c-*erb*B in avian leukosis virus-induced erythroblastosis. *J. Virol.,* **54**, 295–303.
109. Nilsen,T.W., Maroney,P.A., Goodwin,R.G., Rottman,F.M., Crittenden,L.B., Raines,M.A. and Kung,H.-J. (1985) c-*erb*B activation in ALV-induced erythroblastosis: novel RNA processing and promoter insertion result in expression of an amino-truncated EGF receptor. *Cell,* **41**, 719–726.

109a. Raines,M.A., Maihle,N.J., Moscovici,C., Crittenden,L. and King,H.-J. (1988) Mechanism of c-*erbB* transduction: newly released transducing viruses retain poly(A) tracts of *erbB* transcripts and encode c-terminally intact *erbB* proteins. *J. Virol.*, **62**, 2437–2443.

110. Klempnauer,K.-H., Gonda,T.J. and Bishop,J.M. (1982) Nucleotide sequence of the retroviral leukemia gene *v-myb* and its cellular progenitor *c-myb*: the architecture of a transduced oncogene. *Cell*, **31**, 453–463.

111. Perbal,B., Cline,J.M., Hillyard,R.L. and Baluda,M.A. (1983) Organization of chicken DNA sequences homologous to the transforming gene of avian myeloblastosis virus. *J. Virol.*, **45**, 925–940.

112. Castle,S. and Sheiness,D. (1985) Structural organization of the mouse proto-myb gene. *Biochem. Biophys. Res. Commun.*, **132**, 688–695.

113. Gonda,T.J., Gough,N.M., Dunn,A.R. and de Blaquiere,J. (1985) Nucleotide sequence of cDNA clones of the murine *myb* proto-oncogene. *EMBO J.*, **4**, 2003–2008.

114. Rosson,D. and Reddy,E.P. (1986) Nucleotide sequence of chicken c-*myb* complementary DNA and implications for *myb* oncogene activation. *Nature*, **319**, 604–606.

115. Mushinski,J.F., Potter,M., Bauer,S.R. and Reddy,E.P. (1983) DNA rearrangement and altered RNA expression of the c-*myb* oncogene in mouse plasmacytoid lymphosarcomas. *Science*, **220**, 795–798.

116. Shen-Ong,G.L.C., Potter,M., Mushinski,J.F., Lavu,S. and Reddy,E.P. (1984) Activation of the c-*myb* locus by viral insertional mutagenesis in plasmacytoid lymphosarcomas. *Science*, **226**, 1077–1080.

117. Lavu,S. and Reddy,E.P. (1986) Structural organization and nucleotide sequence of mouse c-*myb* oncogene: activation in ABPL tumors is due to viral integration in an intron which results in the deletion of the 5′ coding sequences. *Nucleic Acids Res.*, **14**, 5309–5321.

118. Shen-Ong,G.L.C., Morse,H.C., Potter,M. and Mushinski,J.F. (1986) Two modes of c-*myb* activation in virus-induced mouse myeloid tumors. *Mol. Cell. Biol.*, **6**, 380–392.

119. Weinstein,Y., Ihle,J.N., Lavu,S. and Reddy,E.P. (1986) Truncation of the c-*myb* gene by a retroviral integration in an interleukin 3-dependent myeloid leukemia cell line. *Proc. Natl. Acad. Sci. USA*, **83**, 5010–5014.

120. Gonda,T.J., Cory,S., Sobieszczuk,P., Holtzman,D. and Adams,J.M. (1987) Generation of altered transcripts by retroviral insertion within the c-*myb* gene in two murine monocytic leukemias. *J. Virol.*, **61**, 2754–2763.

121. Rosson,D., Dugan,D. and Reddy,E.P. (1987) Aberrant splicing events that are induced by proviral integration: implications for *myb* oncogene activation. *Proc. Natl. Acad. Sci. USA*, **84**, 3171–3175.

122. Weinstein,Y., Cleveland,J.L., Askew,D.S., Rapp,U.R. and Ihle,J.N. (1987) Insertion and truncation of c-*myb* by murine leukemia virus in a myeloid cell line derived from cultures of normal hematopoietic cells. *J. Virol.*, **61**, 2339–2343.

123. Kanter,M.R., Smith,R.E. and Hayward,W.S. (1988) Rapid induction of B-cell lymphomas: insertional activation of c-*myb* by avian leukosis virus. *J. Virol.*, **62**, 1423–1432.

124. Chen,S.J., Holbrook,N.J., Mitchell,K.F., Vallone,C.A., Greengard,J.S., Crabtree,G.R. and Lin,Y. (1985) A viral long terminal repeat in the interleukin 2 gene of a cell line that constitutively produces interleukin 2. *Proc. Natl. Acad. Sci. USA*, **82**, 7284–7288.

125. Ymer,S., Tucker,W.Q.J., Sanderson,C.J., Hapel,A.J., Campbell,H.D. and Young,I.G. (1985) Constitutive synthesis of interleukin-3 by leukaemia cell line WEHI-3B is due to retroviral insertion near the gene. *Nature*, **317**, 255–258.

126. Kuff,E.L., Smith,L.A. and Lueders,K.K. (1981) Intracisternal A-particle genes in *Mus musculus*: a conserved family of retrovirus-like elements. *Mol. Cell. Biol.*, **1**, 216–227.

127. Lueders,K.K., Fewell,J.W., Kuff,E.L. and Koch,T. (1984) The long terminal repeat of an endogenous intracisternal A-particle gene functions as a promoter when introduced into eucaryotic cells by transfection. *Mol. Cell. Biol.*, **4**, 2128–2135.

128. Lueders,K.K. and Kuff,E.L. (1977) Sequences associated with intracisternal A-particles are reiterated in the mouse genome. *Cell*, **12**, 963–972.

129. Rechavi,G., Givol,D. and Canaani,E. (1982) Activation of a cellular oncogene by DNA rearrangement: possible involvement of an IS-like element. *Nature*, **300**, 607–611.

130. Canaani,E., Dreazan,O., Klar,A., Rechavi,G., Ram,D., Cohen,J.B. and Givol,D. (1983)

Activation of the c-*mos* oncogene in a mouse plasmacytoma by insertion of an endogenous intracisternal A-particle genome. *Proc. Natl. Acad. Sci. USA,* **80**, 7118–7122.

131. Cohen,J.B., Unger,T., Rechavi,G., Canaani,E. and Givol,D. (1983) Rearrangement of the oncogene c-*mos* in mouse myeloma NSI and hybridomas. *Nature,* **306**, 797–799.
132. Gattoni-Celli,S., Hsiao,W.-L.W. and Weinstein,I.B. (1983) Rearranged c-*mos* locus in a MOPC 21 murine myeloma cell line and its persistence in hybridomas. *Nature,* **306**, 795–796.
133. Kuff,E.L., Feenstra,A., Lueders,K.K., Rechavi,G., Givol,D. and Canaani,E. (1983) Homology between an endogenous viral LTR and sequences inserted in an activated cellular oncogene. *Nature,* **302**, 547–548.
134. Horowitz,M., Luria,S., Rechavi,G. and Givol,D. (1984) Mechanism of activation of the mouse c-*mos* oncogene by the LTR of an intracisternal A-particle gene. *EMBO J.,* **3**, 2937–2941.
135. Blair,D.G., Oskarsson,M., Wood,T.G., McClements,W.L., Fischinger,P.J. and Vande Woude,G.F. (1981) Activation of the transforming potential of a normal cell sequence: a molecular model for oncogenesis. *Science,* **212**, 941–943.
136. Wood,T.G., McGeady,M.L., Baroudy,B.M., Blair,D.G. and Vande Woude,G.F. (1984) Mouse c-*mos* oncogene activation is prevented by upstream sequences. *Proc. Natl. Acad. Sci. USA,* **81**, 7817–7821.
137. Kuff,E.L., Feenstra,A., Lueders,K., Smith,L., Hawley,R., Hozumi,N. and Shulman,A. (1983) Intracisternal A-particle genes as movable elements in the mouse genome. *Proc. Natl. Acad. Sci. USA,* **80**, 1992–1996.
138. Shen-Ong,G.L. and Cole,M.D. (1984) Amplification of a specific set of intracisternal A-particle genes in a mouse plasmacytoma. *J. Virol.,* **49**, 171–177.
139. Greenberg,R., Hawley,R. and Marcu,K.B. (1985) Acquisition of an intracisternal A-particle element by a translocated c-*myc* gene in a murine plasma cell tumor. *Mol. Cell. Biol.,* **5**, 3625–3628.
140. Katzir,N., Rechavi,G., Cohen,J.B., Unger,T., Simoni,F., Segal,S., Cohen,D. and Givol,D. (1985) 'Retrotransposon' insertion into the cellular oncogene c-*myc* in canine transmissible venereal tumor. *Proc. Natl. Acad. Sci. USA,* **82**, 1054–1058.
141. Westaway,D., Papkoff,J., Moscovici,C. and Varmus,H.E. (1986) Identification of a provirally activated c-Ha-*ras* oncogene in an avian nephroblastoma via a novel procedure: cDNA cloning of a chimaeric viral-host transcript. *EMBO J.,* **5**, 301–309.
142. George,D.L., Glick,B., Trusko,S. and Freeman,N. (1986) Enhanced c-Ki-*ras* expression associated with Friend virus integration in a bone marrow-derived mouse cell line. *Proc. Natl. Acad. Sci. USA,* **83**, 1651–1655.
142a. Baumbach,W.R., Colston,E.M. and Cole,M.D. (1988) Integration of the BALB/c ecotropic provirus into the colony-stimulating factor locus in a *myc* retrovirus-induced murine monocyte tumor. *J. Virol.,* **62**, 3151–3155.
143. Bingham,P.M., Lewis,R. and Rubin,G.M. (1981) Cloning of DNA sequences from the *white* locus of D. melanogaster by a novel and general method. *Cell,* **25**, 693–704.
144. Moore,D.H., Long,C.A., Vaidya,A.B., Sheffield,J.B., Dion,A.S. and Lasfargues,E.Y. (1979) Mammary tumor viruses. *Adv. Cancer Res.,* **29**, 347–418.
145. Peters,G. and Dickson,C. (1988) On the mechanism of carcinogenesis by mouse mammary tumor virus. In *Cellular and Molecular Biology of Mammary Cancer*, Medina,D., Kidwell,W., Heppner,G. and Andereson,E. (eds.), Plenum, New York, pp. 307–319.
146. Nusse,R. (1988) The activation of cellular oncogenes by proviral insertion in murine mammary cancer. In *Breast Cancer: Cellular and Molecular Biology*, Lippman,M.E. and Dickson,R. (eds.), Martinus Nijhoff, Boston, pp. 283–306.
147. Nusse,R. and Varmus,H.E. (1982) Many tumors induced by the mouse mammary tumor virus contain a provirus integrated in the same region of the host genome. *Cell,* **31**, 99–109.
148. Peters,G., Brookes,S., Smith,R. and Dickson,C. (1983) Tumorigenesis by mouse mammary tumor virus: evidence for a common region for provirus integration in mammary tumors. *Cell,* **33**, 367–377.
149. Gallahan,D. and Callahan,R. (1987) Mammary tumorigenesis in feral mice: identification of a new *int* locus in mouse mammary tumor virus (Czech II) - induced mammary tumors. *J. Virol.,* **61**, 66–74.

150. Garcia,M., Wellinger,R.,Vessaz,A. and Diggelmann,H. (1986) A new site of integration for mouse mammary tumor virus proviral DNA common to BALB/cf(C3H) mammary and kidney adenocarcinomas. *EMBO J.*, **5**, 127–134.
151. Schuermann,M. and Michalides,R. (1987) A rare common integration site of proviruses of the mouse mammary tumor virus in P-type mammary tumors of mouse strain GR. *Virology*, **156**, 229–237.
152. Escot,C., Hogg,E. and Callahan,R. (1986) Mammary tumorigenesis in feral *Mus cervicolor popaeus*. *J. Virol.*, **58**, 619–625.
153. Pathak,V.K., Strange,R., Young,L.J.T., Morris,D.W. and Cardiff,R.D. (1987) Survey of *int* region DNA rearrangements in C3H and BALB/cfC3H mouse mammary tumor system. *J. Natl. Cancer Inst.*, **78**, 327–331.
154. Popko,B.J. and Pauley,R.J. (1985) Mammary tumorigenesis in C3Hf/Ki mice: examination of germinal mouse mammary tumor viruses and the *int*-1 and *int*-2 putative proto-oncogenes. *Virus Res.*, **2**, 231–243.
155. Peters,G., Lee,A.E. and Dickson,C. (1986) Concerted activation of two potential proto-oncogenes in carcinomas induced by mouse mammary tumour virus. *Nature*, **320**, 628–631.
156. Nusse,R., van Ooyen,A., Cox,D., Fung,Y.K.T. and Varmus,H.E. (1984) Mode of proviral activation of a putative mammary oncogene (*int*-1) on mouse chromosome 15. *Nature*, **307**, 131–136.
157. Peters,G., Kozak,C. and Dickson,C. (1984) Mouse mammary tumor virus integration regions *int-1* and *int-2* map on different mouse chromosomes. *Mol. Cell. Biol.*, **4**, 375–378.
158. Gallahan,D., Kozak,C. and Callahan,R. (1987) A new common integration region (*int-3*) for mouse mammary tumor virus on mouse chromosome 17. *J. Virol.*, **61**, 218–220.
159. Dickson,C., Smith,R., Brookes,S. and Peters,G. (1984) Tumorigenesis by mouse mammary tumor virus; proviral activation of a cellular gene in the common integration region *int-2*. *Cell*, **37**, 529–536.
160. van Ooyen,A. and Nusse,R. (1984) Structure and nucleotide sequence of the putative mammary oncogene *int*-1; proviral insertions leave the protein-encoding domain intact. *Cell*, **39**, 233–240.
161. Moore,R., Casey,G., Brookes,S., Dixon,M., Peters,G. and Dickson,C. (1986) Sequence, topography and protein coding potential of mouse *int-2*: a putative oncogene activated by mouse mammary tumor virus. *EMBO J.*, **5**, 919–924.
162. Fung,Y.K.T., Shackleford,G.M., Brown,A.M.C., Sanders,G.S., Varmus,H.E. (1985) Nucleotide sequence and expression in vitro of cDNA derived from mRNA of *int*-1, a provirally activated mouse mammary oncogene. *Mol. Cell. Biol.*, **5**, 3337–3344.
163. Smith,R., Peters,G. and Dickson,C. (1988) Multiple RNAs expressed from the *int-2* gene in mouse embryonal carcinoma cell lines encode a protein with homology to fibroblast growth factors. *EMBO J.*, **7**, 1013–1022.
164. van't Verr,L.J., van Kessel,A.D., van Heerikhuizen,H., van Ooyen,A. and Nusse,R. (1984) Molecular cloning and chromosomal assignment of the human homolog of *int*-1, a mouse gene implicated in mammary tumorigenesis. *Mol. Cell. Biol.*, **4**, 2532–2534.
165. van Ooyen,A., Kwee,V. and Nusse,R. (1985) The nucleotide sequence of the human *int*-1 mammary oncogene; evolutionary conservation of coding and non-coding sequences. *EMBO J.*, **4**, 2905–2909.
166. Casey,G., Smith,R., McGillivray,D., Peters,G. and Dickson,C. (1986) Characterization and chromosome assignment of the human homolog of *int-2*, a potential proto-oncogene. *Mol. Cell. Biol.*, **6**, 502–510.
167. Rijsewijk,F., Schuermann,M., Wagenaar,E., Parren,P., Weigel,D. and Nusse,R. (1987) The drosophila homolog of the mouse mammary oncogene *int*-1 is identical to the segment polarity gene wingless. *Cell*, **50**, 649–657.
168. Brown,A.M.C., Papkoff,J., Fung,Y.K.T., Shackleford,G.M. and Varmus,H.E. (1987) Identification of protein products encoded by the proto-oncogene *int*-1. *Mol. Cell. Biol.*, **7**, 3971–3977.
169. Papkoff,J., Brown,A.M.C. and Varmus,H.E. (1987) The *int*-1 proto-oncogene products are glycoproteins that appear to enter the secretory pathway. *Mol. Cell. Biol.*, **7**, 3978–3984.
170. Dickson,C. and Peters,G. (1987) Potential oncogene product related to growth factors. *Nature*, **326**, 833.

171. Cuypers,H.T.M., Selten,G.C., Zijlstra,M., de Goede,R.E., Melief,C.J. and Berns,A.J. (1986) Tumor progression in murine leukemia virus-induced T-cell lymphomas: monitoring clonal selections with viral and cellular probes. *J. Virol.*, **60**, 230–241.

172. Tsichlis,P.N., Strauss,P.G. and Hu,L.F. (1983) A common region for proviral DNA integration in MoMuLV-induced rat thymic lymphomas. *Nature*, **302**, 445–449.

173. Lemay,G. and Jolicoeur,P. (1984) Rearrangement of a DNA sequence homologous to a cell-virus junction fragment in several Moloney murine leukemia virus-induced rat thymomas. *Proc. Natl. Acad. Sci. USA*, **81**, 38–42.

174. Tsichlis,P.N., Lohse,M.A., Szpirer,C., Szpirer,J. and Levan,G. (1985) Cellular DNA regions involved in the induction of rat thymic lymphomas (*Mlvi-1, Mlvi-2, Mlvi-3* and c-*myc*) represent independent loci as determined by their chromosomal map location in the rat. *J. Virol.*, **56**, 938–942.

175. Jolicoeur,P., Villeneuve,L., Rassart,E. and Kozak,C. (1985) Mouse chromosomal mapping of a murine leukemia virus integration region (*Mis-1*) first identified in rat thymic leukemia. *J. Virol.*, **56**, 1045–1048.

176. Vijaya,S., Steffen,D.L., Kozak,C. and Robinson,H.L. (1987) *Dsi-1*, a region with frequent proviral insertions in Moloney murine leukemia virus-induced rat thymomas. *J. Virol.*, **61**, 1164–1170.

177. Cuypers,H.T., Selten,G., Quint,W., Zijlstra,M., Robanus-Maandag,E., Boelens,W., Van Wezenbeek,P., Melief,C. and Berns,A. (1984) Murine leukemia virus-induced T-cell lymphomagenesis: integration of proviruses in a distinct chromosomal region. *Cell*, **37**, 141–150.

178. Graham,M., Adams,J.M. and Cory,S. (1985) Murine T lymphomas with retroviral inserts in the chromosomal 15 locus for plasmacytoma variant translocations. *Nature*, **314**, 740–743.

179. Silver,J. and Kozak,C. (9186) Common proviral integration region on mouse chromosome 7 in lymphomas and myelogenous leukemias induced by Friend murine leukemia virus. *J. Virol.*, **57**, 526–533.

180. Sola,B., Fichelson,S., Bordereaux,D., Tambourin,P.E. and Gisselbrecht,S. (1986) *fim-1* and *fim-2*: two new integration regions of Friend murine leukemia virus in myeloblastic leukemias. *J. Virol.*, **60**, 718–725.

181. Bordereaux,D., Fichelson,S., Sola,B., Tambourin,P.E. and Gisselbrecht,S. (1987) Frequent involvement of the *fim-3* region in Friend murine leukemia virus-induced mouse myeloblastic leukemias. *J. Virol.*, **61**, 4043–4045.

182. Mucenski,M.L., Taylor,B.A., Ihle,J.N., Hartley,J.W., Morse,H.C.III, Jenkins,N.A. and Copeland,N.G. (1988) Identification of a common ecotropic viral integration site, *Evi-1*, in the DNA of AKXD murine myeloid tumors. *Mol. Cell. Biol.*, **8**, 301–308.

183. Moreau-Gachelin,F., Tavitian,A. and Tambourin,P. (1988) *Spi*-1 is a putative oncogene in virally induced murine erythroleukaemias. *Nature*, **331**, 277–280.

184. Selten,G., Cuypers,H.T. and Berns,A. (1985) Proviral activation of the putative oncogene *Pim*-1 and MuLV-induced T-cell lymphomas. *EMBO J.*, **4**, 1793–1798.

184a. Buchberg,A.M., Bedigian,H.G., Taylor,B.A., Brownell,E., Ihle,J.N., Nagata,S., Jenkins,N.A. and Copeland,N.G. (1988) Localization of *Evi-2* to chromosome 11: linkage to other proto-oncogene and growth factor loci using interspecific back-cross mice. *Oncogene Res.*, **2**, 149–164.

184b. Poirier,Y., Kozak,C. and Jolicoeur,P. (1988) Identification of a common helper provirus integration site in Abelson murine leukemia virus-induced lymphoma DNA. *J. Virol.*, **62**, 3985–3992.

184c. Villemur,R., Monczak,Y., Rassart,E., Kozak,C. and Jolicoeur,P. (1987) Identification of a new common provirus integration site in Gross passage A murine leukemia virus-induced mouse thymoma DNA. *Mol. Cell. Biol.*, **7**, 512–522.

185. Selten,G., Cuypers,H.T., Boelens,W., Robanus-Maandag,E., Verbeek,J., Domen,J., van Beveren,C. and Berns,A. (1986) The primary structure of the putative oncogene *pim*-1 shows extensive homology with protein kinases. *Cell*, **46**, 603–611.

186. Casnellie,J.E., Harrison,M.L., Hellstrom,K.E. and Krebs,E.G. (1983) A lymphoma cell line expressing elevated levels of tyrosine protein kinase activity. *J. Biol. Chem.*, **258**, 10738–10742.

187. Marth,J.D., Peet,R., Krebs,E.G. and Perlmutter,R.M. (1985) A lymphocyte-specific protein-tyrosine kinase gene is rearranged and overexpressed in the murine T cell lymphoma LSTRA. *Cell*, **43**, 393–404.

188. Voronova,A.F. and Sefton,B.M. (1986) Expression of a new tyrosine protein kinase is stimulated by retrovirus promoter insertion. *Nature*, **319**, 682–685.

189. Voronova,A.F., Adler,H.T., Sefton,B.M. (1987) Two *lck* transcripts containing different 5' untranslated regions are present in T cells. *Mol. Cell. Biol.*, **7**, 4407–4413.

189a. Adler,H.T., Reynolds,P.J., Kelley,C.M. and Sefton,B.M. (1988) Transcriptional activation of *lck* by retrovirus promoter insertion between two lymphoid-specific promoters *J. Virol.*, **62**, 4113–4122.

190. Villeneuve,L., Rassart,E., Jolicoeur,P., Graham,M. and Adams,J.M. (1986) Proviral integration site *Mis-1* in rat thymomas corresponds to the *pvt-1* translocation breakpoint in murine plasmacytomas. *Mol. Cell. Biol.*, **6**, 1834–1837.

191. Cory,S., Graham,M., Webb,E., Corcoran,L. and Adams,J.M. (1985) Variant (6;15) translocations in murine plasmacytomas involve a chromosome 15 locus at least 72 kb from the c-*myc* oncogene. *EMBO J.*, **4**, 675–681.

192. Graham,M. and Adams,J.M. (1986) Chromosome 8 breakpoint far 3' of the c-*myc* oncogene in a Burkitt's lymphoma 2;8 variant translocation is equivalent to the murine *pvt*-1 locus. *EMBO J.*, **5**, 2845–2851.

193. Mengle-Gaw,L. and Rabbitts,T.H. (1987) A human chromosome 8 region with abnormalities in B cell, HTLV-1[+] T cell and c-*myc* amplified tumours. *EMBO J.*, **6**, 1959–1965.

194. Tsichlis,P.N., Strauss,P.G. and Kozak,C.A. (1984) Cellular DNA region involved in induction of thymic lymphomas (*Mlvi-2*) maps to mouse chromosome 15. *Mol. Cell. Biol.*, **4**, 997–1000.

195. Kozak,C.A., Strauss,P.G. and Tsichlis,P.N. (1985) Genetic mapping of a cellular DNA region involved in induction of thymic lymphomas (*Mlvi-1*) to mouse chromosome 15. *Mol. Cell. Biol.*, **5**, 894–897.

196. Gisselbrecht,S., Fichelson,S., Sola,B., Bordereaux,D., Hampe,A., Andre,C., Galibert,F. and Tambourin,P. (1987) Frequent c-*fms* activation by proviral insertion in mouse myeloblastic leukaemias. *Nature*, **329**, 259–261.

197. Sherr,C.J., Rettenmier,C.W., Sacca,R., Roussel,M.F., Look,A.T. and Stanley,E.R. (1985) The c-*fms* proto-oncogene product is related to the receptor for the mononuclear phagocyte growth factor, CSF-1. *Cell*, **41**, 665–676.

198. Mucenski,M.L., Taylor,B.A., Copeland,N.G. and Jenkins,N.A. (1988) Chromosomal location of *Evi-1*, a common site of ecotropic viral integration in AKXD murine myeloid tumors. *Oncogene Res.*, **2**, 219–233.

199. Sola,B., Simon,D., Mattei,M., Fichelson,S., Bordereaux,D.,Tambourin,P.E., Guenet,J. and Gisselbrecht,S. (1988) *Fim-1*, *Fim-2/c-fms*, and *Fim-3*, three common integration sites of Friend murine leukemia virus in myeloblastic leukemias, map to mouse chromosomes 13, 18, and 3, respectively. *J. Virol.*, **62**, 3973–3978.

200. Morishita,K., Parker,D.S., Mucenski,M., Jenkins,N.A., Copeland,N.G. and Ihle,J.N. (1988) Retroviral activation of a novel gene encoding a zinc finger protein in IL-3 dependent myeloid cell lines. *Cell*, **54**, 831–840.

201. Mowat,M., Cheng,A., Kimura,N., Bernstein,A. and Benchimol,S. (1985) Rearrangements of the cellular p53 gene in erythroleukaemic cells transformed by Friend virus. *Nature*, **314**, 633–624.

202. Chow,V., Ben-David,Y., Bernstein,A., Benchimol,S. and Mowat,M. (1987) Multistage Friend erythroleukemia: independent origin of tumor clones with normal or rearranged p53 cellular oncogenes. *J. Virol.*, **61**, 2777–2781.

203. Ben David,Y., Prideaux,V.R., Chow,V., Benchimol,S. and Bernstein,A. (1988) Inactivation of the p53 oncogene by internal deletion or retroviral integration in erythroleukemic cell lines induced by Friend leukemia virus. *Oncogene*, **3**, 179–185.

204. Klein,G. (1987) The approaching era of tumor suppressor genes. *Science*, **238**, 1539–1545.

205. Tiollais,P., Pourcel,C. and Dejean,A. (1985) The hepatitis B virus. *Nature*, **317**, 489–495.

206. Dejean,A., Bougueleret,L., Grzeschik,K.H. and Tiollais,P. (1986) Hepatitis B virus DNA integration in a sequence homologous to v-*erb-A* and steroid receptor genes in a hepatocellular carcinoma. *Nature*, **322**, 70–72.

207. Pasquinelli,C., Garreau,F., Bougueleret,L., Cariani,E., Grzeschik,K., Thiers,V., Croissant,O., Hadchouel,M., Tiollais,P. and Brechot,C. (1988) Rearrangement of a

common cellular DNA domain on chromosome 4 in human primary liver tumors. *J. Virol.*, **62**, 629–632.

208. De The,H., Marchio,A., Tiollais,P. and Dejean,A. (1987) A novel steroid thyroid hormone receptor-related gene inappropriately expressed in human hepatocellular carcinoma. *Nature*, **330**, 667–670.

209. Brand,N., Petkovich,M., Krust,A., Chambon,P., De The,H., Marchio,A., Tiollais,P. and Dejean,A. (1988) Identification of a second human retinoic acid receptor. *Nature*, **332**, 850–853.

210. Hsu,T., Moroy,T., Etiemble,J., Louise,A., Trepo,C., Tiollais,P. and Buendia,M.A. (1988) Activation of c-*myc* by woodchuck hepatitis virus insertion in hepatocellular carcinoma. *Cell*, **55**, 627–635.

211. Stewart,T.A., Pattengale,P.K. and Leder,P. (1984) Spontaneous mammary adenocarcinomas in transgenic mice that carry and express MTV/*myc* fusion genes. *Cell*, **38**, 627–637.

3

Molecular pathology of chromosomal abnormalities and cancer genes in human tumors

T.H.Rabbitts and P.H.Rabbitts

1. Introduction

1.1 Definition of chromosomal abnormality

The subject of this chapter is the type of chromosomal abnormalities which are concerned with tumor etiology. These are somatic changes which occur in cells destined to become overt tumors. We will only describe those abnormalities which are gross changes in chromosomal structure, such as translocation. Other important tumorigenic somatic changes do occur (such as point mutations, particularly well documented as single base changes observed in the *ras* oncogene family in a variety of tumor types; see Chapter 4) but these are not considered as chromosomal abnormalities for the purpose of this chapter. The three major types of chromosomal abnormality found in cancer cells are illustrated in *Figure 1*.

1.1.1 Translocation

This involves the exchange of material between chromosomes (*Figure 1A*). It can occur between homologous [e.g. t(14;14) (p11;q32)] or non-homologous chromosomes [e.g. t(8;14) (q24;q32)] and can be a balanced, reciprocal event or an unbalanced event in which material is lost from one or both chromosomal junctions.

1.1.2 Inversion

Inversions involve two breaks within one chromosome. These can occur either on the same arm of the chromosome (paracentric inversion, *Figure 1Bi*), or on both arms, that is on either side of the centromere (pericentric inversion, *Figure 1Bii*). The inversion of the intervening material is illustrated in *Figure 1B* by theoretical breaks between regions 1–2 and 3–4 creating a new genomic order of 1, 3, 2, 4.

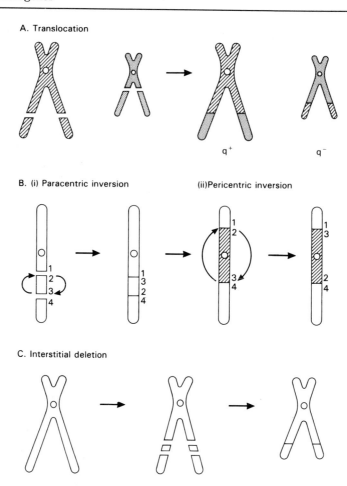

Figure 1.   The three major chromosomal abnormalities detected by cytogenetic techniques in cancer cells. The short arms of chromosomes are designated p and long arms q. (**A**) Representation of a balanced reciprocal translocation between non-homologous chromosomes involving both long arms. Different sized chromosomes result, one with an elongated (q^+) and one with a shortened (q^-) long arm. (**B**) Representation of chromosome inversion which results in no net change of chromosome size. (i) Inversion of the short arm of one chromosome (paracentric); (ii) pericentric inversion involving breaks on either side of the centromere: note the change in location of the centromere giving rise to an altered chromosomal morphology. (**C**) Interstitial deletion. Various sized pieces of chromosomes are lost resulting in the creation of shortened forms.

1.1.3 Interstitial deletion

This also involves two breaks occurring on one chromosome but with the intervening material being subsequently lost (*Figure 1C*).

   Other types of karyotypic abnormalities are also found in tumor cells, notably amplifications exemplified by double minute chromosomes or homogeneously staining regions, ring chromosomes, dicentric chromosomes and trisomies. Of these latter changes, only amplifications will be discussed in detail below.

1.2 Why study chromosomal abnormalities?

In its most simple form, the assumption is that proto-oncogenes are activated (i.e. they become activated oncogenes) by their proximity to the breakpoints of chromosomal aberrations. There are two levels at which chromosome alterations can be examined: the cytogenetic level and the DNA level. The former type of study has been crucial in identifying those chromosomal abnormalities which appear consistently in tumors of a certain type, for example the Philadelphia (Ph') chromosome in chronic myeloid leukemia (CML) (1). The consistent appearance of a particular chromosome abnormality in leukemic cells has thus led to the very important concept that they are involved in the etiology of the tumor carrying the abnormality. The rationale is simply that it is unlikely that consistent chromosomal abnormalities are inadvertent features of leukemia. Thus, we can look upon chromosomal abnormalities as natural mutations and their study should provide an insight into the nature of potential new oncogenes occurring at the junctions of such abnormalities and how such genes are disrupted by the changes. Furthermore, the mechanism of formation of the chromosome abnormalities can frequently be elucidated by studies on sequences at the junctions of chromosomal aberrations. Therefore, the approach of molecular biologists to chromosomal abnormalities has been to study them at the DNA level, particularly at the junctions between the rearranged chromosomal elements. This new study of chromosomal alterations has verified many of the predictions of cancer cytogenetics and, as discussed below, has raised many interesting questions.

1.3 The significance of chromosomal abnormalities to cancer

As stated above, the concept of oncogene activation resulting from chromosomal aberration underlies all the molecular biology work on such aberrations. Although oncogene activation undoubtedly occurs as the result of some chromosomal abnormalities, it is perhaps a misconception in many instances, for example where gene deletion seems to play a role, such as appears to be the case in Wilms tumor (2). We prefer therefore the use of the term cancer genes as a general name for genes involved in tumor pathogenesis, whether these are positively activated proto-oncogenes [e.g. c-*myc* in Burkitt's lymphoma (BL)] or putative recessive genes (e.g. as in Wilms tumor or retinoblastoma).

The molecular biology work on chromosome abnormalities is aimed at identifying new cancer genes at or near the junctions of chromosomal

Table 1.   Tumor-specific chromosome abnormalities studied at the DNA level

Disease	Translocation	Breakpoint cloned	Known proto-oncogene identified	Novel RNA transcript identified	Reference
Chronic myeloid leukemia	t(9;22)(q34;q11)	+	c-*abl*		1
Burkitt's lymphoma	t(8;14)(q24;q32)	+	c-*myc*		3,4
	t(2;8)(p12;q24)	+	c-*myc*		5
	t(8;22)(q24;q11)	+	c-*myc*		5
Follicular lymphoma	t(14;18)(q32;q21)	+		+	6–9
B cell prolymphocytic leukemia	t(11;14)(q13;q32)	+		–	10
T chronic lymphocytic leukemia	inv14(q11;q32)	+		–	11–13
T acute lymphocytic leukemia	t(11;14)(p15;q11)	+		+	14–16
T acute lymphocytic leukemia	t(11;14)(p13;q11)	+		–	17–21

aberrations, elucidating the mechanisms of the abnormality and gaining insight into mechanisms of gene regulation, deregulation and function. A number of consistent tumor-specific abnormalities have been the subject of study at the DNA level and these are summarized in *Table 1*. Details of these molecular analyses are given below. In those cases where a known cellular proto-oncogene has been identified near to the junction of an abnormality (e.g. in BL) a causative role of the chromosomal abnormality in tumor formation is very strongly suggested, mediated by activation of the proto-oncogene (c-*myc* in this case). In those cases where no known gene is identified, the observation of an RNA transcription unit near the chromosomal junction is consistent with the view that this is a new cancer gene affected by the chromosomal abnormality. Although this seems a reasonable assumption at present, it cannot be too strongly stated that no direct evidence has yet been produced to show that this is true. Clearly, the cloning of the various junctional RNAs will open the way to direct experimental tests for oncogenicity of these newly identified genes.

A further note of caution may be drawn from the observation that there are two types of chromosomal abnormality: those that appear consistently in tumors of a particular type and those which are idiopathic to individual tumors. In the latter cases, cytogenetic breakpoints of chromosomal abnormalities often correlate with known cytogenetic locations of cellular proto-oncogenes. *Table 2* is a list of proto-oncogenes whose human chromosomal locations are known. However, the chromosomal banding techniques usually applied to tumor samples and the level of sensitivity with which genes can be localized to chromosome bands (e.g. by *in situ* hybridization) makes such coincidence of cytogenetic breakpoints and oncogene localization speculative. An example of this implied association is the localization of the c-*ets* 'proto-oncogene' to band 11q23(22) and possible association with t(14;11)(q21;q23) in various types of lymphoid tumors. Although recent papers show that the c-*ets* sequences have translocated with the tip of 11q in some tumors (23), no sign of re-arrangement of c-*ets* can be detected in conventional Southern filter hybridization. Since a chromosome band could consist of 10^6 bp or more, breakpoints and genes are not necessarily at the same position. Thus c-*ets* may reside at band 11q23 coincidentally to the 11q23 breakpoint but need not necessarily play a role in the etiology of tumors with t(4;11).

Similarly, any putative association of fragile sites with oncogene localization is speculative at the present time. Fragile chromosomal sites are simply regions where there is a preponderance of chromosome breakage, as judged by the frequency of cytogenetically observed fractures at a given position. Oncogene localization probably has little to do with the most fragile sites.

Table 2.   Chromosomal location of known human proto-oncogenes

Oncogene	Location
src2	1p34-31
L-myc	1p32
N-ras	1p
ski	1q22-qter
N-myc	2p24
rel	2
raf-1	3p25
raf-2	4
fms	5q34
K-ras	6p23
myb	6q15-q24
yes2	6
mos	8q11
myc	8q24
abl	9q34
H-ras	11p15.5
ets-1	11q23
mit-1	12pter-q14
K-ras2	12p12
fos	14q21-q31
akt1	14q32
fes	15q25-26
ng1	17q21-q22
yes1	18q21.3
src1	20q12-q13
sis	22q12.3-q13.1
H-ras2	Xpter-q28

2. The Philadelphia chromosome and the c-abl proto-oncogene

2.1 Cytogenetics of the Philadelphia chromosome

The Ph′ chromosome was first described in 1960 by Nowell and Hungerford as a small chromosome present in two male patients with chronic myelocytic leukemia (1). The Ph′ chromosome has subsequently been found predominantly in CML, although around 10% of acute lymphocytic leukemias (ALL) carry this abnormality. It is known that the Ph′ chromosome is usually a reciprocal translocation between chromosome 9 and 22 (q34;q11). Extensive molecular biology studies have now established that the c-abl proto-oncogene is involved in this translocation, moving from normal chromosome 9q32 to the Ph′ chromosome itself.

2.2 c-abl and bcr genes in the Ph′ chromosome of CML

In Ph′-positive CML, the c-abl oncogene is invariably translocated to chromosome 22 (24,25). The breakpoint with respect to the c-*abl* gene appears to vary and can occur as much as 101 kb from the most 5' exon of c-*abl* which bears homology to v-*abl* sequences (26). The orientation of c-*abl* on the Ph' chromosome is centromere-5' – c-*abl* – 3'-telomere (i.e. the translocation breakpoint on chromosome 9q occurs upstream of c-*abl*). The region of chromosome 22 which is involved is of considerable interest in itself since a gene occurs at this position. A cluster of breakpoints in more than 30 Ph'-positive CML patients has been found to occur with a region of 5.8 kb; this region has been named the breakpoint cluster region (*bcr*) (27). The interesting feature of *bcr* is that it is a gene transcribed in normal cells, giving mRNAs of 4 and 6.5 kb (28). The normal function of the *bcr* is unknown. No homology of *bcr* with other genes has yet been demonstrated, but genomic structural studies have shown that multiple exons exist and that Ph' breakpoints occur within introns of the *bcr* gene (28,29). The orientation of the *bcr* gene on chromosome 22 has been established from these studies and it is similar to that of c-*abl* on chromosome 9 (i.e. centromere-5' – *bcr* – 3'-telomere). This means that after translocation the majority of Ph' chromosomes will produce a new fusion gene whose organization is now centromere-5' – *bcr* – c-*abl* – 3'-telomere.

A summary of this organization and reorganization of *bcr* and c-*abl* after chromosomal translocation is shown in *Figure 2*. Since the translocation breaks within introns of both *bcr* and c-*abl*, and the translocation orientation of the genes is the same, new RNA splicing patterns can take place which generate a fusion RNA of about 8.5 kb which is a chimera of *bcr* (5' end)

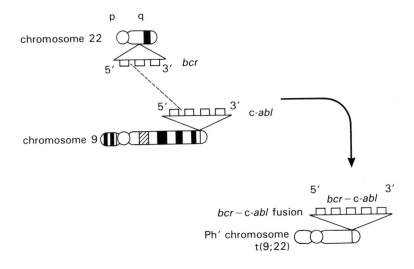

Figure 2. Diagrammatic representation of *bcr* and c-*abl* genes before and after chromosome translocation. The numbers of exons in the *bcr* and c-*abl* genes are reduced for simplicity.

and c-*abl* (3' end) utilizing the *bcr* promoter and c-*abl* poly(A) addition signal. It is assumed that this fusion RNA is crucial for development of CML. This seems a reasonable conclusion given the consistency of the translocation in CML and the observed loss of the reciprocal chromosome $9q^+$ from some tumors. However, a case of CML in blast crisis has also been reported in which the *bcr–abl* fusion gene has been lost (30), suggesting a role of the Ph' fusion in tumor initiation or progression rather than maintenance of the leukemic state.

The 8.5 kb chimeric mRNA discussed above has been shown to yield a 210 kd *bcr–abl* fusion protein which is unique to Ph' CML. The crucial difference between this protein and the normal c-*abl* protein (p150) is that only the former chimeric protein exhibits tyrosine kinase activity (31). (The v-*abl* gene also has such an activity.) It has been proposed that the Ph' chromosome fusion causes an unmasking of the tyrosine kinase activity of c-*abl*, perhaps by creating a new protein fold due to the presence of the *bcr* protein. It is, however, intriguing that the *bcr–abl* fusion protein does not seem able to transform NIH3T3 cells *in vitro* whereas the v-*abl* protein can (32).

2.3 Ph'-positive ALL and c-*abl*

Around 10–20% of patients with ALL have a Ph' chromosome which is cytogenetically similar to that of CML (33,34). However, the CML and ALL diseases are themselves quite distinct. Additionally, a proportion of CML patients develop lymphoid blast transformation which clinically resemble ALL (35). Early work suggested a complex pattern of *bcr* gene involvement which was unlike the 5.8 kb cluster seen in CML and some Ph^+ ALL without *bcr* rearrangement (36,37). A further important difference between CML and ALL with Ph' chromosome is that the latter disease shows the presence of a novel protein of 180–190 kd which may be etiologically related to the ALL (31,38,39). The relationship of *bcr–abl* gene fusions to this novel protein has not yet been clarified.

3. Chromosomal translocation in Burkitt's lymphoma

3.1 Cytogenetic characteristics

BL is a B cell tumor (40) which can be divided into two main forms: endemic (African) BL, which is mainly associated with Epstein–Barr virus (EBV), and sporadic BL, which is usually EVB^-. Recent studies on BL in acquired immunodeficiency syndrome (AIDS) patients have also suggested the involvement of EBV. The significant feature of BL of all types, however, is the occurrence of one of three specific chromosomal translocations involving chromosome 8, band q24. The majority of BL tumors carry t(8;14) (p24;q32) and the minority carry either t(2;8) (p12;q24) or t(8;22) (p24;q11) (3 – 5,41). The common feature of these tumors is the chromosome 8q24 breakpoint and this is the site of the c-*myc* proto-oncogene (42,43). This gene is therefore believed to be the gene involved in the etiology of BL.

Karyotypic studies, together with *in situ* and molecular hybridization, have defined the breakpoints on chromosome 8 and on the other chromosomes involved in BL. Molecular studies have shown that the immunoglobulin (Ig) loci are involved in each of the three translocations. The heavy (H) chain genes are involved in the t(8;14) cases, the kappa ($\varkappa$) light (L) chain genes in t(2;8), and the lambda (λ) L chain genes in t(8;22). The latter two forms of BL are referred to as variant forms since they occupy the minority of cases.

A major difference exists between the t(8;14) and variant BL. This is the difference in breakpoints on chromosome 8 with respect to the c-*myc* gene. The c-*myc* gene has three exons, two of which code for the c-*myc* protein (44) and one which is non-coding (45). [A controversy does exist about this first 'non-coding' exon from which a protein does appear to be made in human cells (46,47) but certainly its open reading frame seems only to have been conserved in man (46).] In t(8;14), breakpoints either occur within c-*myc* itself or on the 5′ side of the gene (42,48 – 51) whereas the variant lymphomas break downstream of the c-*myc* gene (50 – 54). This results in markedly different translocation events, which are summarized in *Figure 3*. All three translocations generally occur in the proximity of the Ig constant (C) region genes (detailed below), although some have been reported within the variable (V) region loci (55). When the break occurs near the C region, the event essentially splits the Ig locus. In the t(8;14) (q24;q32), the translocation moves the tip of chromosome 8q, including the c-*myc* gene, to the tip of 14q to juxtapose the Ig gene with the 5′ region of c-*myc*. Similarly, the Ig V_H locus is moved over the chromosome 8q to the residues of the c-*myc* locus (*Figure 3A*). The variant translocations are different. In these, the Ig C region genes move, with the piece of chromosome 2p or 22q, onto the downstream side of c-*myc* (*Figure 3B* and *C*). The molecular consequences of these varied rearrangements are quite distinct and will be described separately.

3.2 Burkitt's lymphoma (8;14)

3.2.1 Breakpoints in heavy chain genes

In the t(8;14) BL, the breakpoint within the Ig H chain gene (*Igh*) seems to be roughly divisible into two areas (*Figure 4A*). Firstly, the *Igh* chain joining (J) region is the predominant site for breakage in endemic BL (56), presumably translocation taking place during the normal V – D – J joining process (see Section 8.1 for a fuller description of V – D – J joining). Some rare cases have been described in the V_H itself (55). On the other hand, in sporadic BL, breakpoints are frequently encountered in the switch (S)

A. t(8; 14) (q24; q32)

B. t(2; 8) (p12; q24)

C. t(8; 22) (q24; q11)

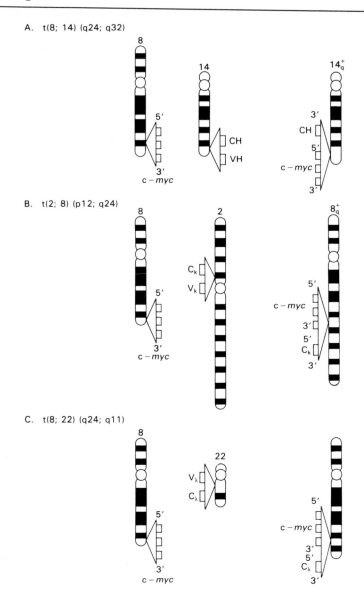

Figure 3.   Diagram of chromosomal rearrangements in various BL translocations.

region located near the H chain C regions (57–59). A noticeable exception to this rule is Raji, a cell line derived from an endemic BL patient, in which the breakpoint occurs in the S region (48,66). These correlations are interesting because they suggest that the translocation takes place in

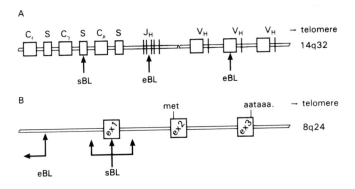

Figure 4. **(A)** Diagram of Igh locus on human chromosome band 14q32. eBL, endemic Burkitt's lymphoma; sBL, sporadic Burkitt's lymphoma; S, switch region. The complexity of V_H, J_H and C_H segments has been simplified in this diagram. See ref. 123 for complete details. **(B)** Diagrammatic representation of the c-*myc* proto-oncogene. The methionine (met) initiation codon occurs at the start of exon two and termination occurs at exon three. Breakpoints of endemic (eBL) and sporadic (sBL) Burkitt's lymphoma are indicated, as is the direction of the telomere of chromosome 8q.

different parts of B cell development. Endemic BL is associated with EBV, which is known to infect B cells and thereby immortalize them. These cells are presumably undergoing $V - D - J_H$ joining and thus the chance for J_H-associated (or V_H-associated) translocation is heightened. Sporadic BL (an EBV$^-$ disease of world-wide occurrence) probably occurs in more mature cells undergoing the Ig class switch (the rearrangement of a $V - D - J$ fused gene to a different C_H segment).

3.2.2 Breakpoints in κ and λ light chain genes

In the variant breakpoints, the L chain C genes seem to be the site for translocation (κ at chromosome 2p12 and λ at chromosome 22q11). This can involve the J regions (at least for the κ chains) (52,53,60) and sometimes $V_κ$ segments (53,60). Importantly, these variant trans-locations bring the $C_κ$ or $C_λ$ behind the c-*myc* gene which remains at 8q32 after translocation.

3.2.3 Breakpoints near c-myc in t(8.14) Burkitt's lymphoma

The c-*myc* proto-oncogene has three exons of which the second two carry protein coding capacity (*Figure 4B*) analogous to the v-*myc* gene (44). Translocation t(8;14) in BL results in the c-*myc* gene being carried over onto the 14q$^+$ chromosome with the piece of translocated chromosome 8. The majority of endemic BL patients do not show DNA rearrangement with c-*myc* probes (61 – 64) even though they have translocated c-*myc* genes. In the t(8;14) BL, there are two main types of c-*myc*-related breakpoints. One cluster, perhaps occurring in pre-B cell precursors (56), appears to occur at a considerable distance upstream of c-*myc* exon 1, and may represent a cluster point for some endemic BL. A few other cell lines from endemic BL show rearrangement of the c-*myc* gene within about 10 kb from the 5′ end. The second cluster of breakpoints is found in the non-endemic BL which have a very high frequency of breakage near to or within a c-*myc* (64); these breakpoints include some in the first exon or intron and some within the region immediately 5′ to exon 1.

3.2.4 Breakpoints near c-myc in variant Burkitt's lymphoma

The variant lymphomas have breakpoints downstream of c-*myc* and the majority do not display rearrangement by the normal filter hybridization procedure. A cluster of breakpoints, designated *pvt*-1-like (67,68) (after the analogous variant translocations studied in mouse), has been studied which include a number of variant (t2;8) BL. The *pvt*-1-like locus at chromosome 8q24 is at least 45 kb from the 3′ end of c-*myc* (68–70) and probably about 300 kb away from it (68). *Figure 5* shows a diagrammatic representation of this downstream region of c-*myc*. An extremely interesting cytogenetic observation has recently been made that perhaps a second, more telomeric cluster of breakpoints may exist for variant t(8;22) translocations (71), which could place such breakpoints at an even greater distance from c-*myc* than the t(2;8) breaks in the *pvt*-1-like region.

Human chromosome band 8q24

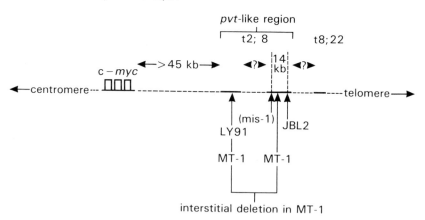

Figure 5. Diagrammatic representation of the human chromosome band 8q24 and *pvt*-like region. The diagram shows the orientation with respect to the centromere and telomere of chromosome 8q. The c-*myc* gene is nearest to the centromere. The *pvt*-like region is split into two regions whose distance apart is unknown. Locations of breakpoints are indicated by vertical arrows. The possibility of a second cluster of *pvt*-like breakpoints due to the t(8;22) translocations is indicated.

Clearly, these apparent long-range effects on the c-*myc* gene are difficult to explain (see below). Such explanations as disruption of chromatin architecture (68) are plausible but fall short of precise molecular explanations for c-*myc* gene activation by chromosomal translocation.

3.2.5 Alterations of the c-myc gene by translocation

Truncation of the c-*myc* first exon by translocation is a readily detectable mutation. Most BL tumors of endemic type do not, however, show any sign of rearrangement within the proximity of the gene. However, mutation in translocated c-*myc* genes is frequent, particularly in the first exon. Mutation in the c-*myc* coding region (72) and non-coding exon 1 (62,63,72) were first shown in endemic BL. This was later confirmed by others (64,73). The high frequency of mutation at a certain site in exon 1 (for example defined by a *Pvu*II site) has been pointed out (64), and this mutation occurs in both t(8;14) (64,72) and t(8;22) translocations (72). The significance of these changes is far from clear, although the low frequency of coding region mutations has been taken to indicate involvement in tumor progression rather than initiation. Future testing of mutation efficacy by transfection studies will allow their importance to be assessed.

3.2.6 Evidence that chromosomal translocation activates the c-myc proto-oncogene

It is assumed from the consistently occurring translocations in Burkitt type lymphoma, that the c-*myc* proto-oncogene is activated by its proximity to one of the immunoglobulin loci after translocation. Direct and indirect evidence in support of this view is derived from both transfection studies and transgenic mouse experiments.

Lymphoblastoid cells can be established in culture by infecting B-cell cultures with EBV. Such lines are non-tumorigenic in nude mice, except after prolonged culture. It has been demonstrated that such lines can be converted to tumorigenic phenotype by transfection with various types of plasmids containing the c-*myc* gene controlled by strong promoter/enhancers, including a truncated c-*myc* gene from the BL line Manca, which included the Ig transcription enhancer (65). Thus, it seems that EBV infection and c-*myc* activation are sufficient for tumorigenic conversion of B cells (74), as apparently occurs in pathogenesis of Burkitt's lymphoma.

Strong support for a role for c-*myc* in B-cell pathogenesis has come from insertion of various forms of c-*myc* genes in the germ-line of mice (transgenes) and subsequent tumor induction (75–78). Particularly effective in generation of transgenic mouse strains in which individuals develop B-cell tumors are c-*myc* gene constructs with strong enhancers. Tissue specific tumor induction of transgenic strains carrying mammary tumor virus LTR-c-*myc* fusion genes demonstrates that inappropriate expression of c-*myc* can result in tumor formation (76), even though synergistic effects of the activity of other oncogenes exacerbates the tumorigenic effect (77).

3.2.7 How does translocation activate the c-myc oncogene?

The data on the variety of chromosomal breakpoints around the c-*myc* gene (upstream and downstream) present a paradox. Can a unifying hypothesis be formulated to explain the consequences of translocation on the c-*myc* gene? The key seems to be transcriptional deregulation rather than increase in the absolute levels of mRNA production in BL compared with, for example, LCLs. A relatively consistent feature of BL translocated c-*myc* genes is the presence of somatic mutations (whether the translocation occurs upstream or downstream of the c-*myc* gene). Such somatic mutations were first shown in a t(8;14) case (the Raji cell line) (63) where the mutations including point mutations, deletions and small duplications, extended through exon 1–intron 1–exon 2 of the translocated c-*myc*. This observation was later extended to cases bearing only exon 1 mutations and to cases (relatively rare compared with exon 1 only changes) with mutations in exon 1 and exon 2 (49,62,63). An extensive study (64) carried out in the light of early reports of somatic mutations within translocated c-*myc* genes, showed the frequent occurrence of mutations within a region near the 3′ end of exon 1. This hot-spot site may be significant in one of two ways. The c-*myc* gene has three exons, and the protein coding reading frame begins with an AUG codon at the beginning of exon 2 (44). A second, non-AUG codon has also been identified near the 3′ end of the otherwise non-coding exon 1 which gives rise to identical protein except for the very N-terminus (79). In BL, the loss of exon 1 by truncation in translocation or by somatic mutation at the end of exon 1 may be involved in the loss of synthesis of the larger of the two myc proteins (79); this may contribute to the oncogenic activation. A different, but not necessarily exclusive mechanism for activation is that the exon 1 mutations correlate with altered transcription control by removal of a transcriptional elongation block (80,81). A number of putative control regions have been identified as DNase I-hypersensitive sites in chromatin and these sites vary according to differentiation states of the c-*myc* gene. Transcriptional control at the level of polymerase II termination has been described in man and mouse. Seemingly, polymerase transcription through the normal c-*myc* can be terminated near the end of exon I (prior to the protein initiation codons) (80), thus normally allowing accumulation of polymerase molecules in the region of exon 1. Since somatic mutations occur in both t(8;14) and variant BL translocations [t(2;8) and t(8;22)], acquisition of somatic mutation which accumulate as a consequence of translocation (63) could be sufficient to be a general activator of translocated c-*myc*.

In relation to loss of transcriptional control, it is important to remember that the concept of c-*myc* activation via the Ig H-chain enhancer is frequently not the case. The human Igh enhancer has been located between J$_H$ and Sμ segments (65,66) and as a result in the BL translocation which involve the switch region sequence (mainly sporadic), the Ig enhancer moves to the reciprocal translocation chromosome and, since the enhancer is a *cis*-acting element, cannot be involved in c-*myc* transcription. In those cases of J$_H$ involvement in translocation, the Igh enhancer can be involved (65).

One intriguing feature of BL is that the normal (non-translocated) allele of c-*myc* is generally silent (49) although there are exceptions (66). A similar effect was observed in EBV+LCLs transfected by activated c-*myc* constructs (74). This feature is reflected in changes in the pattern of DNase I hypersensitivity sites upstream of and within c-*myc*. There are five major sites (82,83) mapping within 2 kb of the 5′ end of the gene. Studies of the pattern of such sites on the normal allele and various translocated alleles have shown that only site I (the most 5′ of the DNase hypersensitivity sites) remains on the silent normal c-*myc* allele (83). Such a situation is also seen in HL60 cells which have been induced to differentiate and which cease to express c-*myc* cytoplasmic mRNA (84). The pattern of hypersensitive sites does not vary between, for example, that seen in an actively growing, non-tumorigenic LCL and that in a translocated c-*myc* gene [except, of course, where translocation breakpoints break near the gene (*Figure 6*)]. No difference is seen between patterns of sites where breakage is far upstream or far downstream of c-*myc* (83).

In conclusion, any explanation for oncogenic activation of c-*myc* must take account of a number of paradoxical findings, principally the great disparity found in the position of the chromosomal translocations with respect to the c-*myc* gene. Loss of control sequences from regions in or around c-*myc* may explain some translocation effects, but cannot be the explanation for many others. On the other hand, acquisition of new transcription control elements may also explain some cases [e.g. the Igh transcription enhancer in some BL (65)]. One very consistent feature of translocated c-*myc* genes in BL, however, is the present of exon 1 mutations (63). These may explain loss of transcriptional control in such genes.

4. t(11;14) translocation and the immunoglobulin heavy chain locus

Translocations between chromosome 14, breaking at q32 (the *Igh* locus), and chromosome 11, breaking at q13, are seen in several different types of B-cell tumors; namely chronic lymphocytic leukemia (CLL) (85), prolymphocytic leukemia (PLL) (86), myeloma (86,87) and plasma cell

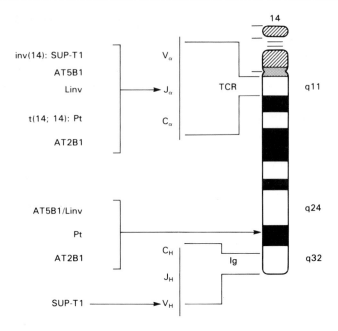

Figure 6. Diagrammatic representation of chromosome 14 with associated breakpoint positions. The various tumors indicated are Linv, Pt AT5B1 and SUP-T1 (118); AT2B1 is from an A-T patient with a clonal t(14;14) but non-malignant.

leukemia (88). The t(11;14) marker chromosome is not diagnostic for any of these diseases, occurring in them only sporadically. However, it is quite often the only (or one of the few) chromosomal abnormalities in the tumors in which it is seen (see e.g. ref. 86), suggesting that it may play an important role in tumor formation.

Erikson *et al.* (89) have prepared somatic cell hybrids using a cell line derived from a patient with CLL carrying the t(11;14) translocation, and have shown that the productive rearrangement of the Ig H chain takes place on the normal chromosome 14, the allelically excluded H chain gene bearing the 11;14 translocation. Further studies by the same group (90) have led to the isolation of a genomic clone which contains the breakpoint region. Sequence studies of the region of chromosome 14 involved in the breakpoint have shown that it occurs at the Ig J_4. Sequence studies of the part of chromosome 11 involved in the translocation reveal a putative 'heptamer – nanomer' recombination signal motif located at the breakpoint on chromosome 11. These sequences (see section 8.1) are thought to be involved in normal $V-D-J$ joining. It has been proposed that the recombinase responsible for $V-D-J$ joining on chromosome 14 is also able to bring together two different chromosomes (90). This might imply that the significance of the translocation was simply the presence of homologous sequences on chromosomes 11 and 14. However, the homology between the chromosome 11 sequences and those of J_H is not complete, and such possible structures occur many times within the human genome (see section 8.2). Since the translocation is frequently the only chromosomal abnormality seen in these B-cell tumors, it seems probable that it positively contributes to malignancy. The affected region of chromosome 11 is often referred to as the 'bcl-1 gene' (10), although so far no transcribed region has been described. The expectation is, however, that a transcribed region will be found, revealing a gene showing similar deregulation to the c-myc gene as a consequence of its juxtaposition to the Igh gene.

DNA probes have been isolated from chromosome 11 in the region of the breakpoint. These allow a study of the occurrence of rearrangement in 11q13 in B cells to be made without depending on cytogenetic analysis. Ince et al. (91) have made such a study using genomic DNA isolated from 32 B-cell tumors and have found rearrangements in only three of them. However, in the cases where breakpoints have been studied by cloning no consistent position has been found. CLL and diffuse B-cell lymphoma (90), although identical cytogenetically, have chromosomal breakpoints that are not exactly the same at the molecular level. Furthermore, a study of a patient with B-cell PLL (B-PLL) with t(11;14) (q13;q32) translocation demonstrated (92) a different breakpoint (close to that of B-CLL). This breakpoint in the B-PLL was separated by about 35 kb from an 11q13 lesion in a plasma cell tumor (91). Therefore, a range of probes may be necessary to detect all B-cell tumors that have breakpoints in the 11q13 cluster.

5.   t(14;18) translocation in follicular lymphomas

Cytogenetic analysis has detected translocations between chromosome 14 (band q32) and chromosome 18 (band q21) as a very frequent event (>80%) in follicular lymphomas (93,94). The involvement of the Igh gene at the breakpoint 14q32 has been shown by the isolation of genomic clones containing DNA from the breakpoint of the translocation (7,9,95). Transcribed sequences have been identified from chromosome 18 at the breakpoint often referred to as the bcl-2 gene. Using these sequences as probes it has been possible to characterize the normal bcl-2 gene on chromosome 18. The gene consists of two exons, separated by at least 50 kb. The second exon has a very long (>5 kb) untranslated region and it is within this 3' region that most breaks occur in the translocation (8,96). The disruption of the bcl-2 gene in cells carrying the 14;18 translocation results in different size transcripts and, although the protein made would be the same from both the normal and translocated genes, it is possible that there may be differences in the processing of these transcripts resulting in altered regulation of protein production. The protein sequence predicted from the nucleic acid sequence suggests a protein without a leader sequence, with no protein kinase activity and without homology to any known oncogenes. There is, however, some homology with a protein predicted to be encoded by the EBV genome.

The use of probes from the breakpoint on chromosome 18 has shown that there are two distinct breakpoint clusters in follicular lymphoma (96,97). Using these probes (98) it has been shown that rearrangement of 18q21 occurs not only in follicular lymphomas but also in large cell lymphomas, and small non-cleaved cell lymphomas.

5.1 Molecular analysis of *bcl-2* translocations

5.1.1 *Molecular junctions of translocation t(14;18)*

The cytogenetic breakpoint at 14q32 was investigated via Igh H chain probes because the 14q32 breakpoint corresponds cytogenetically to this locus. Indeed, the Igh locus is involved in this translocated with cloned breakpoints again occurring within the J_H cluster (7–9), generally precisely at the 5′ end of the J_H segments. In these tumors, unlike t(11;14)(q13;q32) where only one J_H has been found to be involved so far, at least two different J_H segments have been shown at translocation junctions. Thus, FL translocations arise from breakages within the immunoglobulin H chain locus and at positions adjacent to the normal site of DNA rearrangement associated with V–D–J joining. This is one reason for believing that the V–D–J recombinase is aberrantly involved in the process of translocation.

5.1.2 *The bcl-2 gene at the chromosome 18 junction of t(14;18)*

A gene has been identified close to the chromosome 18 junction of the FL t(14;18) (7–9,95) and this gene, *bcl-2*, is thereby implicated in the pathogenesis of those follicular lymphomas carrying the translocation. Detailed analysis of the *bcl-2* gene has been carried out, revealing a gene with three exons (7–9,90). There is a long 5′ untranslated first exon, a small (220 bp) intron between exons I and II, and a very large intron (370 bp) between exons II and III (99). The promoter elements of *bcl-2* contain TATA and CCAAT motifs (99) as well as multiple potential Sp1 binding sites. Somatic mutations resulting in coding region changes, like those seen in translocated c-*myc*, have been found in translocated *bcl-2* (99).

Two chromosomal breakpoints regions have been described for the *bcl-2* gene on chromosome 18q21. The majority of these occur within the 3′ non-coding region of the third exon (7–9). However, a second region of breakage near the 5′ end of *bcl-2* has been found to occur at a reasonable frequency (97). In one such case, an unexpected order of the involved genes was observed (96), being 3′ $C\gamma S\gamma/S\mu$ J_H 5′:5′ *bcl-2* 3′. For this tumor translocation to have occurred, inversion must have occurred to generate the new order; this inversion could be either of *bcl-2* or of the

Igh locus. Interestingly, a stretch of ~ 170 nucleotides was found between the J_H6 element and the part of *bcl-2* locus involved in the translocation (96). The origin or meaning of this fragment is not known.

5.1.3 *Mechanism of t(14;18) translocation*

The precise mechanism of this chromosomal translocation remains controversial. The fact that translocation breakpoints have so far all been found within the J_H cluster obviously argues for errors of V–D–J recombinase activity in the creation of the translocation (100). However, there are two distinct viewpoints. One set of data argues for the existence of heptamer-like and nonamer-like signal sequences at the junction of the translocations on chromosome 18 (100,101). These sequences are suggested as the target for V–D–J recombinase resulting in an interchromosomal rearrangement (i.e. translocation) instead of the normal intrachromosomal V–D–J joining. In contrast to this, other data (102) argue that chromosome 18 cuts do not occur at recombinase signals. A model has been proposed in which illegitimate pairing occurs between D_H and J_H ends from chromosome 14 and staggered double-stranded breaks from chromosome 18, and resulting translocation. Clearly, these two models differ significantly in the role of recombinase.

5.1.4 *Effect of translocation of bcl-2 to the immunoglobulin chain locus*

The various translocation t(14;18) junctions which have been studied so far do not appear to truncate in any way the coding region of the *bcl-2*, and the only alterations reported are some point somatic mutations (99). The best hypothesis at the moment appears to be deregulation via transcription perturbation. The *bcl-2* mRNA has a fairly short half-life (99,103) and it displays transcriptional regulated expression in T and B cells in response to mitogens. The translocated *bcl-2* transcription product displays a similar half-life to normal *bcl-2* mRNA (99) and, therefore, the higher level of mRNA in cells with translocations must reflect higher levels of transcription or post-transcriptional processes. Interestingly, it seems that the untranslocated *bcl-2* allele in FL is transcriptionally silent (99) again suggesting as in the case of the c-*myc* oncogene, that tight controls are exerted on the gene.

The precise nature of the transcriptional deregulation remains uncertain. Many translocation breakpoints do occur in the 3′ untranslated region of *bcl-2* which might remove control elements. On the Igh side, all the breakpoints of the t(14;18) have been found in J_H segments, which leaves the H-chain transcription enhancer close to the translocation junction and therefore potentially active in *bcl-2* transcription. However, since most breakpoints occur at the 3′ end of *bcl-2* and since the *bcl-2* gene is about 370 kb long, the enhancer would need to operate over this large span if it is involved in *bcl-2* activation.

5.1.5 The protein product of the bcl-2 gene

Analysis of genomic and cDNA clones has identified two possible *bcl-2* open reading frames which could yield two proteins, one of 239 amino acids (designated α) and one of 205 amino acids (designated β). The putative α and β chains derive from overlapping reading frames and would be identical except at the carboxy terminus (8,104,105). No homologies with any known protein have been found. Antibodies were raised against the protein product of *bcl-2* (made in a bacterial expression vector system) and subcellular fractionation indicates that the protein is associated with the cell membrane (105) but the absence of a transmembrane domain makes it unlikely to span the cell membrane and behave as a receptor or like a protein (105).

Obviously, these studies are preliminary and much work needs to be done to determine the function the *bcl-2* protein and the role of this protein in tumorigenesis. One recent study bears directly on the latter problem (105). A retrovirus vector was constructed containing the *bcl-2* sequence and this was introduced into normal mouse bone marrow cells or in bone marrow derived from a strain of mice transgenic for the c-*myc* gene controlled by the Igh transcriptional enhancer. The latter cells developed a low number of colonies in soft agar (6.7×10^{-3} per cent of cells plated) whereas normal cells infected with *bcl-2* construct gave no colonies. However, none of the colonies proliferated in culture. Further experiments on growth factor dependence in the presence of the *bcl-2* construct were taken to mean that *bcl-2* provides a survival signal to cells and thus might contribute to neoplasia by allowing a clone to persist in an individual [after the translocation t(14;18)] until other oncogenes become activated by other means (e.g. c-*myc*) (106). This conclusion complements the observed co-existence of t(14;18) and t(8;14) in acute B-cell leukemias affecting *bcl-2* and c-*myc* respectively (107,108)

6. Chromosome band 14q11 in human T-cell leukemias

A number of human T-cell leukemias have abnormalities of the long arm of chromosome 14, principally involving bands q11 and q32. As described below, these bands correspond to the cytogenetic location of the T-cell receptor (TCR) α chain gene (14q11) (109,110) and the *Igh* gene (14q32) (111). The association of the breakpoints with these rearranging genes is obviously not coincidence but is related to these rearrangement processes (this is discussed in a later section).

6.1 Chromosome 14 inversion and translocation t(14;14)

6.1.1 Ataxia telangiectasia-associated abnormalities

Ataxia telangiectasia (A-T) is an autosomal recessive disease which is characterized by a number of disorders such as, for example, the apparent

DNA repair defect and a propensity for the development of T-cell leukemia (112). Cytogenetic studies on the white cells of these patients has shown that a number of different chromosomal abnormalities can be seen, usually in a small proportion of the circulatory T cells of a patient (113 – 115). These aberrations include t(14;14)(q11;q32), inv(14)(q11;q32), t(X;14), inv(7) (p15;q35), t(7;7)(p15;q35) and t(7;14)(p15;q11) or (q35;q11) (116). However, only t(14;14) and inv(14) abnormalities appear in tumor cells of A-T patients (94) or in otherwise normal patients (11,117 – 121). This clearly implicates sequences at 14q32 (or at 14q11) in the genesis of the tumors carrying these abnormalities.

In one A-T patient, an apparent clone of T cells carrying inv(14) (q11;q32) (initially observed at about 10% of total white cells) has been studied from time to time when no leukemic symptoms were visible in the patient, until the emergence of overt T-cell chronic lymphocytic leukemia (T-CLL), in which the T cells still contained the inv(14) chromosome (115). This developmental picture implies that inv(14) has a role in T-cell tumor pathogenesis but that it is itself insufficient to convert cells into the full tumorigenic state. Such a conversion presumably, therefore, requires secondary genetic alterations in other oncogenes. The location of the breakpoints of this A-T-associated T-CLL inv(14) has recently been studied (13). The chromosome 14q11 breakpoint occurs in the TCRα chain locus, immediately adjacent to the J$_\alpha$ segment (i.e. at the position where a V gene would normally join), and about 53 kb upstream of the C$_\alpha$ gene (13). Material from band 14q32 is fused to the J$_\alpha$ at this point and this material originates from a segment of 14q32 which occurs downstream (i.e. on the centromeric side) of the *Igh* locus. This breakpoint locus, variously called the 14q32.1 locus (12,13) or *tcl-1* (88), is thereby implicated in T cell pathogenesis.

6.1.2 The 14q32.1 locus in T-cell chronic lymphocytic leukemia and T-cell prolymphocytic leukemia

The inv(14) chromosome (q11;q32) and the analogous translocation t(14;14)(q11;q32) chromosome appears to be prevalent in T-CLL and T-PLL. Molecular cloning studies have demonstrated that patients with T-cell leukemias bearing these abnormal chromosomes have 14q11 breakpoints within the TCRα locus and precisely at the junction with J$_\alpha$ segments (12,13). The rearranged DNA segment at 14q32 [like that described for the T-CLL in the A-T patient (13)] is on the centromeric side of the *Igh* locus at this chromosome band. This strongly implies that the 14q32.1 locus harbors a gene whose new environment, after the chromosomal aberration has occurred, causes it to behave as a cancer gene and contribute to tumor pathogenesis. This conclusion has been strengthened by the observation that the breakpoint at 14q32.1 in the T-CLL from the A-T patient described above is only about 2100 bp away from that present in an inversion 14 chromosome from a non-A-T patient

(122). Thus, the locus putatively involved in the long-term evolution of the tumor in the T-CLL of the A-T patient is also disrupted by an inversion breakpoint in a non-A-T T-CLL patient.

6.1.3 Sporadic inv(14) chromosome

It has been known for some years that the stimulation of normal T cells in culture by phytohemagglutinin (PHA) allows the detection of cells carrying inv(14) and t(7;14) chromosomes (123–125). These sporadic abnormalities are probably unrelated to the tumor-associated ones and may represent the products of other events occurring in the division of the T cells in culture. Therefore, they probably are not involved in pathogenetic aspects of T-cell tumors. Cytogenetic studies of t(14;14) and inv(14) from normal and non-malignant but clonal T-cell expansions in A-T have led to the identification of two distinct breakpoints at 14q32: 14q32.3 in the former and 14q32.1 in the latter (126–128). Furthermore, one 14q32.1 breakpoint in a A-T-associated inv(14) developed from a non-malignant clone in an overt T-CLL (115) suggesting, as discussed above, that the 14q32.1 locus is the key feature for pathogenesis. What is the 14q32.3 breakpoint? Studies of one such breakpoint in a cell line derived from a T-cell lymphoma (129–131) showed that breakpoints occurred within the J_α segment and within the Ig V_H region. Thus, this representative of the sporadic inv(14) resulted from the mutual joining of two rearranging genes, similar to that which presumably takes place in PHA-cultured T lymphocytes.

In summary, the inv(14) and t(14;14) chromosomes all seem to have breakpoints with J_α at 14q11 (*Figure 6*) while there are two distinct breakpoints at 14q32. One (the sporadic breakpoint) probably occurs within the Ig V_H locus (at 14q32.3) and the other breakpoint seemingly has nothing to do with the *Igh* locus but breaks downstream (i.e. centromeric) of this locus at 14q32.1. The latter is thought to represent the locus involved in tumor etiology, presumably because a cancer gene is located at this position.

6.2 c-*myc* gene translocation in T cells

The c-*myc* gene is transcribed in many cell types and has been shown to be activated when resting T cells are stimulated in culture (132). Apparently, the activity of the gene is necessary for T cell division. In that case, it might be anticipated that chromosomal translocations occur in which the c-*myc* locus is involved in translocations with the TCR rearranging genes. Recently, several examples of such translocations have been described t(8.14)(q24;q11) in which the breakpoints occur just downstream of the c-*myc* gene (i.e. on the telomeric side) (133–136). As well as these translocations, T-cell lymphomas have been described with rearranged c-*myc* genes, apparently resulting from interstitial deletions downstream of c-*myc*, either close to the 3′ end (60) or about 300 kb downstream (68). Such rearrangements of the c-*myc* locus are analogous to those seen in the variant BL B-cell tumor where breaks are also found downstream of the c-*myc* gene. However, the latter breakpoints, in general, do not occur immediately adjacent to the 3' end of c-*myc* whereas the T-cell tumors do. Thus there may be differences in the way in which the translocations of B cells and T cells affect the c-*myc* gene, and the pathogenesis of these tumors may, therefore, follow different routes.

6.3 Other T-cell acute lymphocytic leukemias (T-ALL) and the T-cell receptor α chain locus at chromosome 14q11

There are at least three other chromosome abnormalities associated with the TCRα locus which are worthy of note. Two of these involve the short arm of chromosome 11 (see *Table 1*) and the other involves the long arm of chromosome 10 (137,138). Molecular cloning has so far been carried out on two of these three translocations t(11;14)(p15;q11) (15) and t(11;14)(p13;q11) (21). In a cell line carrying this translocation, the breakpoint at chromosome 14q11 has been identified joined to the $J_\delta 1$ segment associated with the C_δ gene (~ 12 kb downstream of J_δ). This locus probably represents the end of the J_α locus (the human C_δ is 85 kb upstream of C_α) and the equivalent mouse gene has been shown to rearrange early in mouse T-cell development (139). The translocation mechanism (discussed below) is probably due to the direct intervention of the V – J recombinase in a sequence-specific rearrangement. The 11p15 DNA (the breakpoint is at 11p15 distal to 11p15.1) which joins the TCRα has been shown to contain a gene transcribed in the cell carrying the translocation (15). This gene is thus a strong candidate for the type of cancer gene referred to in this chapter, the expression of which is subject to interference as a result of the chromosomal abnormality. The gene could therefore be involved in the pathology of the T-cell tumor.

The translocation t(11;14)(p13;q11) is a rather consistent abnormality associated with human T-ALL (17). Six tumors have so far been examined at the DNA level which carry this translocation (21), and five of these break within 2 kb of each other at 11p13. This region of chromosome 11p13 has been examined for the presence of transcribed sequences but none has been found so far (21). Like the t(11;14)(p15;q11), the 11p13 translocations occur to the TCR δ locus at 14q11. Molecular studies have shown that the chromosomal breakpoints involve joins with D – D – Jδ and germ-line Dδ configurations (21), thereby again providing evidence that aberrant recombinase activity is involved in the formation of chromosomal aberration. Examination of 11p13 sequences at junctions of translocations revealed only one case of a heptamer-like element (no nonamer) (21).

The other translocation described in T-ALL [t(10;14)(q24;q11)] has been studied by somatic cell genetics and been found to possess breakpoints at 14q11 with the TCRα locus (see *Table 1*). The implication is that the TCR J_δ or J_α segments are involved in a manner similar to t(11;14) (p15;q11) and t(8.14)(q24;q11). In addition, there is a clear implication that putative genes near to the translocation junctions at 10q24 is a potential cancer gene involved in pathogenesis of the afflicted tumors.

6.4 Adult T-cell leukemia

Adult T-cell leukemia (ATL) (otherwise known as ATL/lymphoma or ATLL) is a malignancy of mature T cells which is characterized by the presence of the retrovirus HTLV-I (140) (synonomous with ATLV). The disease is endemic to certain areas of south-western Japan where a high proportion (>10 per cent) of healthy people are seropositive for HTLV-I (141). Epidemiological studies suggest a causative role for HTLV-I in this class of T-cell lymphoma and it has been shown that proviral ATLV DNA is present in most (142) but not all (143) patients presenting with classical symptoms of ATL. The presence of a small group of patients whose clinically diagnosed ATL is ATLV⁻ suggests that other factors are also involved in pathogenesis of ATL. Thus, this disease has many parallels with EBV in BL, where both EBV⁺ and EBV⁻ forms of the disease are found.

The putative secondary factors which are involved in the genesis of ATL are obviously unknown. One clue that gross DNA changes may be involved was the recent observation of a deletion in the DNA of a cell line derived from an ATL patient (68). This deletion was found on chromosome band 8q24 within the human *pvt*-like region (see *Figure 5*), a region which is implicated in the pathogenesis of variant BL (67,68). Thus, the ATL-associated *pvt*-like deletion may be an important activator of the c-*myc* gene in this tumor, even though the involvement of HTLV-I in the deletion seems unlikely (68). More studies of DNA from ATL patients are clearly required to attempt to elucidate other abnormalities which might be present in these tumors.

The roles of HTLV-I in ATL and EBV in BL thus have clear parallels. Both viruses infect progenitors of the tumor cells and in doing so create 'stabilized' clones. Subsequently, these clones can acquire secondary DNA rearrangements (e.g. c-*myc* translocation in BL) which moves the clone toward malignancy. Whether this is sufficient to produce the tumor phenotype is debatable although recent work on conversion of non-tumorigenic EBV⁺ B cell lines into tumorigenic lines by transfection of activated c-*myc* genes suggest that it is sufficient (144).

7. Why are antigen receptor genes involved in lymphoid tumor-associated chromosomal abnormalities?

7.1 Normal rearrangement of antigen receptor genes

The lymphoid cells engaged in antigen-specific immune responses have, on their cell surfaces, antigen receptors for this purpose. These receptors have V regions which are cell-specific since a single receptor will only bind a single antigen. The V regions of these receptors are covalently linked to C regions. The genetic organization of the elements which encode these various segments is complex and their expression involves chromosomal DNA rearrangement during B- or T-cell differentiation (reviewed in ref. 145).

The Ig genes (the genes coding for the antigen receptor of B cells) and the TCR genes (the genes coding for the antigen receptor of T cells) show a very marked degree of conservation in the basic composition of inherited DNA segments and in the mechanism whereby these segments rearrange in the development of the antigen specific cells. All the known rearranging genes (shown in *Table 3*, together with their chromosomal locations) have, in germ line DNA, a set of V region genes, a set of D segments (these only occur in *Igh*, TCRβ and probably TCRα) and a set of J segments separated from one or more C genes by an intervening sequence or intron. The feature of these genes which is different from other genes is that the V – D – J segments are rearranged in the mature B or T cells by an enzyme system called recombinase. This enzyme is probably common to T and B cells and seems to operate by recognition of precise signal sequences near to the elements to be joined (V, D and J) [this is extensively reviewed by Tonegawa (145) for the Ig genes]. These signal sequences occur downstream of V genes, on either side of D segments and upstream of J segments, and consist of heptamer and nanomer sequence separated by 12 or 23 bp. The rule that applies to the recombinase-mediated joining is that heptamer – nanomer motifs must be separated by 12 bp on one side of the join and 23 bp on the other side (145).

In the *Igh* locus, there is a further rearrangement process called the class switch. This is the switching of a joined V gene (i.e. a complete, productive V – D – J) from one C_H gene to another. This is not mediated by heptamer – nanomer signal sequences but rather by an internally

Table 3. Chromosomal location of lymphoid rearranging genes

Gene		Chromosome band
B-cell genes	Igh	14q32
	Ig$\varkappa$	2p12
	Igλ	22q11
T-cell genes[a]	TRGγ	7p15
	TCRβ	7q32
	TCRα	14q11
	TCRδ	14q11

[a]The TCR$\varkappa$ gene recently described in mouse (139) has a human counterpart at the end of the TCRα locus (15). This gene which rearranges early in T-cell ontogeny is the TCRδ gene, the partner of the TRGγ gene product.

repetitive element (called an S sequence) which is repeated upstream of all the C$_H$ genes (146). This switch process seems to be restricted to the *Igh* locus.

7.2   Mechanism of translocation and inversion involving recombinase errors

Chromosomal aberrations may arise either by chance events between chromosomes or they may occur by usurping pre-existing mechanisms for chromosomal rearrangement. Obviously, aberrations found in B or T cells associated with rearranging antigen receptor genes can fall into the latter category. Furthermore, the creation of a chromosomal abnormality utilizing a rearranging gene locus can occur by any of three processes:

(i)   by random association of these latter loci with any other chromosomal region followed by selection of the cell carrying the abnormality by virtue of the consequent growth advantage conferred on the cell;

(ii) by direction to a particular region by the specific use of heptamer and nanomer recombination signals, as recently demonstrated for an inversion 14 chromosome (129–131);

(iii) by use of sequences related to the heptamer and nanomer sequences but which are not themselves apparently normally involved in DNA rearrangement.

Clear examples of the latter two types of situation have been documented. The inv(14) chromosome of the sporadic type [which we believe is not related to tumor pathogenesis (129,131)] is the best example of the specific use of heptamer–nanomer recombination signals. Many examples for tumor breakpoints in which sequences apparently related to signal sequences have been identified. This association of signal-like sequences with, for example, the follicular lymphoma t(14;18) breakpoints has been suggested (7), but there is some dispute about whether this is always the case (147). The cases where no apparent signal-like sequence can be identified presumably places the mechanism for the translocation in the first category mentioned above.

The t(11;14) (p15;q11) seen in the human T-ALL-derived T-cell line RPMI 8402 (14,15) and in one example of t(11;14)(p13;q11) (21) seem to have arisen by the third of these possibilities but only utilizing heptamer-like sequences. A perfect match to a heptamer consensus sequence occurs on chromosome 11, exactly at the borders of the translocation breakpoints, and these heptamers are back-to-back with a heptamer sequence from chromosome 14, which is separated by 12 bp from a nanomer sequence. This back-to-back organization of the reciprocal rearrangements is reminiscent of the residual DNA resulting from normal V–D–J joins. The strong implication of this sequence arrangement is that the recombinase normally involved in V–D–J rearrangement is involved in the chromosomal translocation process by artefactually recognizing the chromosome 11 heptamers as that downstream of a V segment. Translocations thereby result between chromosomes rather than the normal intra-chromosomal rearrangement. The possible presence of a D element at the translocation breakpoints supports this general mechanism. The chromosomal translocation in RPMI 8402 cells takes place near to the $J_\delta 1$ sequence. This is the human equivalent of the mouse element (139) which has been shown to be rearranged and expressed at early times in T-cell development. RPMI 8402 cells do not express CD3, CD4 or CD8 on the cell surface and can be regarded as early thymocytes. The association of the chromosomal translocation t(11;14) with the J_δ segment and the data discussed above suggest that normal rearrangement took place at the J_δ locus in the early thymocyte precursor of the tumor cell perhaps firstly with a $D_\delta - J_\delta$ fusion. Then by a mistake mediated through the conserved chromosome 11 heptamer the recombinase generated an aberrant inter-chromosome rearrangement instead of completing a $V - D_\delta - J_\delta$ fusion.

The idea that chromosome aberrations have some basis in the formation of cancers stems from the consistent observation of these changes in primary tumors. The involvement of the $V - J$ recombinase in the creation of such abnormalities poses some problems if it is assumed that a heptamer sequence (as discussed above) at the translocation breakpoint is involved. Assuming a random occurrence, a given heptamer sequence should appear every 1.6×10^4 bp or 1.8×10^5 times in a haploid genome. Clearly this frequency would result in an enormous error rate if the mere presence of this sequence were necessary to facilitate translocation. However, it seems likely that the chromosomal configuration at such sequences determines their accessibility for the recombination enzyme(s). In addition, the reasonable additional assumption must be made that chromosomal aberrations do occur on a sporadic basis but the vast majority of these are innocuous. However, the rare aberration which occurs in a crucial chromosomal region will provide the afflicted cell with a growth advantage or immortality, allowing it to progress to an overt tumor.

7.3   The $J_\delta - J_\alpha$ region of chromosome 14q11 is a hot spot for chromosomal abnormalities

The $J_\delta - J_\alpha$ locus in man occupies about 100 kb of DNA (15). One end of this region is the C_α gene (telomeric side) and the other is the J_δ gene (centromeric side). A large number of J_α segments occur between the human C_δ and C_α (148). Molecular studies of T-cell leukemias have shown that this 100 kb region is a hot spot for involvement in a number of different chromosomal abnormalities apparently involved in different T-cell leukemias. The positions of identified chromosome aberration breakpoints within this region are illustrated in *Figure 7*. These include T-PLL (12), CLL (12,13) and ALL (15). Other breakpoints have also been identified in the J_α region but precise locations in our map are not

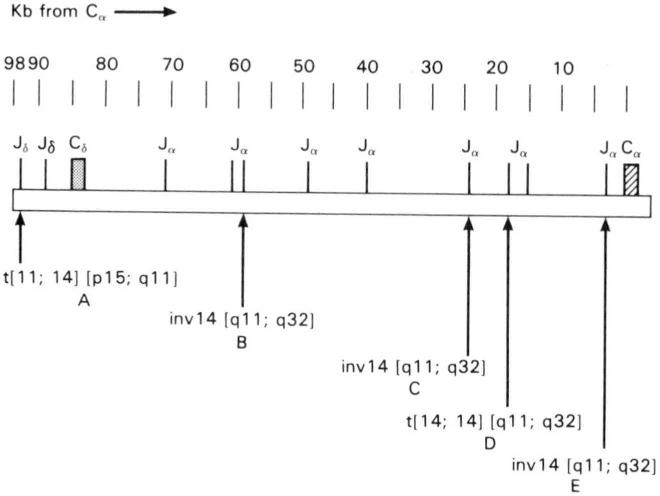

Kb from C_α →

Abnormality	Tumour	Breakpoint	Surface phenotype
t[11; 14] [p15; q11]	T-ALL	A	CD3− CD4− CD8−
inv14 [q11; q32]	T-CLL	B	CD3+ CD4− CD8+
inv14 [q11; q32]	T-CLL	C	CD3+ CD4+ CD8−
t[14; 14] [q11; q32]	T-PLL	D	not known
inv14 [q11; q32]	T lymphoma	E	CD3+ CD4+ CD8+

Figure 7. Map of $C_\delta - C_\alpha$ locus at chromosome band 14q11. The positions of a number of relevant J_α segments are shown together with breakpoints of the various tumors listed.

available (134 – 136,149). The range of T-cell types involved covers a wide spectrum, ranging from very immature cells such as RPMI 8402 (15) to $CD3^+, CD4^+$ (12) and $CD3^+, CD8^+$ cell types (13). Rearrangments at the TCR_α locus take place both at early ($J_\delta - C_\delta$) and late ($J_\alpha - C_\alpha$) stages of T-cell differentiation. To date, molecular cloning has been described in five different chromosomal abnormalities involving chromosome band 14q11. These are inv(14)(q11;q32) (12,13), t(14;14)(q24;q11) (12), t(11;14) (p15;q11) (14), t(11;14)(p13;q11) (21) and t(8;14)(q24;q11) (133 – 136). Each of these breakpoints involves joining with the 5′ end of J segments spread throughout the $J_\delta - J_\alpha$ locus, with events which eliminate the recombination signals at the site of inversion or translocation. Therefore, the involvement of the V – J recombinase in all of these rearrangements seems likely, some with the use of specific recombination signals and some with the use of sequences homologous to recombination signals.

7.4 Involvement of switch region sequences in Burkitt's lymphoma

The *Igh* locus has extra rearrangement capability which is the so-called class switch involving S region sequences. The mediation of translocations in sporadic BL by S regions has been noted above. There seems to be no sequence specificity at the c-*myc* locus responsible for this chromosome translocation; rather it would seem that pre-leukemic precursor B cells undergoing the class switch generate a *trans*-rearrangement (translocation) rather than the usual *cis*-rearrangement ($V_H - D_H - J_H$ switching). The cell carrying this abnormality thereby acquires a growth advantage in those cases where the translocation involves the c-*myc* locus (by virtue of the c-*myc* gene activation). Thus in these tumors, selective cell growth is the driving force for the emergence of tumor cells resulting from random (rather than sequence targeted) translocation.

8. Gene amplification

A very striking difference between normal cells and tumor cells is the occurrence of the chromosomal abnormalities known as double minutes (DM) and homogeneously staining regions (HSRs). These abnormalities only occur in tumor cells, although they can be induced artificially in cell lines which have become resistant to high levels of cytotoxic drugs. DMs are self-replicating, small, paired bodies containing chromosomal material. They have no centromere and therefore segregate randomly at mitosis. HSRs occur within chromosomes and are recognized when the normal banding pattern of a chromosome seen after staining is interrupted by a region of even staining. These regions are often very large and result in chromosomes which are noticeably larger than their homolog. DMs and HSRs are considered together because they are thought to represent the same structure. Cell lines exist in which different clones contain either DMs or HSRs and where the same gene has been localized to both structures (150). Experiments with cell lines made drug resistant *in vitro* (151) indicated that DMs and HSRs were the site of amplified genes (in these cases, the genes responsible for drug resistance), so that when oncogenes were found amplified in tumor cells, and these tumor cells contained DMs, it was a small step to show that the chromosomal abnormalities were often the site of amplified oncogenes.

  The amplification of oncogenes has usually been detected in tumor cell lines rather than tumor biopsy samples. This is partly because oncogenes are more frequently amplified in cell lines (see below) and partly because cell lines are more favored for these studies than biopsy samples. There are many examples of different oncogenes amplified in cell lines isolated from a variety of tumor types. Frequently quoted types are c-*myc*, amplified about 20-fold in the acute prolymphocytic cell line HL60 (152), and c-*abl*, amplified 10-fold in K562, a cell line from a patient with chronic myelogenous leukemia (153). Solid tumors also have examples of amplified genes; c-*erb*B is amplified 15- to 20-fold in the cell line A431, a vulval epidermoid carcinoma, and c-*myc* is amplified about 30-fold in the colon carcinoma cell line COLO 320 (150).

Besides these examples of oncogenes which are amplified sporadically, there are a few tumors where a much larger proportion of samples have particular oncogenes amplified. After the observation that c-*myc* was amplified in a colon carcinoma cell line containing DMs and HSRs, neuroblastoma cell lines were analyzed with c-*myc* probes in a search for amplified sequences because DMs and HSRs are also often observed in these lines (154). This led to the discovery of a gene homologous to c-*myc* called N-*myc* (155) and to the observation that this gene was frequently amplified in neuroblastoma cell lines and in neuroblastoma tumors (155,156). In the Brodeur study (156), 24 out of 63 (39%) untreated patients showed amplification; 100- to 300-fold in 12 cases and 3- to 10-fold in 10 others. In cell lines, the proportion of samples showing amplification is much higher. When the neuroblastoma tumor samples are divided according to clinical staging, then tumors of stages III and IV have N-*myc* amplification in about 50% of cases. Cell lines are only easily established from patients with advanced disease so it is not surprising that studies of cell lines show a higher proportion of N-*myc* amplification than tumor samples.

The observation that N-*myc* is more often amplified in stages III and IV (>50%, 156) and not at all in stages I and II (156) has led to the belief that N-*myc* amplification is a prognostic factor in neuroblastoma since patients with stage I or II disease usually have a 75–90% two-year survival whereas patients with stage III or IV disease have a 10–30% two-year survival (157). Seeger *et al.* (158) analyzed the N-*myc* copy number in 89 untreated primer neuroblastoma patients and looked at progression-free survival after 18 months. Estimated progression-free survival was 70% with unamplified N-*myc*, 30% with 3–10 N-*myc* copies and 5% for patients with more than 10 copies, that is, the more copies of N-*myc*, the more likely that during the 18 months follow up, the patients' disease will have progressed.

Noteworthy were the patients with stage IVS disease. This, like stage IV, has distant metastases but the primary tumor is small, as seen in stages I and II. These patients usually have a good prognosis, tumors often regressing spontaneously (159); in this study all five patients had single copies of the N-*myc* gene and all five were alive 18 months after initial staging. However, the very important observation has been made that for any one patient progression of the disease does not result in increased N-*myc* copy number with time (160).

Lung tumors, in particular those of the small-cell type (SCLC), have members of the *myc* family amplified. c-*myc*, N-*myc* and L-*myc*, the latter an oncogene recently discovered because of its amplification in small-cell lung tumors, are all found amplified. One sampling of 31 cell lines found L-, c- and N-*myc* amplified in 14 of them (161). L-*myc* amplification was not tested in this particular study. However, in another study of 44 cell lines, 13 were amplified for c-, N- and L-*myc* (162). SCLC cell lines are divided into two groups (classic and variant) according to various phenotypic parameters. These two forms are also seen in patients; the variant form having a poorer prognosis. It is assumed that the two forms are equivalent *in vitro* and *in vivo*. The observation has been made that the variant cell lines often carry amplified c-*myc* genes whereas the classics rarely do (163), so that it has been suggested that amplified c-*myc* is a prognostic indicator of the more aggressive form of the disease analogous to N-*myc* amplification in neuroblastoma. However, *myc* genes seem to be amplified much less frequently in tumor samples from patients with small-cell carcinoma of the lung than they are in cell lines. Wong *et al.* (164) examined DNA from 45 patients extracted from paraffin-embedded samples and found evidence of *myc* gene amplification in only five of them. Of these five, two had a good response to therapy but the authors do not state whether these patients had N- or c-*myc* amplified.

Recently, it has been shown that *myc* gene amplification is more often seen in cell lines isolated from patients after chemotherapy than in cell lines from previously untreated patients (162). DNA samples have been isolated from 38 small cell tumors and examined for *myc* family gene amplification. Six of these tumor DNAs had amplified L- or N-*myc* sequences (none had c-*myc* amplification) and these were all from patients who had previously received chemotherapy. Thus it seems likely that in SCLC, amplification of *myc* genes represents a response to selection, that is, the ability to grow in tissue culture or ability to grow in presence of cytotoxic drugs. So far, therefore, there is no confirmation of the *in vitro* studies which suggest that c-*myc* is a prognostic indicator of the more aggressive form of small-cell lung cancer.

The oncogene called HER-2/*neu* or c-erbB2 has been found amplified in 58 of 257 human adenocarcinomas (165 – 167). In the work on breast cancer, 26% of 189 tumors showed amplification of c-erbB2. Amplification correlated with poor prognosis and was a stronger predictor of disease progression than the presence of estrogen or progesterone receptors (previously shown to have some prognostic value). The amplification is associated with over-expression of the protein which can be detected by immunohistochemical staining, a technique which may prove useful in predicting tumor progression in the clinic (168).

Proto-oncogenes can be activated to behave as oncogenes in several different ways. When the structure of amplified oncogenes is compared with their single copy counterpart it is infrequently rearranged as would be expected if a new chromosomal position was responsible for activation. Where the sequence of the amplified oncogene and the unamplified oncogene has been compared by restriction enzyme mapping, no differences have been detected. For one oncogene, N-*myc*, a complete sequence comparison between a single copy N-*myc* gene and one of the copies from an array of amplified N-*myc* genes is possible. No sequence differences have been found (169). Therefore the activation event is most probably over-production of normal protein.

9. Deletion and recessive cancer genes in tumors

Oncogenes, identified by transfection assays and acting at the site of translocations, are thought to act dominantly in a genetic sense, but there is evidence that 'recessive' oncogenes also exist. Oncogene is probably a misnomer as these genes may act by suppressing genes (or the product of genes) whose aberrant activity leads to uncontrolled cellular proliferation and invasiveness. The best evidence that tumor-suppressing genes exist comes from analysis of the tumorigenicity of hybrids formed between normal and tumor cells (179,180). A recent refinement of this is the observation that when normal chromosome 11 is introduced into Wilms tumor cells (which have a deletion at 11p13 thought to be the site of a repressor gene), these cells lose their capacity for growth in nude mice (181)—currently the best indicator of tumorigenicity. A variety of human tumors exhibit chromosomal deletions and these deletions are expected to represent the location of such genes. *Table 4* lists some common deletions and others have been recently reviewed (178). Originally deletions were identified by cytogenetic analysis of tumor biopsy specimens, but more recently detection of allelic loss has been demonstrated by reduction to homozygosity of an allele for which the patient is constitutionally heterozygous.

Apart from those chromosomal deletions discussed in detail below, recent molecular analysis of RFLP has illustrated that meningioma (generally benign intracranial tumors) is characterized by loss of part or all of chromosome 22 (177). In an analogous type of analysis, reduction to homozygosity was elucidated for regions of chromosome 11p in breast tumors (176) and in insulinoma (182). Using mini-satellite DNA, loss of material from chromosome 1p was found in endocrine neoplasia of an inherited cancer syndrome, multiple endocrine neoplasia type 2 (183).

Work on other familial cancer occurrence has located genes important for tumor origin and/or progression. Particularly, recent studies have been

Table 4. Chromosomal deletions/allelic loss and human tumors

Tumor	Locations of deletion	Reference
Neuroblastoma	del 1p (p31-p36)	170
Small-cell lung cancer	del 3p (p14-p23)	171
Retinoblastoma	del 13q (q13)	172
Wilms tumor	del 11p (p13)	2
Renal cell carcinoma	del 3p	173
Transitional cell carcinoma of the bladder	del 11p	174
Colorectal carcinoma	del 5q	175
Breast	del 11p	176
Meningioma	del 22	177

concerned with colorectal tumors, such as familial adenomatous polyposis or familial polyposis coli (FAP). Several studies of allelic loss have identified chromosome 5 as crucial, probably in the area of chromosome 5q21-q22 (184–189). Other studies of colorectal tumors implicate in addition chromosome 17p and chromosome 18 allelic loss (190).

9.1 Small-cell lung cancer and 3p deletion

The history of the deletion in the short arm of chromosome 3 in SCLC illustrates the advantages of the newer approach to molecular biology. Using only cytogenetic analysis of cell lines, two large studies disagreed about the importance of the deletions in SCLC. Whang Peng et al. (171) found a deletion in all 20 samples examined (mainly cell lines), but another group, working with different cell lines of their own, found chromosome 3p deletions in only three out of the 19 lines examined (191). The few cytogenetic analyses of biopsy material (as opposed to cell lines) all showed a deletion, but this material is a very poor source of chromosomes. The discrepancies were resolved by the comparison of DNA isolated from paired samples of normal and tumor tissue of patients with SCLC analyzed with probe pH_3H_2, a probe for a restriction fragment length polymorphism (RFLP). All patients who are heterozygous for the RFLP constitutionally, show reductions to homozygosity at the locus detected by this probe (192). Since deletions have been frequently observed in chromosome 3 in cell lines isolated from patients with this disease, the RFLP results presumably reflect this mechanism. The probe pH_3H_2 has been shown by *in situ* hybridization to localize to the p21 band on chromosome 3. The deletions seen in SCLC are large and this band is almost invariably included in the deletion. So far, no candidate genes have been identified from the deleted region which might be involved in tumor pathogenesis.

Using three RFLP probes for chromosome 3 (173), 11 patients with renal cell carcinoma were evaluated and it was found that in all cases there was reduction to homozygosity for at least one of the probes. The probe which showed reduction to homozygosity was always the one nearest to the deletion in SCLC, suggesting a common lesion. There is also cytogenetic evidence from other studies of renal cell carcinoma that the region of chromosome 3 involved may be the same as in SCLC (193). The observation of a possible common genetic lesion in distinct tumor types has also been made for chromosome 11, deletions of which are thought to be involved in Wilms tumor and several embryonal tumors (194).

9.2 Wilms tumor and 11p13 deletion

Wilms tumor is a malignancy of embryonal kidney in which there is a very strong correlation to a deletion of the chromosome band 11p13 (2). The tumor is associated with a number of other defects, indicating the existence of a complex locus at 11p13 whose activity is important in development. Wilms tumor is both familial and sporadic, the suggestion being that the tumor arises by a two-hit process (195). Thus, in familial Wilms patients, a pre-existing constitutional deletion affects the band 11p13 and a second putative event affects the 'normal' allele of 11p13 to generate the tumor. In sporadic Wilms tumors, it must be assumed that both 11p13 loci are affected by random mutations, to allow tumor formation.

No gene has yet been identified which might be involved in the pathogenesis of this tumor..Clearly, this is a major goal of molecular biology at the present time.

Much work has gone into derivation of probes from chromosome 11p13 which will help to locate the Wilms predisposition gene. In this way, the gene has been placed between the genes for catalase (CAT) and follicule stimulating hormone β subunit, FSH. The occurrence of Wilms tumor is frequently associated with aniridia (AN) (lack of iris development) and genito-urinary dysplasia (GU) (the so-called Wilms triad), and these associations are in turn due to 11p13 deletions. Since these deletions vary markedly in size, it has been possible to establish the order of these loci by studying somatic cell hybrids derived from tumors with such heterogeneous deletions. The order is – (centromere) – CAT – Wilms – AN2 – GU – FSH – (telomere).

Recently, a candidate gene from the 11p13 area, identified via a translocation in T-cell leukemia, has been placed between CAT and Wilms (196). The possibility that this 'T-cell gene' represents part of the Wilms predisposition locus remains open.

9.3 Retinoblastoma

Retinoblastoma occurs in a sporadic and hereditary form. In the hereditary form, susceptibility is inherited as an autosomal dominant trait. After the deletion of band q14 in chromosome 13 was observed in retinoblastoma samples, examination of pedigrees of families suffering from this disease showed a very close linkage of the phenotype and defective genotype. The sporadic form of the disease also shows chromosomal abnormalities of chromosome 13. It was proposed (197) that inactivation of both alleles of a gene (called RB) located in band q14 on chromosome 13 was necessary for development of the disease and that in the inherited form, inactivation of one allele was inherited requiring only one more mutation (in the other allele) for disease development (197). Using RFLP probes for chromosome 13, it was shown that reduction to homozygosity occurred in the tumors of patients who were constitutionally heterozygous (198). Tumor samples which showed no visible deletions showed this reduction to homozygosity, indicating the presence of microdeletions. Using a probe for the esterase D gene, which is known to be closely linked to the RB gene, it has been possible to isolate the RB gene by chromosome walking (199,200). The gene consists of 12 exons scattered over more than 100 kb of DNA and produces a transcript of 4.6 kb which is expressed in placenta and fetal retina. In the six retinoblastoma samples tested, none showed a normal transcription pattern. Some of these tumor samples had genes which appeared normal on Southern blot analysis, suggesting that point mutations were responsible for the altered transcripts.

Studies of the sequence of the retinoblastoma cDNAs and immuno-precipitation of protein show that the gene product is a nuclear phosphoprotein (201) apparently with DNA binding properties. Interesting observations followed, showing that the product of retinoblastoma gene associates with the transforming protein EIA of adenovirus (202). This interaction of oncogene and anti-oncogene products raises many speculations about functions of such proteins as growth and/or differentiation. The recent observation of structural abnormalities in the 'retinoblastoma' gene in small-cell lung carcinoma (203) and in breast carcinoma (204) implies a more general role for the gene product.

10. Other consistent chromosomal cancer genes in tumors

A number of other consistently occurring chromosomal abnormalities have been described in various tumors, but these have not yet been amenable to study because of the lack of suitable probes near to the respective chromosome breakpoints. *Table 5* lists a few of the most common of such abnormalities. Obviously, this is not a totally exhaustive list of chromosome abnormalities but merely serves to further illustrate the point that many specific tumor types have chromosomal changes and that these changes are most likely involved in tumor pathogenesis. It is a very important issue now that genes identified at or near to the junctions of chromosomal aberrations should be examined carefully for their ability to act as cancer genes, using *in vitro* and *in vivo* tests. If, as should be expected, these various genes are indeed shown to be capable of transforming cells, then the long-standing assumption of cancer cyto-genetics that chromosomal abnormalities are crucial to tumor pathogenesis will be validated and the molecular basis of tumor formation will be better understood.

Table 5.   Other consistent chromosomal abnormalities

Tumor type	Abnormality
Pre-B cell	t(1;19)(q23;p13.3)
B cell	t(4;11)(q21;q23)
Acute myelogenous leukemia	t(8;21)(q22;q22)
Acute non-lymphocytic leukemia	t(5;21)(q13;q22)
	t(15;17)(q22;q21)
Ewings sarcoma	t(11;22)(q24;q12)

11. References

1. Nowell,P.C. and Hungerford,D.A. (1960) Chromosome studies on normal and leukaemic human leykocytes. *J. Natl. Cancer. Inst.*, **25**, 85.
2. Riccardi,V.M., Sujansky,E., Smith,A.C. and Francke,U. (1978) Chromosomal imbalance in the aniridia-Wilms tumour association: 11p interstitial deletion. *Pediatrics*, **61**, 604.
3. Manolov,G. and Manolova,Y. (1972) Marker band in one chromosome 14 from Burkitt lymphoma. *Nature*, **237**, 33.
4. Zech,L., Haglund,U., Nilsson,K. and Klein,G. (1976) Characteristic chromosome abnormalities in biopsies and lymphoid-cell lines from patients with Burkitt and non-Burkitt lymphomas. *Int. J. Cancer*, **17**, 47.
5. Bernheim,A., Berger,R. and Lenoir,G. (1981) Cytogenetic studies on African Burkitt's lymphoma cell lines: t(8;14) t(2;8) and t(8;22) translocations. *Cancer Cytogenet.*, **3**, 307.
6. Yunis,J.J. (1983) The chromosomal basis of human neoplasia. *Science*, **221**, 227.
7. Tsujimoto,Y., Finger,L.R., Yunis,J., Nowell,P.C. and Croce,C.M. (1984) Cloning of the chromosome breakpoint of neoplastic B cells with the t(14;18) chromosome translocation. *Science*, **226**, 1097.
8. Cleary,M.L., Smith,S.D. and Sklar,J. (1986) Cloning and structural analysis of cDNAs for *bcl-2* and a hybrid *bcl-2/immunoglobulin* transcript resulting from the t(14;18) translocation. *Cell*, **47**, 19.
9. Bakhshi,A., Jensen,J.P., Goldman,P.G., Wright,J.J., McBridge,O.W., Epstein,A.L. and Korsmeyer,S.J. (1985) Cloning the chromosomal breakpoint of t(14;18) human lymphomas: clustering around J_H on chromosome 14 and near a transcriptional unit on 18. *Cell*, **41**, 889.
10. Tsujimoto,Y., Yunis,J., Onorato-Showe,L., Erikson,J., Nowell,P.C. and Croce,C.M. (1984) Molecular cloning of the chromosomal breakpoint of B-cell lymphomas and leukaemis with the t(11;14) chromosome translocation. *Science*, **224**, 1403.
11. Zech,L., Gahrton,G., Hammarström,L., Juliusson,G., Mellstedt,H., Robert,K.H. and Smith,C.I.E. (1984) Inversion of chromosome 14 marks human T cell chronic lymphocytic leukaemia. *Nature*, **308**, 858.
12. Mengle-Gaw,L., Willard,H.F., Smith,C.I.E., Hammarström,L., Fischer,P., Sherrington,P., Lucas,G., Thompson,P.W., Baer,R. and Rabbitts,T.H. (1987) Human T cell tumours containing chromosome 14 inversion or translocation with breakpoints proximal to immunoglobulin joining regions at 14q23. *EMBO J.*, **6**, 2273.
13. Baer,R., Heppell,A., Taylor,A.M.R., Rabbitts,P.H., Boullier,B. and Rabbitts,T.H. (1987) The breakpoint of an inversion chromosome 14 in a T cell leukaemia; sequences downstream of the immunoglobulin heavy chain locus implicated in tumorigenesis. *Proc. Natl. Acad. Sci. USA*, **84**, 9069.
14. LeBeau,M.M., McKeithan,T.W., Shima,E.A., Goldman-Leikin,E., Chan,S.J., Bell,G.I., Rowley,J.D. and Diaz,M.O. (1986) T cell receptor α chain gene is split in a human T cell leukaemia cell line with a t(11;14)(p15;q11). *Proc. Natl. Acad. Sci. USA*, **83**, 9744.
15. Boehm,T., Baer,R., Lavenir,I., Forster,A., Waters,J.J., Nacheva,E. and Rabbitts,T.H. (1988) A chromosomal translocation t(11;14) in a human leukaemia involving the T cell receptor Cδ locus on chromosome 14q11 and a transcriptionally active region of chromosome 11p15. *EMBO J.*, **7**, 385.
16. Takasaki,N., Kaneko,Y., Maseki,N., Sakurai,M., Shimawura,K. and Takayama,S. (1987) Hemophagocyte syndrome: T cell acute lymphoblastic leukaemia with a novel t(11;14)(p15;q11) chromosome translocation. *Cancer Res.*, **59**, 424.
17. Williams,D.L., Look,A.T., Melvin,S.L., Roberson,P.K., Dahl,G., Flake,T. and Stass,S. (1984) New chromosomal translocations correlate with specific immunophenotypes of childhood acute lymphoblastic leukaemia. *Cell*, **35**, 101.
18. White,L., Meyer,P.R. and Benedict,W.F. (1984) Establishment and characterisation of a human T-cell leukaemia line (LALW-2) in nude mice. *J. Natl. Cancer Inst.*, **72**, 1029.
19. Lewis,W.H., Michalopoulos,E.E., Williams,D.L., Minden,M.D. and Mak,T.W. (1985) Breakpoints in the human T-cell antigen receptor α chain locus in two T-cell leukaemia patients with chromosomal translocations. *Nature*, **317**, 544.
20. Erikson,J., Williams,D.L., Finan,J., Nowell,P.C. and Croce,C.M. (1985) Locus of the

α-chain of the T cell receptor is split by chromosome translocation in T-cell leukaemias. *Science,* **229**, 784.

21. Boehm,T., Buluwela,L., Williams,D., White,L. and Rabbitts,T.H. (1988) A cluster of chromosome 11p13 translocations found via distinct D–D and D–D–J rearrangemements of the human T cell receptor δ chain gene. *EMBO J.,* **7**, 2011.
22. de Taisne,C., Gegonne,A., Stehelin,D., Bernheim,A. and Berger,R. (1984) Chromosomal localisation of the human proto-oncogene c-*ets*. *Nature,* **310**, 581.
23. Sacchi,N., Watson,D.K., Guerts van Kessel,A.H.M., Hagemeijer,A., Kersey,J., Drabkin,H.D., Patterson,D. and Papas,T.S. (1986) Hu-*ets*-1 and Hu-*ets*-2 genes are transposed in acute leukaemias with (4;11) and (8;21) translocations. *Science,* **231**, 379.
24. de Klein,A., Guerts van Kessel,A., Grosveld,G., Bartram,C.R., Hagemeijer,A., Bootsma,D., Spurr,N.K., Heisterkamp,N., Groffen,J. and Stephenson,J.R. (1982) A cellular oncogene is translocated to the Philadelphia chromosome in chronic myelocytic leukaemia. *Nature,* **300**, 765.
25. Bartram,C.R., de Klein,A., Hagemeijer,A., van Agthoven,T., Guerts van Kessel, A.-G., Bootsma,D., Grosveld,G., Ferguson-Smith,M.A., Davies,T., Stone,M., Heisterkamp,N., Stephenson,J.R. and Groffen,J. (1983) Translocation of c-*abl* oncogene correlates with the presence of a Philadelphia chromosome in chronic myelocytic leukaemia. *Nature,* **306**, 277.
26. Grosveld,G., Verwoerd,J., van Agthoven,T., de Klein,A., Ramachandran,K.L., Heisterkamp,N., Stam,K. and Groffen,J. (1986) The chronic myelocytic cell line K562 contains a breakpoint in *bcr* and produces a chimeric *bcr/c-abl* transcript. *Mol. Cell. Biol.,* **6**, 607.
27. Groffen,J., Stephenson,J.R., Heisterkamp,N., de Klein,A., Bartram,C. and Grosveld,G. (1984) Philadelphia chromosome breakpoints are clustered within a limited region *bcr* on chromosome 22. *Cell,* **36**, 93.
28. Heisterkamp,N., Stam,K., Groffen,J., de Klein,A. and Grosveld,G. (1985) Structural organisation of the *bcr* gene and its role in the Ph′ translocation. *Nature,* **315**, 758.
29. Shtivelman,E., Lifshity,B., Gale,R.P., Roe,B.A. and Canaani,E. (1986) Alternative splicing of RNAs transcribed from the human *abl* gene and from the *bcr*–*abl* fused gene. *Cell,* **47**, 277.
30. Bartram,C.R., Janssen,J.W.G., Becher,R., de Klein,A. and Grosveld,G. (1986) Persistence of chronic myelocytic leukaemia despite deletion of rearranged *bcr*–*c-abl* sequences in blast crisis. *J. Exp. Med.,* **164**, 1389.
31. Clark,S.S., McLaughlin,J., Crist,W.M., Champlin,R. and Witte,O.N. (1987) Unique forms of the abl tyrosine kinase distinguish Ph′-positive CML from Ph′-positive ALL. *Science,* **235**, 85.
32. Daley,G.Q., McLaughlin,J., Witte,O.N. and Baltimore,D. (1987) The CML-specific P210 bcr/abl protein, unlike v-abl, does not transform NIH-3T3 fibroblasts. *Science,* **237**, 532.
33. Catovsky,D. (1979) Ph′ positive acute leukaemia and chronic granulocyte leukaemia: one or two diseases. *Br. J. Haematol.,* **42**, 493.
34. Rowley,J.D. (1973) A new consistent chromosomal abnormality in chronic myelogenous leukaemia identified by quinacrine fluorescence and Giemsa staining. *Nature,* **243**, 290.
35. Koeffler,H.P. and Golde,D.W. (1981) Chronic myelogenous leukaemia—new concepts. *N. Engl. J. Med.,* **304**, 1201.
36. de Klein,A., Hagemeijer,A., Bartram,C.R., Houwen,R., Hoefsloot,L., Carbonell,F., Chan,L., Barnett,M., Greaves,M., Kleihauer,E., Heisterkamp,N., Groffen,J. and Grosveld,G. (1986) bcr rearrangement and translocation of the c-*abl* oncogene in Philadelphia positive acute lymphoblastic leukaemia. *Blood,* **68**, 1369.
37. Erikson,J., Griffin,C.A., Ar-Rushdi,A., Hoxie,J., Finan,J., Emanuel,B.S., Rovera,G., Nowell,P.C. and Croce,C.M. (1986) Heterogeneity of chromosome 22 breakpoints in Philadelphia positive acute lymphoblastic leukaemia. *Proc. Natl. Acad. Sci. USA,* **83**, 1807.
38. Kurzrock,R., Shtalrid,M., Romera,P., Kloetzer,W.S., Talpas,M., Trujillo,J.M., Blick,M., Bevan,M. and Gutterman,J.U. (1987) A novel c-abl protein product in Philadelphia positive acute lymphoblastic leukaemia. *Nature,* **325**, 631.
39. Chan,L.C., Karhi,K.K., Rayter,S.I., Heisterkamp,N., Eridani,S., Powles,R., Lawler,S.D., Groffen,J., Foulkes,J.G., Greaves,M.F. and Wiedemann,L.M. (1987) A novel abl protein expressed in Philadelphia chromosome positive acute lymphoblastic leukaemia. *Nature,* **325**, 635.

40. Klein,E., Klein,G., NadKarni,J.S., NadKarni,J.H., Wigzell,H. and Clifford,P. (1968) Surface IgM kappa specificity on a Burkitt lymphoma cell *in vivo* and in derived culture lines. *Cancer Res.*, **28**, 1300.

41. van der Berghe,H., Parloir,C., Gosseye,S., Englebienne,V., Cornu,G. and Sokal,G. (1979) Variant translocation in Burkitt lymphoma. *Cancer Cyt. Genet.*, **1**, 9.

42. Dalla-Favera,R., Bregni,M., Erikson,J., Patterson,D., Gallo,R.C. and Croce,C.M. (1982) Human c-*myc* oncogene is located on the region of chromosome 8 that is translocated in Burkitt lymphoma cells. *Proc. Natl. Acad. Sci. USA*, **79**, 7824.

43. Neel,B.G., Jhanwar,S.C., Chaganti,R.S.K. and Hayward,W.S. (1982) Two human c-*onc* genes are located on the long arm of chromosome 8. *Proc. Natl. Acad. Sci. USA*, **79**, 7842.

44. Colby,W.W., Chen,E.Y., Smith,D.H. and Levinson,A.D. (1983) Identification and nucleotide sequence of a human locus homologous to the v-*myc* oncogene of avian myelocytomatosis virus MC29. *Nature*, **301**, 722.

45. Watt,R., Stanton,L.W., Marcu,K., Gallo,R.C., Croce,C.M. and Rovera,G. (1983) Nucleotide sequence of cloned cDNA of human c-*myc* oncogene. *Nature*, **303**, 725.

46. Gazin,C., Dupont de Dinechin,S., Hampe,A., Masson,J.M., Martin,P., Stehelin,D. and Galibert,F. (1984) Nucleotide sequence of the human c-*myc* locus: provocation open reading frame within the first exon. *EMBO J.*, **3**, 383.

47. Gazin,C., Rigolet,M., Briand,J.P., van Regenmortel,M.H.V. and Galibert,F. (1986) Immunochemical detection of proteins related to the human c-*myc* exon 1. *EMBO J.*, **5**, 2241.

48. Hamlyn,P.H. and Rabbitts,T.H. (1983) Translocation joins c-*myc* and immunoglobulin γ genes in a Burkitt lymphoma revealing a third exon in the c-*myc* oncogene. Nature, **304**, 135.

49. Leder,P., Battey,J., Lenoir,G., Moulding,C., Murphy,W., Potter,H., Steward,T. and Taub,R. (1983) Translocation among antibody genes in human cancer. *Science, 222*, 765.

50. Davis,M., Malcolm,S. and Rabbitts,T.H. (1984) Chromosome translocation can occur on either side of the c-*myc* oncogene in Burkitt lymphoma cells. *Nature*, **308**, 286.

51. Taub,R., Kirsch,I., Morton,C., Lenoir,G., Swan,D., Tronick,S., Aaronson,S. and Leder,P. (1982) Translocation of the c-*myc* gene into the immunoglobulin heavy chain locus in human Burkitt's lymphoma and murine plasmacytoma cells. *Proc. Natl. Acad. Sci. USA, 79*, 7837.

52. Taub,R., Kelly,K., Battey,J., Latt,S., Lenoir,G.M., Tantravahi,U., Tu,Z. and Leder,P. (1984) A novel alteration in the structure of an activated c-*myc* gene in a variant t(2;8) Burkitt lymphoma. *Cell, 37*, 511.

53. Erikson,J., Nishikura,K., Ar-Rushdi,A., Finan,J., Emanuel,B., Lenoir,G., Nowell,P.C. and Croce,C.M. (1983) Translocation of an immunoglobulin *x* locus to a region of 3' of an unrearranged c-*myc* oncogene enhances c-*myc* transcription. *Proc. Natl. Acad. Sci. USA, 80*, 7581.

54. Croce,C.M., Thierfelder,W., Erikson,J., Nishikura,K., Finan,J., Lenoir,G.M. and Nowell,P.C. (1983) Transcriptional activation of an unrearranged and untranslocated c-*myc* oncogene by translocation of a Cλ locus in Burkitt lymphoma cells. *Proc. Natl. Acad. Sci., 80*, 6922.

55. Erikson,J., Finan,J., Nowell,P.C. and Croce,C.M. (1982) Translocation of immuno-globulin V_H genes in Burkitt lymphoma. *Proc. Natl. Acad. Sci. USA, 79*, 5611.

56. Haluska,F.G., Finger,S., Tsujimoto,Y. and Croce,C.M. (1986) The t(8;14) chromosomal translocation occurring in B-cell malignancies results from mistakes in V–D–J joining. *Nature, 324*, 158.

57. Gelmann,E.P., Psallidopoulos,C.M., Papas,T.S. and Dalla-Favera,R.D. (1983) Identification of reciprocal translocation sites within the c-*myc* oncogene and immunoglobulin μ locus in a Burkitt lymphoma. *Nature, 306*, 799.

58. Wiman,K.G., Clarkson,B., Hayday,A.C., Saito,H., Tonegawa,S. and Hayward,W.S. (1984) Activation of translocated c-*myc* gene: role of structural alterations in the upstream region. *Proc. Natl. Acad. Sci. USA, 81*, 6798.

59. Showe,L.C., Ballantine,M., Nishikura,K., Erikson,J., Kaji,H. and Croce,C.M. (1985) Cloning and sequencing of a c-*myc* oncogene in a Burkitt's lymphoma cell line that is translocated to a germ line alpha switch region. *Mol. Cell Biol., 5*, 501.

60. Rabbitts,T.H., Baer,R., Buluewla,L., Mengle-Gaw,L., Taylor,A.M. and Rabbitts,P.H.

(1986) Molecular genetics of antigen receptors and associated chromosomal abnormalities in human leukaemias. *Cold Spring Harbor Symp. Quant. Biol.*, **51**, 923.

61. Bernard,O., Cory,S., Gerondakis,S., Webb,E. and Adams,J.M. (1983) Sequence of the murine and human cellular *myc* oncogenes and two modes of *myc* transcription resulting from chromosome translocation in B lymphoid tumours. *EMBO J.*, **2**, 2375.

62. Battey,J., Moulding,C., Taub,R., Murphy,W., Stewart,T., Potter,H., Lenoir,G. and Leder,P. (1983) The human c-*myc* oncogene: structural consequence of translocation into the IgH locus in Burkitt lymphoma. *Cell*, **34**, 779.

63. Rabbitts,T.H., Forster,A., Hamlyn,P. and Baer,R. (1984) Effect of somatic mutation within translocated c-*myc* genes in Burkitt's lymphoma. *Nature*, **309**, 592.

64. Pelicci,P.-G., Knowles,D.M., Magrath,I.T. and Dalla-Favera,R. (1986) Chromosomal breakpoints and structural alterations of the c-*myc* locus differs in endemic and sporadic forms of Burkitt lymphoma. *Proc. Natl. Acad. Sci. USA*, **83**, 2984.

65. Hayday,A.C., Gillies,S.D., Saito,J., Wood,C., Wiman,K., Hayward,W.S. and Tonegawa,S. (1984) Activation of a translocated human c-*myc* gene by an enhancer in the immunoglobulin heavy chain locus. *Nature*, **307**, 334.

66. Rabbitts,T.H., Forster,A., Baer,R. and Hamlyn,P.H. (1983) Transcription enhancer identified near the human Cμ immunoglobulin heavy chain gene is unavailable to the translocated c-*myc* gene in a Burkitt lymphoma. *Nature*, **306**, 806.

67. Graham,M. and Adams,J.M. (1986) Chromosome 8 breakpoint far 3′ of the c-*myc* oncogene in a Burkitt's lymphoma 2;8 variant translocation is equivalent to the murine *pvt*-1 locus. *EMBO J.*, **5**, 2845.

68. Mengle-Gaw,L. and Rabbitts,T.H. (1986) A human chromosome 8 region with abnormalities in B cell, HTLV-I$^+$ T cell and c-*myc* amplified tumours. *EMBO J.*, **6**, 1959.

69. Sun,L.K., Showe,L.C. and Croce,C.M. (1986) Analysis of the 3′ flanking region of the human c-*myc* gene in lymphomas with t(8;22) and t(2;8) chromosomal translocations. *Nucleic Acids Res.*, **14**, 4037.

70. Rappold,G.A., Hameister,H., Cremer,T., Adolph,S., Henglein,B., Freese,U.-K., Lenoir,G.M. and Bornkamm,G.W. (1984) c-*myc* and immunoglobulin $\varkappa$ light chain constant genes are on the 8q$^+$ chromosome of three Burkitt lymphoma lines with t(2;8) translocations. *EMBO J.*, **3**, 2951.

71. Manolov,G., Manolova,Y., Klein,G., Lenoir,G. and Levan,A. (1986) Alternative involvement of two cytogenetically distinguishable breakpoints on chromosome 8 in Burkitt's lymphoma associated translocation. *Cancer Genet. Cytogenet.*, **20**, 95.

72. Rabbitts,T.H., Hamlyn,P.H. and Baer,R. (1983) Altered nucleotide sequence of a translocated c-*myc* in Burkitt lymphoma. *Nature*, **306**, 760.

73. Murphy,W., Sarid,J., Taub,R., Vasicek,T., Battey,J., Lenoir,G. and Leder,P. (1986) A translocated human c-*myc* oncogene is altered in a conserved coding sequence. *Proc. Natl. Acad. Sci. USA*, **83**, 2939.

74. Lombardi,L., Newcomb,E.W. and Dalla-Favera,R. (1987) Pathogenesis of Burkitt lymphoma: Expression of an activated c-*myc* oncogene causes the tumorigenic conversion of EBV-infected human B lymphoblasts. *Cell*, **49**, 161.

75. Adams,J.M., Harris,A.W., Pinkert,C.A., Corcoran,L.M., Alexander,W.S., Cory,S., Palmiter,R.D. and Brinster,R.L. (1985) The c-*myc* oncogene driven by immunoglobulin enhancers induces lymphoid malignancy in transgenic mice. *Nature*, **318**, 533.

76. Leder,A., Pattengale,P.K., Kuo,A., Stewart,T.A. and Leder,P. (1986) Consequences of widespread deregulation of the c-*myc* gene in transgenic mice: multiple neoplasms and normal development. *Cell*, **45**, 485.

77. Sinn,E., Muller,W., Pattengale,P., Tepler,I., Wallace,R. and Leder,P. (1987) Coexpression of MMTV/v-Ha-*ras* and MMTV/c-*myc* genes in transgenic mice: synergistic action of oncogenes *in vivo*. *Cell*, **49**, 465.

78. Harris,A.W., Pinkert,C.A., Crawford,M., Langdon,W.Y., Brinster,R.L. and Adams,J.M. (1988) The Eμ-*myc* transgenic mouse. *J. Exp. Med.*, **167**, 353.

79. Hann,S.R., King,M.W., Bentley,D.L,. Anderson,C.W. and Eisenman,R.N. (1988) A non-AUG translational initiation in c-*myc* exon 1 generates an N-terminally distinct protein whose synthesis is disrupted in Burkitt's lymphoma. *Cell*, **52**, 185.

80. Bentley,D. and Groudine,M. (1988) Sequence requirements for premature termination of transcription in the human c-*myc* gene. *Cell*, **53**, 245.

81. Cesarman,E., Dalla-Favera,R., Bentley,D. and Groudine,M.(1987) Mutations in the

first exon are associated with altered transcription of c-*myc* in Burkitt lymphoma. *Science*, **238**, 1272.

82. Siebenlist,U., Henninghausen,L., Battey,J. and Leder,P. (1984) Chromatin structure and protein binding in the putative regulatory region of the c-*myc* gene in Burkitt lymphoma. *Cell*, **37**, 381.

83. Dyson,J. and Rabbitts,T.H. (1985) Chromatin structure around the c-*myc* gene in Burkitt lymphomas with upstream and downstream translocation points. *Proc. Natl. Acad. Sci. USA*, **82**, 1984.

84. Dyson,J., Littlewood,T.D., Forster,A. and Rabbitts,T.H. (1985) Chromatin structure of transcriptionally active and inactive human c-*myc* alleles. *EMBO J.*, **4**, 2885.

85. Vahdati,M., Graafland,H. and Emberger,J.M. (1983) Karyotype analysis of B-lymphocytic leukaemia. *Hum. Genet.*, **63**, 372.

86. Melo,J.V., Brito-Babapulle,V., Foroni,L., Robinson,D.S.V., Luzzatto,L. and Catovsky,D. (1986) Two new cell lines from B-prolymphocytic leukaemia: characterization by morphology, immunological markers, karyotype and Ig gene rearrangement. *Int. J. Cancer*, **38**, 531.

87. Van der Berghe,H., Vermaelen,K., Louwagie,A., Criel,A., Mecucc,C. and Vaerman,J.-P. (1984) High incidence of chromosome abnormalities in IgG3 myeloma. *Cancer Genet. Cytogenet.*, **11**, 381.

88. Gahrton,G., Zech,L., Nilsson,K., Lonnqvist,B. and Carlstrom,A. (1980) Two translocations t(11;14) and t(1;6) in a patient with a plasma cell leukaemia and two populations of plasma cells. *Scand. J. Haematol.*, **24**, 42.

89. Erikson,J., Finan,J., Tsujimoto,Y., Nowell,P.C. and Croce,C.M. (1984) The chromosome 14 breakpoint in neoplastic B cells with the t(11;14) translocation involves the immunoglobulin heavy chain locus. *Proc. Natl. Acad. Sci. USA*, **81**, 4144.

90. Tsujimoto,Y., Jaffe,E., Cossman,J., Gorham,J., Nowell,P.C. and Croce,C.M. (1985) Clustering of breakpoints on chromosome 11 in human B-cell neoplasms with the t(11;14) chromosomal translocation. *Nature*, **315**, 340.

91. Ince,C., Vahdan,S., Selvanayagam,P., Cabanillas,F., Gutterman,J. and Blick,M. (1987) The *bcl*-1 sequences located at chromosome 11q13 are rearranged in uncultured malignant lymphomas. *Proc. ASCO*, **6**, 14.

92. Rabbitts,P.H., Forster,A., Stinson,M.A., Catovsky,D. and Rabbitts,T.H. (1988) Chromosome abnormalities at 11q13 in B cell tumors. *Oncogene*, **2**, 99–103.

93. Bloomfield,C.D., Arthur,D.C., Frizzera,G., Levine,G.G., Peterson,B.A. and Gajl-Peczalska,K.J. (1983) Non-random chromosome abnormalities in lymphoma. *Cancer Res.*, **47**, 2975.

94. Yunis,J.J., Oken,M.M., Kaplan,M.E., Ensrud,K.M., Howe,R.R. and Theologides,A. (1982) Multiple recurrent genomic defects in follicular lymphoma: a possible model for cancer. *N. Engl. J. Med.*, **307**, 1231.

95. Cleary,M.L. and Sklar,J. (1985) Nucleotide sequence of a t(14;18) chromosomal breakpoint in follicular lymphoma and demonstration of a breakpoint cluster region near a translocationally active locus on chromosome 18. *Proc. Natl. Acad. Sci. USA*, **82**, 7439.

96. Tsujimoto,Y., Bashir,M.M., Givol,I., Cossman,J., Jaffe,E. and Croce,C.M. (1987) DNA rearrangements in human follicular lymphoma can involve the 5' or 3' region of the *bcl*-2 gene. *Proc. Natl. Acad. Sci. USA*, **84**, 1329.

97. Cleary,M.L., Galili,N. and Sklar,J. (1986) Detection of a second t(14;18) breakpoint cluster region in human follicular lymphomas. *J. Exp. Med.*, **164**, 315.

98. Lee,M.S., Blick,M.B., Pathak,S., Trujillo,J.M., Butler,J.J., Katz,R.L., McLaughlin,P., Hagemeister,F.B., Valasquez,W.S., Goodacre,A., Cork,A., Gutterman,J.U. and Cabanillis,C. (1987) The gene located at chromosome 18 band q21 is rearranged in uncultured diffuse lymphomas as well as follicular lymphomas. *Blood*, **70**, 90.

99. Seto,M., Jaeger,U., Hocket,R.D., Graninger,W., Bennett,S., Goldman,P. and Korsmeyer,S.J. (1988) Alternative promoters and exons, somatic mutations and deregulation of the Bcl-2-Ig fusion gene in lymphoma. *EMBO J.*, **7**, 123.

100. Tsujimoto,Y., Gorham,J., Cossman,J., Jaffe,E. and Croce,C.M. (1985) The t(14;18) chromosome translocations involved in B-cell neoplasms result from mistakes in VDJ joining. *Science*, **229**, 1390.

101. Ysujimoto,Y., Louie,E., Bashir,M.M. and Croce,C.M. (1988) The reciprocal partners of both the t(14;18) and the t(11;14) translocations involved in B-cell neoplasms are rearranged by the same mechanism. *Oncogene*, **2**, 347.

102. Bakhshi,A., Wright,J.J., Graninger,W., Seto,M., Owens,J., Cossman,J., Jensen,J.P., Goldman,P. and Korsmeyer,S.J. (1987) Mechanism of the t(14;18) chromosome translocation: Structural analysis of both derivative 14 and 18 reciprocal partners. *Proc. Natl. Acad. Sci. USA*, **84**, 2396.

103. Reed,J.C., Tsujimoto,Y., Alpers,J.D., Croce,C.M. and Cowell,P.C. (1987) Regulation of *bcl-2* proto-oncogene expression during normal human lymphocyte proliferation. *Science*, **236**, 1295.

104. Tsujimoto,Y. and Croce,C.M. (1986) Analysis of the structure, transcripts and protein product of *bcl-2*, the gene involved in human follicular lymphoma. *Proc. Natl. Acad. Sci. USA*, **83**, 5214.

105. Tsujimoto,Y., Ikegki,N. and Croce,C.M. (1987) Characterisation of the protein product of *bcl-2*, the gene involved in human follicular lymphoma. *Oncogene*, **2**, 3.

106. Vaux,D.L., Cory,S. and Adams,J.M. (1988) *Bcl-2* gene promotes haemopoietic cell survival and cooperates with c-*myc* to immortalize pre-B cells. *Nature*, **335**, 440.

107. Pegoraro,L., Palumbo,A., Erikson,J., Falda,M., Giovanazzo,B., Emanuel,B.S., Rovera,G., Nowell,P.C. and Croce,C.M. (1984) A 14;18 and an 8;14 chromosome translocation in a cell line derived from an acute B-cell leukemia. *Proc. Natl. Acad. Sci. USA*, **81**, 7166.

108. Gauwerky,C.E., Hoxie,J., Nowell,P.C. and Croce,C.M. (1988) Pre-B-cell leukemia with a t(8;14) and a t(14;18) translocation is preceded by follicular lymphoma. *Oncogene*, **2**, 431.

109. Croce,C.M., Isobe,M., Palumbo,A., Puck,J., Ming,J., Tweardy,D., Erikson,J., Davis,M. and Rovera,G. (1985) Gene for α chain of human T cell receptor: location on chromosome 14 region involved in T cell neoplasms. *Science*, **227**, 1044.

110. Caccia,N., Bruns,G.A., Kirsch,I.R., Hollis,G.F., Bertress,V. and Mak,T.W. (1985) T cell receptor α chain genes are located on chromosome 14 at 14q11-q12 in humans. *J. Exp. Med.*, **161**, 1255.

111. Kirsch,I.R., Morton,C.C., Nakahara,K. and Leder,P. (1982) Human immunoglobulin heavy chain genes map to a region of translocations in malignant B lymphocytes. *Science*, **216**, 301.

112. Spector,B.D., Filipovich,A.H., Perry,G.S. and Kersey,J.H. (1982) Epidemiology of cancer in ataxia-telangiectasia. In *Ataxia-telangiectasia—A Cellular and Molecular Link Between Cancer, Neuropathology and Immune Deficiency*. B.A.Bridges and D.G.Harnden (eds), Wiley, New York, p. 103.

113. Oxford,J.M., Harnden,D.G., Parrington,J.M. and Delhanty,J.D.A. (1975) Specific chromosome aberrations in ataxia telangiectasia. *J. Med. Genet.*, **12**, 251.

114. Taylor,A.M.R., Oxford,J.M. and Metcalf,J.A. (1981) Spontaneous cytogenetic abnormalities in lymphocytes from 13 patients with ataxia telangiectasia. *Int. J. Cancer*, **27**, 311.

115. Taylor,A.M.R. and Butterworth,S.V. (1986) Clonal evolution of T cell chronic lymphocytic leukaemia in a patient with ataxia telangiactasia. *Int. J. Cancer*, **37**, 511.

116. Aurias,A. and Dutrillaux,B. (1986) Probable involvement of immunoglobulin superfamily genes in most recurrent chromosomal rearrangements from ataxia telangiactasia. *Hum. Genet.*, **72**, 210.

117. Zech.L., Godal,T., Hammarström,L., Mellstedt,H., Smith,C.I.E., Tötterman,T. and Went,M. (1986) Specific chromosome markers involved with chronic T lymphocyte tumours. *Cancer Genet. Cytogent.*, **21**, 67.

118. Hecht,F., Morgan,R., Kaiser-McCaw Hecht,B. and Smith,S.D. (1984) Common region on chromosome 14 in T-cell leukaemia and lymphoma. *Science*, **226**, 1445.

119. Ueshima,Y., Rowley,J.D., Variakojis,D., Winter,D. and Gordon,L. (1984) Cytogenetic studies on patients with chronic T cell leukaemia/lymphoma. *Blood*, **63**, 1028.

120. Sadamori,N., Kusano,M., Nishino,K., Tagawa,M., Yao,E., Yamada,Y., Amagasaki, T., Kinoshita,Y. and Ichimaru,M. (1985) Abnormalities of chromosome 14 at band 1411 in Japanese patients with adult T cell leukaemia. *Cancer Genet. Cytogenet.*, **17**, 279.

121. Clare,N., Baldt,D., Messerschmidt,G., Zeltzer,P., Hansen,K. and Manhoff,L. (1986) Lymphocyte malignancy and chromosome 14 structural observations involving band q11. *Blood*, **67**, 704.

122. Mengle-Gaw,L., Albertson,D.G., Sherrington,P.D. and Rabbitts,T.H. (1988) Analysis of a T-cell tumor-specific breakpoint cluster at human chromosome 14q32. *Proc. Natl. Acad. Sci. USA*, **85**, 9171–9175.

123. Welch,J.P. and Lee,C.L.Y. (1975) Non-random occurrence of 7-14 translocations in human lymphocyte cultures. *Nature*, **255**, 241.
124. Beatty-DeSana,J.W., Haggard,M.J. and Cooledge,J.W. (1975) Non-random occurrence of 7-14 translocations in human lymphocytic cultures. *Nature*, **255**, 242.
125. Hecht,F., McCaw,B.K., Peakman,D. and Robinson,A. (1975) Non-random occurrence of 7-14 translocations in human lymphocyte cultures. *Nature*, **255**, 243.
126. Aurias,A., Dutrillaux,B. and Griscelli,C. (1983) Tandem translocation t(14;14) in isolated and clonal cells in ataxia telangiectasia are different. *Hum. Genet.*, **63**, 320.
127. Aurias,A., Couturier,J., Dutrillaux,A.M., Durtillaux,B., Herpin,F., Lamoliate,E., Lombard,M., Muleris,M., Paravatou,M., Prieur,M., Prod'homme,M., Sportes,M., Viegas-Pequignot,E. and Volobouev,V. (1985) Inversion 14; the most frequently acquired rearrangement in lymphocytes. *Hum. Genet.*, **71**, 19.
128. Aurias,A., Croquette,M.F., Nuyts,J.P., Griscelli,C. and Dutrillaux,B. (1986) New data on clonal anomalies of chromosome 14 in ataxia telangiectasia:t(14;14) and inv(14). *Hum. Genet.*, **72**, 22.
129. Baer,R., Chen,K.-C., Smith,S.D. and Rabbitts,T.H. (1985) Fusion of an immunoglobulin variable gene and a T cell receptor constant gene in the chromosome 14 inversion associated with T cell tumours. *Cell*, **43**, 705.
130. Denny,C.T., Yoshikai,Y., Mak,T.W., Smith,S.D., Hollis,G.F. and Kirsch,I.R. (1986) A chromosome 14 inversion in a T cell lymphoma is caused by site-specific recombination between immunoglobulin and T cell receptor loci. *Nature*, **320**, 549.
131. Baer,R., Forster,A. and Rabbitts,T.H. (1987) The mechanism of chromosome 14 inversion in a human T cell lymphoma. *Cell*, **50**, 97.
132. Kelly,K., Cochran,B.H., Stiles,C.D. and Leder,P. (1983) Cell specific regulation of the c-*myc* gene by lymphocyte mitogens and platelet-derived growth factor. *Cell*, **35**, 603.
133. Mathieu-Mahul,D., Caubet,J.F., Bernheim,A., Mauchauffe,M., Palmer,E., Berger,R. and Larsen,C.-J. (1985) Molecular cloning of a DNA fragment from human chromosome 14 (14q11) involved in T-cell malignancies. *EMBO J.*, **4**, 3427.
134. McKeithan,T.W., Shima,E.A., LeBeau,M.M., Minowada,J., Rowley,J.D. and Diaz,M.D. (1986) Molecular cloning of the breakpoint junction of a human chromosomal 8;14 translocation involving the T-cell receptor α-chain gene and sequences on the 3′ side of *myc*. *Proc. Natl. Acad. Sci. USA*, **83**, 6636.
135. Finger,L.R., Harvey,R.C., Moore,R.C.A., Showe,L.C. and Croce,C.M. (1986) A common mechanism of chromosomal translocation in T- and B-cell neoplasmia. *Science*, **234**, 982.
136. Erikson,J., Finger,L., Sun,L., Ar-Rushdi,A., Nishikura,K., Minowada,J., Finan,J., Emanuel,B.S., Nowell,P.C. and Croce,C.M. (1986) Deregulation of c-*myc* by translocation of the α-locus of the T-cell receptor in T-cell leukaemias. *Science*, **232**, 884.
137. Smith,S.D., Morgan,R., Link,M.P., McFall,P. and Hecht,F. (1986) Cytogenetic and immunophenotypic analysis of cell lines established from patients with T cell leukaemia/lymphoma. *Blood*, **67**, 650.
138. Dube,I.D., Raimondi,S.C., Pi,D. and Kalousek,D.K. (1986) A new translocation t(10;14) (q24;q11) in T cell neoplasia. *Blood*, **67**, 1181.
139. Chien,Y.-H., Iwashima,M., Kaplan,K.B., Elliott,J.F. and Davis,M.M. (1987) A new T cell receptor gene located within the alpha locus and expressed early in T cell differentiation. *Nature*, **327**, 677.
140. Sugamura,K. and Hinuma,Y. (1985) Human retrovirus in adult T cell leukaemia/lymphoma. *Immunol. Today*, **6**, 83.
141. Hinuma,Y. (1986) Adult T-cell leukemia virus: the present picture. In *The Cytobiology of Leukaemias and Lymphomas*. D.Quaglino and F.R.G.Hayhoe (eds), Elsevier, Amsterdam, p. 435
142. Yoshida,M., Seiki,M., Yamaguchi,K. and Takatsuki,K. (1984) Monoclonal integration of human T-cell leukaemia provirus in all primary tumours of adult T-cell leukaemia suggests causative role of human T-cell leukaemia virus in the disease. *Proc. Natl. Acad. Sci. USA*, **81**, 2534.
143. Shimoyama,M., Kagami,Y., Shimotohno,K., Miwa,M., Minato,K., Tobinai,K., Suemasu,K. and Sugimura,T. (1986) Adult T-cell leukaemia/lymphoma not associated with human T-cell leukaemia virus type I. *Proc. Natl. Acad. Sci. USA*, **83**, 4524.

144. Lombardi,L., Newcomb,E.W. and Dalla-Favera,R. (1987) Pathogenesis of Burkitt lymphoma: expression of an activated c-*myc* oncogene causes the tumorigenic conversion of EBV-infected human B lymphoblasts. *Cell*, **49**, 161.
145. Tonegawa,S. (1983) Somatic generation of antibody diversity. *Nature*, **302**, 575.
146. Honjo,T. (1983) Immunoglobulin genes. *Annu. Rev. Immunol.*, **1**, 499.
147. Bakhshi,A., Wright,J.J., Graninger,W., Seto,M., Owens,J., Cossman,J., Jensen,J.P., Goldman,P. and Korsmeyer,S.J. (1987) Mechanism of the t(14;18) chromosomal translocation: structural analysis of both derivature 14 and 18 reciprocal partners. *Proc. Natl. Acad. Sci. USA*, **84**, 2396.
148. Yoshikai,Y., Clark,S.P., Taylor,S., Sohn,U., Wilson,B.I., Minden,M.D. and Mak,T.W. (1985) Organisation and sequences of the variable, joining and constant regions of the human T cell receptor α chain. *Nature*, **316**, 837.
149. Kagan,J., Finan,J., Letoksky,J., Besa,E.C., Nowell,P.C. and Croce,C.M. (1987) α chain locus of the T cell receptor in the t(10;14) chromosome translocation of T cell acute lymphocytic leukaemia. *Proc. Natl. Acad. Sci. USA*, **84**, 4543.
150. Alitalo,K., Schwab,M., Lin,C.C., Varmus,H.E. and Bishop,M. (1983) Homogeneously staining chromosomal regions contain amplified copies of an abundantly expressed cellular oncogene (c-*myc*) in malignant neuroendocrine cells from a human colon carcinoma. *Proc. Natl. Acad. Sci. USA*, **80**, 1707.
151. Biedler,J.L. and Spengler,B.A. (1976) Metaphase chromosome anomaly: association with drug resistance and cell-specific products. *Science*, **191**, 185.
152. Collins,S. and Groudine,M. (1982) Amplification of endogenous myc-related DNA sequences in a human myleoid leukaemia cell line. *Nature*, **298**, 679.
153. Collins,S.J. and Groudine,M.T. (1983) Rearrangement and amplification of c-*abl* sequences in the human chronic myelogenous leukaemia cell line K-562. *Proc. Natl. Acad. Sci. USA*, **80**, 4813.
154. Biedler,J.L., Meyers,M.B. and Spengler,B.A. (1983) Homogeneously staining regions and double minute chromosomes, prevalent cytogenetic abnormalities of human neuroblastoma cells. *Adv. Cell Neurobiol.*, **4**, 268.
155. Schwab,M., Alitalo,K., Klempnauer,K.-H., Varmus,H.E., Bishop,J.M., Gilbert,F., Brodeur,G., Goldstein,M. and Trent,J. (1983) Amplified DNA with limited homology to *myc* cellular oncogene is shared by human neuroblastoma cell lines and a neuroblastoma tumour. *Nature*, **305**, 245.
156. Brodeur,G.M., Seeger,R.C., Schwab,M., Varmus,H.E. and Bishop,J.M. (1984) Amplification of N-*myc* in untreated human neuroblastomas correlates with advanced disease stage. *Science*, **224**, 1121.
157. Pochedly,C. (1982) *Neuroblastoma*. Elsevier Biochemical Press, New York.
158. Seeger,R.C., Brodeur,G.M., Sather,H., Dalton,A., Siegel,S.E. Wong,K.Y. and Hammond,D. (1985) Association of multiple copies of the N-*myc* oncogene with rapid progression of neuroblastoma. *N. Engl. J. Med.*, **313**, 1111.
159. Evans,A.E., Baum,E. and Chard,R. (1981) Do infants with stage IV-S neuroblastoma need treatment? *Arch. Dis. Child.*, **31**, 2098.
160. Brodeur,G., Hayes,A., Green,A., Casper,J., Lee,H. and Seeger,R. (1986) Consistent N-*myc* copy number in simultaneous or consecutive neuroblastoma samples from a given patient's tumour. *Proc. ASCO*, **5**, No. 52.
161. Nau,M.M., Brooks,B.J., Carney,D.N., Gazdar,A.F., Battey,J.F., Sausville,E.A. and Minna,J.D. (1986) Human small-cell lung cancers show amplification and expression of the N-*myc* gene. *Proc. Natl. Acad. Sci. USA*, **83**, 1092.
162. Johnson,B.E., Makuch,R.W., Simmons,A.D., Gazdar,A.F., Burch,D. and Cashesll,A.M. (1988) *myc* family DNA amplification in small cell lung cancer patients' tumors and corresponding cell lines. *Cancer Res.*, **48**, 5163.
163. Little,C.D., Nau,M.M., Carney,D.N., Gazdar,A.F. and Minna,J.D. (1983) Amplification and expression of the c-*myc* oncogene in human lung cancer cell lines. *Nature*, *306*, 194.
164. Wong,A.J., Ruppert,J.M., Eggleston,J., Hamilton,S.R., Baylin,S.B. and Vogelstein,B. (1986) Gene amplification of c-*myc* and N-*myc* in small cell carcinoma of the lung. *Science*, **233**, 461.
165. Slamon,D.J., Clark,G.M., Wong,S.G., Levin,W.J., Ullrich,A. and McGuire,W.L. (1987) Human breast cancer: correlation of relapse and survival with amplification of the HER-2/*neu* oncogene. *Science*, **235**, 177.
166. Yokato,J., Yamamoto,T. and Toyoshima,K. (1986) Amplification of c-*erb* B-2 oncogene in human adenocarcinomas *in vivo*. *Lancet*, **i**, 765.

167. Semba,K., Kamata,N., Toyoshima,K. and Yamamoto,T. (1985) A v-*erb* B related protooncogene, c-*erb*-2, is distinct from the c-*erb* B-1/EGF receptor gene and is amplified in a human salivary gland adenocarcinoma. *Proc. Natl. Acad. Sci. USA,* **82**, 6497.

168. Venter,D.J., Tuzi,N.L., Kumar,S. and Gullick,W.J. (1987) Overexpression of the c-*erb* B-2 oncoprotein in human breast carcinomas: immunohistological assessment correlates with gene amplification. *Lancet,* **i**, 69.

169. Ibson,J.M. and Rabbitts,P.H. (1988) Sequence of a germ-line N-*myc* gene and amplification as a mechanism of activation. *Oncogene,* **2**, 399.

170. Gilbert,F. (1985) Human neuroblastomas and abnormalities of chromosome 1 and 17. *Cancer Res.,* **44**, 5444.

171. Whang-Peng,J., Bunn,P.A., Kao-Shan,C.S., Lee,E.C., Carney,D.N., Gazdar,A.F. and Minna,J.D. (1982) A non-random chromosomal abnormality del 3p(14-23) in human small cell lung cancer (SCLC). *Cancer Genet. Cytogenet.,* **6**, 119.

172. Yunis,J.J. and Ramsay,N. (1978) Retinoblastoma and subband deletion of chromosome 13. *Am. J. Dis. Child.,* **132**, 161.

173. Zbar,B., Brauch,H., Talmadge,C. and Linehan,M. (1987) Loss of alleles of loci on the short arm of chromosome 3 in renal cell carcinoma. *Nature,* **327**, 721.

174. Fearon,E.R., Feinberg,A.P., Hamilton,S.H. and Vogelstein,B. (1985) Loss of genes on the short arm of chromosome 11 in bladder cancer. *Nature,* **318**, 377.

175. Solomon,E., Voss,R., Hall,V., Bodmer,W., Jass,J.R., Jeffreys,A.F., Lucibello,F.C., Patel,I. and Rider,S.H. (1987) Chromosome 5 allele loss in human colorectal carcinomas. *Nature,* **328**, 616.

176. Ali,I.U., Lidereau,R., Theillet,C. and Callahan,R. (1987) Reduction to homozygosity of genes on chromosome 11 in human breast neoplasia. *Science,* **238**, 185.

177. Dumanski,J.P., Carlbom,E., Collins,V.P. and Nordenskjöld,M. (1987) Deletion mapping of a locus on human chromosome 22 involved in the oncogenesis of meningioma. *Proc. Natl. Acad. Sci. USA,* **84**, 9275.

178. Ponder,B. (1988) *Nature,* **335**, 400.

179. Harris,H. (1979) Some thoughts about genetics, differentiation and malignancy. *Somatic Cell. Genet.,* **5**, 923.

180. Stanbridge,E.J. (1967) Suppression of malignancy in human cells. *Nature,* **260**, 17.

181. Weissman,B.E., Saxon,P.J., Pasquale,S.R., Jones,G.R., Geiser,A.G. and Stanbridge, E.J. (1987) Introduction of a normal human chromosome 11 into a Wilms tumour cell line controls its tumorigenic expression. *Science,* **236**, 175.

182. Larsson,C., Skogseid,B., Öberg,K., Nakamura,Y. and Nordenskjöld,M. (1988) Multiple endocrine neoplasia type 1 gene maps to chromosome 11 and is lost in insulinoma. *Nature,* **332**, 85.

183. Mathew,C.G.P., Smith,B.A., Thorpe,K., Wong,Z., Royale,N.J., Jeffreys,A.J. and Ponder,B.A.J. (1988) Deletion of genes on chromosome 1 in endocrine neoplasia. *Nature,* **328**, 524.

184. Cannon-Albright,L.A., Skolnick,M.H., Bishop,D.T., Lee,R.G. and Burt,R.W. (1988) Common inheritance of susceptibility to colonic adenomatous polyps and associated colorectal cancers. *N. Engl. J. Med.,* **319**, 533.

185. Bodmer,W.F., Bailey,C.J., Bodmer,J., Bussey,H.J.R., Ellis,A., Gorman,P., Lucibello,F.C., Murday,V.A., Rider,S.H., Scambler,P., Sheer,D., Solomon,E. and Spurr,N.K. (1987) Localisation of the gene for familial adenomatous polyposis on chromosome 5. *Nature,* **328**, 614.

186. Solomon,E., Voss,R., Hall,V., Bodmer,W.F., Jass,J.R., Jeffreys,A.J., Lucibello,F.C., Patel,I. and Rider,S.H. (1987) Chromosome 5 allele loss in human colorectal carcinomas. *Nature,* **328**, 616.

187. Vogelstein,B., Fearon,E.R., Hamilton,S.R., Kern,S.E., Preisinger,A.C., Leppert,M., Nakamura,Y., White,R., Smits,A.M.M. and Bos,J.L. (1988) Genetic alterations during colorectal-tumor development. *N. Engl. J. Med.,* **319**, 526.

188. Law,D.J., Olschwang,S., Monpezat,J.-P., Lefrancois,D., Jagelman,D., Petrelli,N.J., Thomas,G. and Feinberg,A.P. (1988) Concerted nonsyntenic allelic loss in human colorectal carcinoma. *Science,* **241**, 961.

189. Okamoto,M., Sasaki,M., Sugio,K., Sato,C., Iwama,T., Ikeuchi,T., Tonomura,A., Sasuzuki,T. and Miyaki,M. (1988) Loss of constitutional heterozygosity in colon carcinoma from patients with familial polyposis coli. *Nature,* **331**, 273.

190. Fearon,E.R., Hamilton,S.R. and Vogelstein,B. (1987) Clonal analysis of human colorectal tumors. *Science*, **238**, 193.
191. Wurster-Hill,D.H., Cannizzaro,L.A., Pettengill,O.S., Sorenson,G.D., Cate,C.C. and Maurer,L.H. (1984) Cytogenetics of small cell carcinoma of the lung. *Cancer Genet. Cytogenet.*, **13**, 303.
192. Kok,K., van der Hout,A.H., Osinga,J., Beredsen,H.H., Carritt,B. and Buys,C.H.C.H. (1987) Molecular analysis shows a common deletion in the major types of lung cancer. *Eur. Assoc. Cancer* ∎ *es. Abstracts*, p. 44.
193. Drabkin,H.A., Bradley,C., Hart,I., Bleskan,L., Li,F.P. and Patterson,D. (1985) Translocation of c-*myc* in the hereditary renal cell carcinoma associated with a t(3;8) (p14.2;q24.13) chromosomal translocation. *Proc. Natl. Acad. Sci. USA*, **82**, 6980.
194. Koufos,A., Hansen,M.F., Copeland,N.G., Jenkins,N.A., Lampkin,B.C. and Cavenee,W.K. (1985) Loss of heterozygosity in three embryonal tumours suggests a common pathogenetic mechanism. *Nature*, **316**, 330.
195. Knudson,A.G. and Strong,L.C. (1972) Mutation and cancer: a model for Wilms tumour of the kidney. *J. Natl. Cancer Inst.*, **48**, 313.
196. Boehm,T., Lavenir,I., Forster,A., Wadey,R.B., Cowell,J.K., Harbott,J., Lampert,F., Waters,J., Sherrington,P., Couillin,P., Azoulay,M., Junien,C., van Heyningen,V., Porteous,D.J., Hastie,N.D. and Rabbitts,T.H. (1988) The T-ALL specific t(11;14) (p13;q11) translocation breakpoint cluster region is located near to the Wilms' tumor predisposition locus. *Oncogene*, **3**, 691–695.
197. Benedict,W., Murphee,A.L., Banerjee,A., Spina,C.A., Sparkes,M.C. and Sparkes,R.S. (1983) Patient with 13 chromosome deletion: evidence that the retinoblastoma gene is a recessive cancer gene. *Science*, **219**, 973.
198. Cavanee,W.K., Dryja,T.P., Phillips,R.A., Benedict,W.F., Godbout,R., Gallie,B.L., Murphee,A.L., Strong,L.C. and White,R.L. (1983) Expression of recessive alleles by chromosomal mechanisms in retinoblastoma. *Nature*, **305**, 779.
199. Lee,W.H., Bookstein,R., Hong,F., Young,L.-J., Shew,J.-Y. and Lee,E.Y.H.-P. (1987) Human retinoblastoma susceptibility gene: cloning, identification and sequence. *Science*, **235**, 1394.
200. Friend,S.H., Bernards,R., Rogel,S., Weinberg,R.A., Rapaport,J.M., Albert,D.M. and Dryja,T.P. (1986) A human DNA segment with properties of the gene that predisposes to retinoblastoma and osteosarcoma. *Nature*, **323**, 643.
201. Lee,W.-H., Shew,J.-H., Hong,F.D., Sery,T.W., Donoso,L.A., Young,L.-J., Booksein,R. and Lee,E.Y.-H.P. (1987) The retinoblastoma susceptibility gene encodes a nuclear phosphoprotein associated with DNA binding activity. *Nature*, **329**, 642.
202. Whyte,P., Buchkovich,K.J., Horowitz,J.M., Friend,S.H., Raybuck,M., Weinberg,R.A. and Harlow,E. (1988) Association between an oncogene and an anti-oncogen: the adenovirus E1A proteins bind to the retinoblastoma gene product. *Nature*, **334**, 124.
203. Harbour,J.W., Lai,S.-H., Whang-Peng,J., Gazard,A.F., Minna,J.D. and Kaye,F.J. (1988) Abnormalities in structure and expression of the human retinoblastoma gene in SCLC. *Science*, **242**, 162.
204. T'Ang,A., Varley,J.M., Chakraborty,S., Murphee,A.L. and Fung,Y.-K.T. (1988) Structural rearrangement of the retinoblastoma gene in human breast carcinoma. *Science*, **242**, 263.

4

Biochemical functions of oncogenes

Sydonia I.Rayter, Kenneth K.Iwata,
Richard W.Michitsch, John M.Sorvillo,
David M.Valenzuela and J.Gordon Foulkes

1. Introduction

In this chapter we review the known biochemical functions of oncogene-encoded proteins and elaborate the concept that transformation is an aberration of signal transduction. Before discussing these ideas in more detail, however, the following introduction deals with mechanisms of oncogene activation, and then briefly reviews the major signal transduction pathways which regulate the normal cell's physiology. It will be helpful throughout the chapter to refer to *Figures 1* and *2* which summarize the biochemical roles of oncogenes and depict signal transduction pathways mediated by growth factors.

1.1 Activation of proto-oncogenes

In the last decade, it has become apparent that the inappropriate expression of a number of cellular genes can lead to malignant transformation. Approximately 50 such genes, termed oncogenes, have been described thus far, and it is estimated that at least 200–300 will be identified (for general reviews, see 1–4).

The generation of oncogenes from their non-transforming homologs (termed proto-oncogenes) can occur in a variety of ways.

(i)   In many animal species retroviruses have been found to transduce proto-oncogenes from their host cell during viral replication, thereby placing the cellular gene under control of the strong retroviral promoter. This can result in either the constitutive overproduction of the normal protein, or in a mutational activation of the protein as a consequence of the transduction event.

(ii)   Spontaneous or evoked gene amplification can also lead to the increased concentration of a proto-oncogene product. For example, amplification of the *neu* proto-oncogene has been reported to occur in approximately 30 per cent of human breast carcinomas (5).

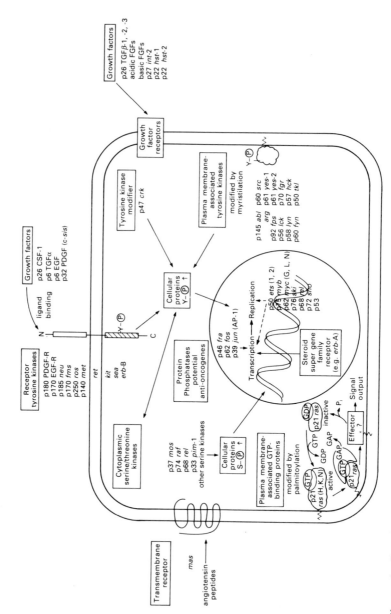

Figure 1. Functional characteristics of oncogenes and growth factors. Molecular weights displayed are for the human proteins.

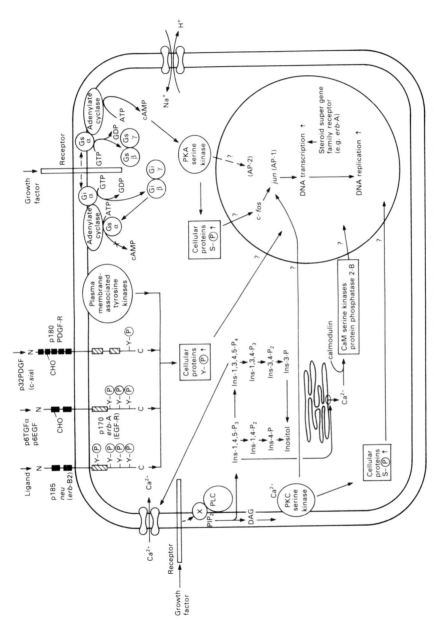

Figure 2. Growth factor-mediated signal transduction pathways involved in mitogenesis.

(iii)   Proto-oncogenes can become activated following chromosomal translocation, the best characterized example being the reciprocal translocation of the human c-*abl* proto-oncogene from chromosome 9 to 22, observed in over 95 per cent of patients with chronic myelogenous leukemia (CML) (6,7). This translocation results in the expression of a novel fusion protein and the constitutive activation of the c-*abl* protein's enzymatic activity (8,9). This mechanism of oncogene activation is comprehensively reviewed in Chapter 3.

(iv)   Oncogenes can be generated by mutations within coding regions, the paradigm being the human *ras* oncogene, in which a single point mutation is sufficient to produce a transforming protein (10,11). *In vitro* experiments using animal models with chemical-induced tumors suggest that *ras* mutations are directly responsible for the initiation of such tumors and are not merely secondary consequences of the transformed phenotype (12–16). Point mutations have also been identified in growth factor receptors (in the transmembrane region of rat *neu* and extracellular domain of human CSF-1), which are sufficient to activate the receptor's intrinsic protein–tyrosine kinase activity and transforming ability (see section 4.1 and reference 17).

1.2   Oncogenes and signal transduction

To date, although the biochemical functions of oncogene-encoded proteins have only been partially defined, numerous observations suggest their intimate involvement in signal transduction. Aspects of these mechanisms were introduced in Chapter 1. For instance, several oncogenes have been shown to encode growth-stimulating hormones, the first example being the oncogene of the simian sarcoma virus (*sis*), found to be derived from the cellular gene encoding the β chain of platelet-derived growth factor (PDGF) (18). The *hst* oncogene, identified by transfection experiments as a transforming gene present both in human stomach cancer (19) and in Kaposi's sarcoma (20), shares greater than 40 per cent homology to basic fibroblast growth factor (FGF). In addition, the protein encoded by the murine *int*-2 oncogene also shows homology to FGF (21,22). Next, in many transformed cells, oncogenes activate the expression of normally quiescent growth factor genes such as PDGF and the transforming growth factor alpha (TGF-α), which act subsequently to stimulate tumor cell growth (23). Third, 30 per cent of all known oncogenes described to date have been found to encode protein–tyrosine kinases, an activity shared with the plasma membrane receptors for several growth-stimulating hormones. These include the epidermal growth factor (EGF) receptor, the kinase domain of which is encoded by the *erb*B oncogene (24), and the receptor for colony stimulating factor-1, which is encoded by the *fms* proto-oncogene (25). Fourth, the *ras* oncogenes of Harvey and Kirsten murine sarcoma viruses show homology to G-proteins, a family of polypeptides which serve to couple extracellular stimuli to intracellular effector systems including adenylate cyclase and phospholipase C (26). The p21*ras* proteins display both GTP binding and GTPase activities, but lack the two subunits involved in the regulation of normal G-proteins (reviewed in 26). Oncogenes can also act to modulate signal transduction pathways via the constitutive activation of phosphoinositide (PI) turnover. Finally, the human proto-oncogene c-*jun*, defined originally as the oncogene of an avian sarcoma virus (ASV-17), encodes a transcription factor of the AP-1 fraction, believed to be responsible for mediating gene expression in response to activation of protein kinase C (27,28).

Most oncogenes, therefore, appear to transform cells by mimicking growth factors, hormone receptors, G-proteins, intracellular effector systems, or transcription factors which control the expression of key regulatory genes. In the following section, we briefly review the major signal transduction pathways which regulate cell growth, before discussing the biochemical functions of oncogene-encoded proteins in detail.

1.3 The role of receptors in signal transduction

Living cells respond to their environment by binding external ligands to specific receptors and then translating this information into internal signalling systems. In general, most receptors are localized to the plasma membrane and transmit the information via changes in protein phosphorylation. The second class of receptors are intracellular and respond to hydrophobic ligands which enter the cell via passive diffusion and act primarily via changes in gene expression.

1.3.1 Plasma-membrane receptors involving G-proteins

A large number of plasma membrane receptors lack intrinsic enzymatic activity and transmit extracellular signals to inside the cell via activation of guanine nucleotide binding proteins (G-proteins) (26,29). Formation of the ligand receptor complex stimulates the release of GDP from the G-protein and the formation of the activated GTP:G-protein complex (reviewed in 26). In turn, G-proteins can interact with second messenger generating systems localized on the cytoplasmic surface of the plasma membrane. G-proteins possess an intrinsic GTPase activity which is involved in the termination of the signal.

The main second-messenger systems are generated by two enzymes, adenylate cyclase and phospholipase C (30–32). cAMP, produced in response to hormonal activation of adenylate cyclase, serves to specifically activate the cAMP-dependent protein kinase (reviewed in 30,31). The majority of the hormones which act via cAMP utilize the G_s and G_i proteins to either activate or inhibit adenylate cyclase, respectively. In addition, a number of hormones, such as insulin, appear to alter cAMP levels via modulation of a cyclic nucleotide phosphodiesterase (33). Phospholipase C cleaves the minor membrane phospholipid, phosphatidyl-inositol-4,5- bisphosphate (PIP2) into two second messengers (34):

1,2-diacylglycerol (DAG) which activates protein kinase C (35), and inositol-1,4,5-trisphosphate (IP3) which acts to mobilize Ca^{2+} from the endoplasmic reticulum (36). The rise in intracellular Ca^{2+} triggered by IP3 is transient, but suffices to elicit a number of rapid metabolic changes when tissues are exposed to Ca^{2+}-mobilizing agents. In several cell types, however, receptor stimulation can also give rise to a second prolonged phase of Ca^{2+} mobilization, which appears to involve an Ins-1,3,4,5-P4 (IP4)-mediated entry of extracellular Ca^{2+} across the plasma membrane (36,37). Although an increasing body of evidence suggests a role for G-proteins in the regulation of PI metabolism, their formal involvement remains to be demonstrated.

As G-proteins are intimately involved in the control of a wide variety of metabolic processes, one can readily visualize mechanisms whereby abrogation of their coupling to either receptor or effector molecules could induce neoplasia. G-proteins are composed of α, β and γ subunits. The GTP-α complex represents the active moiety, while β/γ act to inhibit the α subunit. The *ras* family of oncogenes are associated with the cytoplasmic surface of the cell membrane and exhibit limited biochemical and sequence homology to the Gα subunit. Unlike normal G-proteins, p21*ras* has not been demonstrated to associate with a β/γ complex. Research by Houslay and colleagues (38) had initially suggested that the product of the N-*ras* proto-oncogene could be a G-protein involved in the PI pathway. However, this work remains to be confirmed, and other interpretations of these data are more probable (see section 3.7.3).

1.3.2 Plasma-membrane receptors encoding protein–tyrosine kinases

A second class of receptors comprises a group of polypeptides which span the plasma membrane and transmit an intracellular signal directly following ligand binding to their extracellular domains. The mechanism by which ligand binding results in receptor activation is unknown, but is presumed to be a consequence of either receptor aggregation or a conformational change in the receptor. This class of receptors possess an intrinsic protein–tyrosine kinase activity localized to an intracellular cytoplasmic domain. Ligands which activate such receptors include: EGF (39); PDGF (40); insulin-like growth factor-1 (41); insulin (42); and colony stimulating factor-1 (CSF-1) (43). The importance of these receptors in transformation is illustrated by the finding that the cellular homologs of a number of oncogenes correspond to known members of this receptor class (see *Table 1*). The biochemical properties of these receptors are discussed in detail in section 4.

1.3.3 Plasma-membrane ion channels

A third class of membrane receptors is utilized by certain neuro-transmitters, e.g. acetylcholine and gamma-aminobutyric acid, in which the receptors function as ion channels. Following ligand binding, these ion channels open to allow cations to flow into the cell (44).

A number of mitogens induce a rapid cytoplasmic alkalinization prior to the onset of DNA synthesis, and current evidence suggests this involves activation of the Na^+/H^+ exchange (44). This observation led to the idea that H^+ ions could act as a second-messenger system. There are very few enzymes, however, which exhibit significant activity changes in response to such small changes in either pH (typically 0.2 pH units) or Na^+ ion concentration. We believe an alternative explanation relates to the fact that most, if not all, growth stimulatory signals activate glycolysis to provide the necessary energy to drive DNA synthesis. Cytoplasmic alkalinization may have evolved, therefore, as a necessary cellular adjustment to prevent an excessive decrease in pH due to the stimulation of glycolysis, rather than as a second-messenger system *per se*. Since to date no oncogene products have been described which are members of this class of receptors, these proteins are not discussed further in this review.

1.3.4 Intracellular hormone receptors

The final class of receptors are located in either the cytoplasm [e.g. progesterone (45,46)] or the nucleus [e.g. thyroid hormone and estrogen (47–49)]. In this case, lipophilic ligands rapidly penetrate the cell membrane and bind to their cognate receptors. Binding of the ligand – receptor complex to the appropriate DNA regulatory sequences activates or inhibits transcription of the corresponding genes (50). As with other receptor systems, these intracellular receptors also appear to be intimately involved with the transformation process. The oncogene product of the avian erythroblastosis virus, p75 *gag/erb*A, is a mutated form of the receptor for the thyroid hormone, triiodothyronine (T3) (51,52). The v-*erb*A product does not bind T3 and appears to represent a constitutively active form of the receptor.

1.4 The role of intracellular effector systems in signal transduction and growth control

Since the discovery of the cAMP-dependent protein kinase some two decades ago (53), it has now become apparent that reversible protein phosphorylation is the major mechanism for the regulation of protein function in eukaryotic cells (30,31,54). At least five groups of protein kinases involved in growth regulation have been described:

(i)   the cAMP-dependent protein kinase (PKA);
(ii)   protein kinases which are Ca^{2+}/calmodulin dependent;
(iii)   the Ca^{2+}/phospholipid-dependent protein kinase C (PKC);
(iv)   protein – tyrosine kinases intrinsic to plasma membrane receptors; and
(v)   those protein – serine kinases for which neither regulatory second

messengers nor important physiological substrates have been identified, although clear evidence exists that they play key roles in cell transformation.

1.4.1 The cAMP-dependent protein kinase

Although originally discovered as an activator of phosphorylase kinase (53), it soon became apparent that PKA had a much wider role in eukaryotic metabolism (55). The great diversity of action of the numerous hormones which act via the single second messenger, cAMP, is achieved by the tissue specific expression of different hormone receptors (55) and by the existence of distinct substrates for PKA in different cell types (30). The PKA holoenzyme consists of two regulatory (R) ($M_r = 43-45$ kd) and two catalytic (C) ($M_r = 41$ kd) subunits, which dissociate when each of the R subunits binds two molecules of cAMP (56). PKA exists in at least two forms, type I and type II, which differ primarily in their R subunits (57,58). In addition, two distinct murine cDNA clones of the C-subunit (Cα and Cβ) have been isolated, which share 91 per cent homology at the amino acid level (59).

Probably all of the effects of PKA involve phosphorylation of substrates by the C subunit, despite the claims of several reports that the free R subunits may also play a direct role in the regulation of cellular functions (60,61). For example, the RII subunit was reported to possess an intrinsic topoisomerase activity (62), but other laboratories have been unable to confirm this observation. Furthermore, despite the existence of limited structural homology between RII and the prokaryotic cAMP-catabolite gene activator protein (CAP) (63) [a factor which is necessary for the activity of RNA polymerase in bacteria (64)], it appears that in eukaryotes the C subunit is exclusively responsible for cAMP-regulated gene transcription (65).

Despite numerous reports detailing the effect of cAMP on the control of cell growth, its mechanism of action in this regard remains completely unknown. A sustained increase in the intracellular concentration of cAMP has been reported to constitute a growth-regulatory signal which can either stimulate or inhibit cell growth (66), and either antagonize (67) or synergize (68) with the transformed phenotype. The availability of full-length C-subunit cDNAs should help to elucidate the role PKA plays in the regulation of cell growth and transformation.

1.4.2 Calcium/calmodulin-dependent protein kinases

Many of the biological actions of a wide variety of neural and hormonal stimuli are mediated by the Ca^{2+}-binding protein, calmodulin (69). In contrast to PKA, most protein kinases activated by calmodulin exhibit a more restricted substrate specificity. The PKA and Ca^{2+}/calmodulin-dependent protein kinase enzymatic pathways are closely interlinked and can act either in a synergistic or antagonistic manner in different tissues (see reference 30).

Ca^{2+}/calmodulin-dependent protein kinases include: phosphorylase kinase (a key regulatory enzyme in the metabolism of glycogen), myosin light chain kinase (which regulates myosin contraction), and three enzymes termed Ca^{2+}/calmodulin-dependent protein kinases I, II and III (reviewed in reference 54). The monomeric kinase I (37–45 kd) is found in high concentrations in the brain and phosphorylates neuron-specific substrates. Kinase II, a multifunctional Ca^{2+}/calmodulin-dependent protein kinase, is a dodecamer (M$_r$=500–700 kd) composed of two stacked hexameric rings. This kinase is found in several different tissues and phosphorylates a variety of substrates. Kinase III is found in skeletal tissue and appears to phosphorylate elongation factor-2 (EF-2) specifically (70). Phosphorylation of EF-2 results in the dissociation of the factor from the ribosomes, thereby inactivating protein synthesis. In fibroblasts, EF-2 undergoes rapid phosphorylation/dephosphorylation in response to exogenous growth signals such as EGF (71), which may allow newly synthesized mRNAs produced shortly after mitogen binding (e.g. *fos, myc*) rapid access to the translation machinery.

1.4.3 Protein kinase C

1,2-Diacylglycerol (DAG) is one of two immediate products of plasma membrane phosphatidylinositol 4,5-bisphosphate (PIP2) hydrolysis. DAG remains in membranes and initiates the activation of a serine and threonine-specific protein kinase, protein kinase C (PKC). PKC requires high levels of Ca^{2+} and a phospholipid (particularly phosphatidylserine) for maximal activity. DAG dramatically increases the affinity of PKC for Ca^{2+}, thereby rendering the enzyme fully active without a net increase in resting Ca^{2+} concentrations (72). PKC can undergo proteolysis to generate a C-terminal 50 kd catalytic fragment which is active in the absence of either Ca^{2+} or phospholipid (73). Recently, it has been demonstrated that there is a family of at least seven distinct PKC gene products, the genes for which map to distinct chromosomal locations (74–77). Like PKA, and perhaps most protein kinases, PKC has a regulatory domain which serves to maintain the catalytic domain in an inactive state until the activating ligand is bound (78–80). The appearance of DAG in membranes is normally transient (disappearing within a few seconds or minutes) (81), implying the PKC holoenzyme may be active for only a short time following stimulation of the PI pathway. Whether the 50 kd fragment of PKC plays a functional role *in vivo* remains to be determined.

One of the major actions of PKC appears to be the phosphorylation of cell surface receptors, a function which in general appears to serve as a feedback or down-regulatory mechanism. PKC substrates include the receptors for transferrin (82), insulin (83), T-cell antigens (84), α agonists (85,86) and EGF (87). In the case of the EGF receptor, phosphorylation by PKC results in inhibition of the receptor's tyrosine kinase activity (87), while PKC mediated phosphorylation of the T-cell antigen receptor increases the rate of receptor internalization (84).

Several observations strongly suggest that PKC plays an important role in the transformation process. In 1977, Diringer and Friis first reported that phosphatidylinositol metabolism was increased in transformed cells (88). This observation received comparatively little attention until the reports published nearly a decade later which claimed that oncogenic protein – tyrosine kinases might actually phosphorylate the PI lipids directly (89). While these latter observations now appear incorrect, it is clear that at least several different oncoproteins, including those encoded by *abl, ras* and *src*, modulate PI metabolism [90,91, and see sections 3.6.3 and 4.2.3(ii)]. By inference, therefore, PKC may play a critical role in transformation by a variety of oncogenes. Secondly, PKC has been shown to be the major receptor for tumor-promoting phorbol esters, such as 12-*O*-tetradecanoyl-phorbol-13-acetate (TPA) (92). TPA activates PKC directly by replacing the requirement for DAG (93). Furthermore, the addition of TPA to cells has been shown to induce the transcriptional activation of several proto-oncogenes, including c-*fos* (94,95), c-*myc* (96), c-*sis* (97) and c-*jun* (27,28). Whether these genes are constitutively transcribed in cells transformed by oncogenes which activate PI metabolism remains to be determined. Finally, the complete primary amino acid structure of PKC α reveals a tandem repeat containing the cysteine-rich sequence $C-X_2-C-X_{13}-C-X_2-C$, a pattern which is observed in a number of DNA-binding proteins (98). Although the current model for TPA induction proposes a PKC mediated activation of the transcription factor AP-1 (27), it is possible that the amino terminal repeat region of PKC could bind directly to promoter/enhancer elements. There is at least one report indicating the presence of PKC in the nucleus (99).

1.4.4 Protein – tyrosine kinases

Protein – tyrosine kinase activity was discovered in 1979, in association with the proteins encoded by the oncogenes of polyoma virus (middle T) (100) and the Rous sarcoma virus (v-*src*) (101,102). In the following year, Cohen's laboratory demonstrated that the EGF receptor also possessed an intrinsic protein – tyrosine kinase activity (39). The demonstration of the *erb*B oncogene as the kinase domain of the EGF receptor (24), the *fms* oncogene as the cellular homolog of the CSF-1 receptor (25) and the *sis* oncogene as the growth factor PDGF (18) (whose receptor also encodes a protein – tyrosine kinase activity), clearly demonstrates the importance of this gene family in transformation. More recently, several new human oncogenes with homology to protein – tyrosine kinases have been detected by either transfection of human tumor DNA into NIH 3T3 cells or as amplified sequences in primary human tumors, e.g. *trk, ret, met, mcf*-3 and *neu* (103 – 108). Recent cloning strategies have been based on the extensive sequence homology of this gene family. Using catalytic-domain probes at lowered stringency, the screening of cDNA libraries has revealed four new *src*-related genes, i.e. *hck, fyn, lyn* and *lck* (109 – 113). In a similar manner, novel *abl*- and *fps*-related genes have been identified by screening genomic libraries, i.e. *arg, eph*, NCP94, TKR11 and TKR16 (114 – 116).

In humans, there is now good evidence for at least two protein – tyrosine kinases playing an important role in cancer. Amplification of the *neu* gene has been observed in over 30 per cent of patients with breast cancer and appears to correlate with a more aggressive metastatic phenotype (5), while the c-*abl* protein – tyrosine kinase has been found to be constitutively activated in over 95 per cent of patients with CML (117,118). Activation of the *src* kinase has been reported to occur in 100 per cent of human colon cancers. The significance of this observation, however, is not clear, and the mechanism whereby the *src* kinase is activated remains to be determined (119). To date, approximately 30 proven or putative protein – tyrosine kinases have been identified (*Table 1*).

1.4.5 Oncogene-encoded protein – serine kinases

The final class of signal effector kinases includes those encoded by the oncogenes *mil/raf, mos* and *pim*-1, all of which are protein – serine/threonine kinases (120 – 122). In contrast to protein – tyrosine kinases, the *mil/raf* and *mos* transforming proteins are cytosolic rather than membrane associated. Activation of the *mil/raf* proto-oncogene is generated by truncation of the 5′ coding region (123,124). The complete primary structure of the *mil* proto-oncoprotein has revealed a cysteine-rich segment with two significant regions of homology to PKC (123). To date, no *in vivo* substrates for these enzymes have been identified.

2. Growth factors and oncogenes

The proliferation and differentiation of normal cells is tightly controlled by exogenous growth factors. In marked contrast, many transformed cells circumvent this requirement in a variety of ways.

(i)   The cell may be transformed by the inappropriate expression of an oncogene which encodes either a growth factor [e.g. the *sis* (18) or *hst* (19) oncogenes, see below] or a constitutively activated growth hormone receptor [e.g. *neu* (108), *erb*B (125) or *fms* (25), see section 4].

(ii)   An activated oncogene can modulate one or more intracellular signal transduction pathways to obviate the need for extracellular control [e.g. the stimulation of phosphoinositide turnover observed in *abl* (90), *src* (126) and *ras* (38,91) transformed cells].

(iii)   An oncogene product may either activate the production of a positive growth factor, or inhibit the expression of a factor whose normal

Table 1. Protein – tyrosine kinases

Mammalian protein – tyrosine kinases

Membrane associated

src gene family
  p60*src* (fibroblast, neuronal forms)
  p62*yes*, p56*lck* (= lsk = tck)
  fgr hck, fyn, lyn proteins

abl gene family
  p145*abl*
  arg protein

fps gene family
  p98*fps*
  NCP94
  fps-related proteins (TKR11 and TKR16)

Growth factor receptors and transmembrane proteins

EGF receptor family
  EGF receptor (*erb*B1 protein)
  neu protein (*erb*B2 protein)

Insulin receptor family
  Insulin receptor
  IGF-1 receptor
  ros, met, trk proteins (mcf3 = truncated *ros*1)

PDGF receptor family
  PDGF receptor
  CSF-1 receptor (*fms* protein)
  kit protein

sea, ret proteins
eph

Others

  p75 (liver)
  p120 (brain)

Drosophila protein – tyrosine kinases
D*src*64B protein
D*src*28C protein
D*ash* protein
fps-related protein
EGF receptor (types 1,11, 111 N terminus)
Insulin receptor
sevenless protein

function is to inhibit cell growth. In the following section we review a number of growth regulatory factors whose expression is modulated in transformed cells.

2.1 Transforming growth factor-α

2.1.1 The role of transforming growth factor-α in human cancer

The initial observation that retrovirally-transformed cells exhibit a decreased requirement for exogenous growth factors led to the concept of 'autocrine' secretion, whereby transformed cells might produce factors which could stimulate their own growth (127). Analysis of the conditioned media in which transformed cells had been cultured resulted in the discovery of two acid, heat-stable peptides which could induce the anchorage independent growth of normal rat fibroblasts (127 – 129). These two distinct peptides were termed transforming growth factor-α (TGF-α) and transforming growth factor-β (TGF-β) (128).

TGF-α, isolated from conditioned media of a variety of transformed cells, exhibits heterogeneity in molecular weights, including 6 kd, 18 kd and 25 kd forms (130). The 6 kd species of both rat and human TGF-α have been purified (131,132) and their cDNAs isolated (133,134). The mature 6 kd TGF-α species is synthesized as a larger precursor of 160 amino acids (25 kd). The 25 kd precursor is predicted to be a transmembrane protein containing a region of hydrophobic amino acids at the C terminus, which is post-translationally palmitoylated on cysteine (135). The hydrophobic segment is believed to be embedded in the plasma membrane, while the extracellular portion of the precursor is cleaved by a specific proteinase to yield a glycosylated 18 kd species and the mature 6 kd TGF-α. The different molecular forms of TGF-α are the result of the proteolytic processing of the 25 kd precursor (136).

Human TGF-α is encoded by a 4.5 – 4.8 kb mRNA (133). Examination of a wide variety of human tumors by Northern hybridization analysis indicates that squamous, renal and mammary carcinomas, as well as melanomas, synthesize TGF-α mRNA (137). As discussed below, the principal receptor for TGF-α appears to be the EGF receptor (130,131) and many of the tumors that contain TGF-α mRNA also express high levels of EGF receptor mRNA. TGF-α has also been detected in the urine of cancer patients (138). Hematopoietic tumors, which lack EGF receptors, do not express TGF-α.

Until recently, TGF-α was thought to be a tumor-specific growth factor. However, TGF-α has now been shown to be present in normal tissues from both the fetus and the adult (139 – 141). In humans, TGF-α is produced in keratinocytes, anterior pituitary cells and the theca-interstitial cells in the ovary (140,141). Although the function of TGF-α in normal cells remains to be defined, it has also been found in the urine of pregnant females (142; J.Sorvillo, unpublished observations).

2.1.2 Biological responses which differentiate between TGF-α and EGF

TGF-α and EGF exhibit similar amino acid sequence homology. Furthermore, both factors bind the same receptor. With the few exceptions described above, however, production of TGF-α is limited to transformed cells while the synthesis of EGF is only observed in normal cells. This suggests that although both molecules bind to the same receptor, the biological response of cells to these two ligands may be different.

In the past several years, large amounts of TGF-α have become available, allowing for a detailed comparison of properties of the two factors (133). Both EGF and TGF-α are equipotent for the induction of anchorage-independent growth of normal rat fibroblasts in soft agar (133), in their effects on newborn mice (143) and in inhibiting gastric acid secretion (144). In some biological systems, however, TGF-α has been shown to be more potent than EGF, i.e. epithelial wound healing of thermally induced lesions is dramatically enhanced by topical application of TGF-α as compared to EGF (145). TGF-α is also about 3- to 10-fold more potent than EGF in stimulating bone resorption, suggesting that TGF-α could play a role in hypercalcemia in malignancy (146). Finally, TGF-α has been reported to be more effective than EGF when tested in an *in vivo* angiogenesis assay (148).

To date, there have been very few investigations into the possible biochemical mechanisms which account for the quantitative differences observed between TGF-α and EGF. As discussed above, both TGF-α and EGF bind to the same receptor. In addition, TGF-α is like EGF in inducing tyrosine phosphorylation of EGF receptors (149), although a detailed comparison of these two factors in this assay remains to be performed. A number of laboratories have suggested that hormone – receptor complexes may be capable of acting within the cell following receptor – ligand internalization. In this regard, it has been reported that methylamine, which inhibits lysosomal proteolysis, completely prevents the breakdown of EGF while having no effect on the processing of TGF-α (147). These data suggest that intracellular processing of the complex formed between TGF-α and the EGF receptor is distinct from that of the EGF – EGF receptor complex. Thus, the species of ligand bound to the EGF receptor may determine transportation to different intracellular sites of action, and could also lead to temporal differences in the activation of the EGF receptor complex.

In Balb/c 3T3 fibroblasts, EGF results in mitogenesis without a measurable increase in the levels of inositol phosphates (153). In A431 human epidermal carcinoma cells, however, EGF has been shown to stimulate phosphatidylinositol metabolism (150,151). Regulation by the growth factor is thought to occur at several points in the pathway. First, EGF has been shown to stimulate the breakdown of PIP2, although it appears this may occur via changes in Ca^{2+} ion flux from extracellular sites, rather than as a direct activation of a phospholipase C (150,151). In addition, EGF causes an increase in the level of phosphatidylinositol 4-phosphate (PIP) as a result of the stimulation of a phosphatidylinositol kinase (150). In rat hepatocytes, EGF has also been reported to stimulate the hydrolysis of PIP2 (152). Treatment of rats with pertussis toxin for 72 h prior to isolating hepatocytes blocked the ability of EGF, but not angiotensin, to increase Ins 1,4,5-P_3 and Ins 1,3,4-P_3. The observation that pertussis toxin selectively abolishes EGF-stimulated inositol lipid breakdown suggests the possibility that EGF may activate phospholipase C via a G-protein dependent mechanism. As it appears unlikely that a distinct membrane receptor system exists for TGF-α, it is interesting to speculate that the contrasting actions of EGF and TGF-α in certain biological systems may be due to differences in temporal activation of PI metabolism.

2.2 Transforming growth factor-β

Although the term transforming growth factors (TGFs) was used initially to describe peptides produced by transformed cells that have the ability to confer the transformed phenotype to normal fibroblasts grown *in vitro*, TGF activity was later found in extracts from a variety of non-neoplastic cells and tissues (128,155). In surveying various sources of TGF activity, a new class of TGFs was discovered which, in contrast to TGF-α, did not compete with EGF binding to the EGF receptor (128). This new TGF activity was termed TGF-β (128).

2.2.1 Biochemical and physical properties

TGF-βs are acid- and heat-stable homodimeric proteins of 25 kd which upon reduction dissociate to two identical monomers of 12.5 kd, consisting of 112 amino acids. Each monomer contains nine cysteine residues. The biological activity of the TGF-βs is inactivated by reducing agents (155), indicating the requirement for disulfide linkages. The cDNA sequence of TGF-β revealed that the 112 amino acid monomer is generated by proteolytic cleavage from the carboxyl-terminal end of a 390 amino acid precursor (156,157).

2.2.2 The TGF-β gene family

Three forms of TGF-β have been reported in porcine platelets: TGF-β1, -β2 and -β1.2 (158). Total sequence identity exists between the human, bovine and porcine forms of TGF-β1, and the mouse sequence has only a single amino acid substitution (156,157). TGF-β1.2 is a heterodimer consisting of one chain of TGF-β1 and one chain of TGF-β2. Recently, we isolated a cDNA for a new form of TGF-β, which we termed TGF-β3 (159). Like TGF-β1 and -β2, TGF-β3 appears to be synthesized as a precursor (in this case, 412 amino acids), which undergoes dimerization and proteolytic cleavage to produce the active C-terminal 112 amino acid homodimer. Northern analysis of cancer tissue has revealed higher levels of TGF-β1 mRNA in tumors compared to adjacent normal tissue (137). TGF-β3 mRNA has also been observed in a number of human tumor cell lines (159). The significance of these observations, however, remains to be determined.

2.3 Bombesin and gastrin-releasing peptides

Bombesin is a 14 amino acid, growth-stimulating peptide isolated originally from amphibian skin (160). Bombesin and its human equivalent, gastrin-releasing peptide (GRP) (a 27 amino acid peptide) promote various physiological responses in mammalian cells, including proliferation of both 3T3 mouse fibroblasts (161) and human bronchial epithelial cells. Bombesin and GRP have nearly identical sequences in their carboxy-terminal regions, which is the domain responsible for receptor recognition and biological activity (160).

GRP has been identified in neuroendocrine cells of the stomach, fetal and adult lung, and in small-cell lung carcinomas (162). Human small-cell lung carcinomas (SCLC) produce GRP-like peptides and express a single class of high-affinity receptors (162). A monoclonal antibody to bombesin which binds the carboxy-terminal region blocks the binding of GRP to its receptor and inhibits both the clonal growth of SCLC *in vitro* and the growth of SCLC xenografts *in vivo* (163). These results suggest that GRP can function as an autocrine growth factor for human SCLC, and that GRP antagonists may be useful in the treatment of this disease.

A recent study has identified a cell surface protein of 115 kd in Swiss 3T3 fibroblasts, which is recognized by phosphotyrosine antibodies following treatment of these cells with bombesin (164). Complexes of the 115 kd protein with the phosphotyrosine antibody were reported to bind labeled bombesin in a specific and saturating manner, suggesting that the 115 kd protein might represent the cell surface receptor for bombesin (164). Whether the bombesin receptor actually possesses a protein – tyrosine kinase activity remains to be clarified, however, since a contrasting study has shown that the addition of bombesin to 3T3 cells resulted in no detectable increase in the levels of phosphotyrosyl-proteins (165).

Addition of bombesin to quiescent Swiss 3T3 fibroblasts causes a rapid stimulation of phosphoinositide turnover (166,167). PIP2 turnover appears to be crucial in eliciting cell proliferation induced by bombesin, since microinjection of antibodies to PIP2 inhibits bombesin-stimulated mitogenesis (168). Interestingly, these antibodies also inhibit DNA synthesis induced by PDGF, but not by FGF, EGF, insulin or serum. Recent studies have shown that overexpression of normal p21 N-*ras* in NIH 3T3 fibroblasts leads to an increased responsiveness of PI turnover in response to bombesin (38), which initially suggested that *ras* could function as the G-protein for the GRP receptor. However, as discussed in the section on *ras*, other interpretations of these data are probable.

2.4 Fibroblast growth factors, *hst* and *int-2*

Recently, a family of oncogenes has been described which encode proteins structurally similar to the fibroblast growth factors (FGFs). FGFs not only stimulate the proliferation of certain cells, but are also potent angiogenic agents (169). Vascularization is essential for solid tumor development and overexpression of FGFs could play an important role in the development of solid tumors. The *hst* oncogene was first detected by the NIH 3T3 transfection assay using DNAs isolated from human stomach cancer (19) and Kaposi's sarcoma (20). DNA sequence determination of the cloned human gene revealed *hst* encodes a 205 amino acid protein with 42 per cent homology to bovine FGF. The *int-2* gene, first identified in tumors induced by mouse mammary tumor virus (21), encodes a protein of 245 amino acids which also exhibits homology to FGF. Expression of this gene has not been detected in normal mammary glands.

To date, little information is available regarding the biochemical properties of the FGF receptor. Basic FGF, along with a wide variety of other growth factors, stimulates the phosphorylation of the 40S ribosomal subunit S6 (170). The FGF-receptor signalling pathway appears not to be coupled to phospholipase C activation, however, as both S6 phosphorylation and FGF-stimulated DNA synthesis can be initiated independently of inositol lipid breakdown (171).

2.5 Platelet-derived growth factor and *sis*

PDGF is the major mitogen in serum for cells of connective tissue origin. The protein is a 30 kd heterodimer consisting of similar sized chains (A and B) which are disulfide linked (172,173). In humans, the genes for the A and B chains of PDGF are located on chromosomes 7 and 22, respectively (174). The transforming protein of the simian sarcoma virus (SSV), p28sis, consists of the N-terminal 109 amino acids of the c-*sis* B-chain (18,175). In SSV-transformed cells the v-*sis* product is dimerized and proteolytically processed to yield a 24 kd homodimer (176). Interestingly, a number of human tumors have been found to express biologically active homodimers of the A-chain (174).

The PDGF receptor is encoded in a single 5.3 kb mRNA with the chromosomal location of the gene mapping to chromosome 5. Sequence analysis of the cloned receptor (177) reveals a 1090 amino acid open reading frame with structural features common to other cell surface growth factor receptors. These include an amino terminal extracellular ligand binding domain, a stretch of hydrophobic amino acids (residues 500–524) which serve as the transmembrane anchor sequence, and an intracellular domain (residues 525–1067) with tyrosine kinase activity. The PDGF receptor is closely related to the v-*kit* oncogene product and to the receptor for CSF-1 (177).

Binding of PDGF to its receptor results in a rapid activation of the receptor's protein–tyrosine kinase activity, as well as the stimulation of PI turnover (178). Treatment of 3T3 cells with PDGF results in the rapid appearance of an 85 kd phosphotyrosyl-protein concomitant with an increase in PI kinase activity (179). Initial purification of the 85 kd phosphoprotein suggests it copurifies with a type 1 PI kinase (179).

2.6 The autocrine model of cell transformation

The autocrine model of cell transformation postulates that the constitutive release of a mitogenic growth factor contributes to the uncontrolled proliferation characteristic of the transformed phenotype (23). Several reports have demonstrated the importance of autocrine stimulation in the transformation process. For example, antibodies to bombesin block the ability of a human small-cell lung carcinoma to form tumors in nude mice (163). The addition of PDGF antibodies to at least certain clones of SSV-transformed human fibroblasts can retard the growth of these cells and result in the appearance of a more normal cell morphology (176). Following binding of PDGF, its receptor is downregulated, rapidly internalized and degraded (172,180). Recent studies have shown that autocrine activation of PDGF receptor in v-*sis* transformed cells can occur in an intracellular compartment (181) and that degradation of the PDGF receptor in v-*sis* transformed cells occurs by a different pathway than that found in normal cells (181). These results may explain, at least in part, the inability of normal PDGF to induce cell transformation. In addition, it has been shown that cells transformed by either the *abl* or H-*ras* oncoproteins secrete a PDGF-like factor which can stimulate phospho-inositide breakdown in NIH 3T3 cells (182). Thus, secretion of PDGF by transformed cells may lead to an autocrine effect on PI turnover.

TGF-α has also been demonstrated to be capable of acting in an autocrine manner (183,184). Introduction of the human TGF-α cDNA clone into rat fibroblasts results in the transformation of these cells *in vitro* (183). Antibodies against TGF-α decreases the growth of these cells in soft agar, indicating that the phenotypic changes resulting from the TGF-α cDNA are at least partially mediated by an autocrine mechanism. Furthermore, Abelson-transformed fibroblasts lacking EGF receptors fail to form tumors in nude mice, implying the importance of an autocrine mechanism for TGF-α in the establishment of v-*abl* induced fibrosarcomas (185). Interestingly, the constitutive overproduction of EGF by introduction of the cloned gene can also lead to autocrine transformation (186).

In summary, increasing evidence suggests that the autocrine action of a number of growth factors plays an important role in tumorigenesis. The ability to block autocrine circuits by using either antibodies to the receptor or receptor antagonists may prove to be clinically useful in the treatment of a variety of human cancers.

3. p21ras proteins

3.1 Introduction

In mammalian species, the *ras* family of proto-oncogenes consists of three closely related members (187–189). Each of the functional *ras* genes, designated H(Harvey)-*ras*-1, K(Kirsten)-*ras*-2 and N-*ras* code for a 21 kd protein, termed p21 (189,190). In addition, several other genes have been identified which show limited homology with those of the *ras* gene family. These include *rho* (191), R-*ras* (192), *ral* (193), YPTI (194) and four *rab* genes (195). The relevance of these *ras*-related genes in cellular transformation, however, remains to be elucidated. Therefore, the remainder of this section will focus specifically on the H-, N- and K-*ras* genes.

Based on the NIH 3T3 transfection assay, mutations in the *ras* genes were originally estimated to occur in 10–30 per cent of human cancers (196). With the recent use of sensitive DNA hybridization technology, however, the frequency of *ras* gene mutations in human tumors now appears to be higher. Activating *ras* mutations have been found in up to 40 per cent of human colon cancers and in 95 per cent of pancreatic cancers (197,198,222). These and other data support a causal role for mutated *ras* oncogenes in human malignancies.

The three functional *ras* proto-oncogenes probably derive from a single ancestral gene, since each contains four translated exons, with intron–exon boundaries at the same sites relative to the p21 coding sequence. The human *ras* genes map to different chromosomes: N-*ras* to the short arm of chromosome 1 (1p22–p32), H-*ras*-1 to the short arm of chromosome 11 (11p15.1–p15.5) and K-*ras*-2 to the short arm of chromosome 12 (12p12.1–pter) (199). The sizes of the three *ras* genes differ substantially due to the varying lengths of the intervening sequences. K-*ras*-2 spans approximately 50 kb, N-*ras*, 10 kb, and H-*ras*, 4.5 kb. Both H- and K-*ras* genes have genetic counterparts in the Harvey and Kirsten strains of the acutely transforming rat sarcoma retroviruses (200). To date, no viral homolog of N-*ras* has been identified, although recombinant retroviral constructs containing an activated human N-*ras* gene display transforming activity (201). The *ras* genes have been highly conserved throughout evolution, and are found in such diverse organisms as man (187–189), chicken (202), insects (203) and yeast (204,205). This high degree of conservation suggests one or more fundamental roles for the normal *ras* genes. The limited homology of p21*ras* to G-proteins suggests that the normal p21*ras* proteins may also function to transduce extracellular information to intracellular effector systems. The following sections detail the prevalence of the different *ras* mutations in human cancer, the biochemistry of the *ras* proteins and how alterations in their biochemical functions relate to tumorigenesis.

3.2 Protein structure and post-translational modifications

Sequence comparison of the mammalian *ras* proteins has defined four regions (189,206). The first is a highly conserved domain, encompassing the amino terminal one-third of the protein. The second and third regions show slightly more divergence, yet retain about 85 per cent homology between any two *ras* genes, while the C-terminal sequences exhibit a high degree of variability (189). Two alternate fourth coding exon regions have been described for the K-*ras*-2 gene, designated IVB and IVA, which result in two K-*ras* proteins of 188 and 189 residues, respectively (190,206).

The *ras* proteins undergo fatty acid acylation at the cysteine residue (position 186) (207). Acylated N-*ras* p21 is found only in the membrane fraction (208). On SDS – PAGE analysis, acylated p21 migrates faster than the cytoplasmic form. Mutant proteins lacking the cysteine residue at position 186 do not undergo acylation, remain in the cytosol, and fail to transform NIH 3T3 cells (209). Fatty acid acylation of p21 appears to be a dynamic process, with a high turnover rate of the palmitate moiety ($t_{1/2}$ ~20 min) (208). Pulse-chase experiments have revealed a second cytosolic form of intermediate mobility which does not correlate with acylation (207). The nature of this modification remains to be defined.

A further post-translational modification has been reported for the K-*ras* protein (210). Treatment of a mouse adrenocortical cell line with TPA induced phosphorylation of K-*ras* p21. The *ras* protein from phorbol ester-treated NIH 3T3 cells expressing activated c-H-*ras* or v-H-*ras*, however, was not phosphorylated. In experiments carried out in our laboratory, however, purified PKC phosphorylated neither human K-*ras* nor H-*ras* p21 effectively *in vitro* (C.Weinstein, unpublished observations). This observation suggests that the phosphorylation of p21 *in vivo* is not mediated by PKC directly, but may involve a protein kinase cascade.

3.3 Activation of *ras* proto-oncogenes

3.3.1 Mutational activation of ras proto-oncogenes in human cancer

Generally, it has been observed that certain cancers exhibit a bias for transforming mutations of a particular member of the *ras* family. H-*ras* mutations have been reported in human bladder and urinary tract carcinomas (10,11,211), while mutant K-*ras* genes are more common in lung and colon carcinomas (212,213). In studies of colorectal cancer, 40 per cent have been found to contain mutated K-*ras* genes (197,198) with a high incidence of glycine to aspartic acid substitutions at codon 12 (197,198). Recent evidence shows that mutations in K-*ras* genes are also found in about 95 per cent of primary pancreatic carcinomas (222). Activated N-*ras* genes predominate in melanomas and hematopoietic malignancies (214,215) and occur in up to 60 per cent of patients with acute myeloid leukemia (216). Furthermore, in three of eight patients with myelodysplastic syndrome (a preleukemic condition), an activated N-*ras* gene with a single nucleotide substitution in codon 13 was noted, with disease in all three patients progressing to a true leukemia within one year (217).

Codon 12 of *ras* proteins is one of two principal regions where a base change can result in oncogenic activation (10,11,218). *In vitro* mutagenesis studies of the H-*ras* gene at codon 12 have shown that the substitution of any amino acid, with the exception of glycine and proline, results in oncogenic activation (218). The second region for *ras* activation has been localized around codon 61 (219). Substitution of glutamine at position 61 by amino acids other than proline or glutamic acid leads to oncogenic activation (220). At amino acid 59, a threonine substitution for alanine also results in transforming activity (218).

Perhaps the most convincing evidence to support a correlation between *ras* gene point mutations and tumorigenesis has come from the chemical carcinogenesis studies of Balmain, Barbacid and others (12–16,221). Sequential treatment of mouse skin with the mutagenic tumor initiator, dimethyl-benzanthracine (DMBA), followed by the tumor promoter, TPA, induces formation of epidermal tumors. By utilizing a restriction site polymorphism (*Xba*I), it was determined that mutations at codon 61 of H-*ras* were present in 90 per cent of the DMBA-initiated benign papillomas (13). Furthermore, when tumors were initiated with a different carcinogen (MNNG), none of the restriction fragments diagnostic of the mutation at codon 61 were observed (14). A study by Bizub *et al.* also demonstrated the presence of a specific mutation (A to T transition) in the second position of codon 61 from a majority of the DNAs isolated from either NIH 3T3 foci transformed with DNA from polycyclic aromatic hydrocarbon-induced mouse skin carcinomas or the chemically-induced primary tumors (221).

A similar line of experimentation has been conducted by Barbacid and co-workers, using the induction of mammary carcinomas in rats by *N*-nitroso-*N*-methylurea (NMU), which specifically induces G to A transitions (15,16). Malignant transformation of NIH 3T3 cells was induced by DNA from 75 per cent of the 58 NMU-induced tumors examined (16). Hybridization analysis demonstrated that the transforming H-*ras*-1 oncogene present in NMU-induced mammary carcinomas is activated by a G to A mutation in the second nucleotide of the twelfth codon. No mutations were observed at codon 61 (15). Furthermore, because of the highly labile nature of NMU, this mutagenic effect achieved by a single dose of the carcinogen must occur within hours following its administration, concomitant with the initiation of carcinogenesis.

3.3.2 Overexpression of ras proto-oncogene in human cancer

A number of reports have implicated a link between overexpression of the normal *ras* gene and neoplastic transformation (223–228). Immunoblotting of human colorectal carcinomas and adjacent normal tissue with anti-*ras* antibodies showed that 9 of 17 primary colon tumors had greater than a three-fold elevation of p21, with a greater than ten-fold elevation of p21 observed in three of these tumors (226). In a study of seven primary human gastric cancers, six tissue samples had increased amounts of *ras* protein when compared to the amount of p21 in human placenta. None of the seven cancers showed p21 with the altered electrophoretic mobility which is characteristic of mutant forms of the protein (227). There is also evidence suggesting elevated levels of p21 occurring in both prostate and bladder carcinomas (223,224), as well as overexpression of the N-*ras* oncogene in human glioblastomas (225). Collectively, these data implicate overexpression of p21*ras* as an alternate mechanism for activation of the *ras* proto-oncogene in human cancers, albeit at a lower frequency.

3.3.3 *Transgenic studies*

Another approach to address the role of the *ras* oncogene in tumor development has been the use of the transgenic mouse model (228). Transgenic mice carrying the v-H-*ras* gene driven by the mouse mammary tumor virus (MMTV) promoter exhibit two distinct pathological changes; nearly all of the animals develop a benign hyperplasia of the Harderian gland, while malignancies of the mammary gland, salivary gland and lymphoid tissues also occur but at a lower frequency (229). The effects of expression of an activated human H-*ras* oncogene in transgenic mice has also been examined (230). Both the normal H-*ras* and activated H-*ras*^{val12} mutant were fused to the rat elastase I promoter. The large majority of the H-*ras* oncogene transgenic mice developed massive tumors involving the pancreatic acinar cells either at birth or shortly thereafter. Hybridization studies revealed that the transgene was expressed specifically in the pancreas. The wild-type *ras* transgene was found to induce some morphological anomalies in the pancreas but not true malignancies. These data demonstrate that the *ras* oncogene product acting alone can elicit both non-neoplastic proliferative effects and, at least in the pancreas, malignant tumor formation. This may help to account for the extremely high frequency of *ras* point mutations in human pancreatic carcinoma (222). In contrast, progression to malignancy in mammary, salivary and lymphoid tissue most likely requires additional somatic events (229,231).

3.4 Functions of p21ras in normal cells

The functions of p21*ras* proteins in normal cells have yet to be elucidated. *ras* proteins are localized to the inner surface of the cell membrane (232–234), bind guanine nucleotides, possess an intrinsic GTP hydrolytic activity (236,237) and share limited sequence homology to G-proteins (238). Various functions have been proposed for the p21*ras* proteins. One is a regulatory role in signal transduction, possibly acting to mediate the hormonal regulation of a phospholipase C (239), a topic which is discussed in detail in section 3.6.2. The *ras* proteins may also play a critical role in the control of cell growth and in maintenance of differentiation. Microinjection of high concentrations of the normal *ras* protein into NIH 3T3 fibroblasts causes both a morphological transformation and initiation of DNA synthesis in the absence of added serum (240). Co-injection of the *ras*-specific monoclonal antibody Y13-259 with p21*ras* has been reported to neutralize the protein's transforming ability (241). Normal NIH 3T3 cells induced to enter the resting stage of the cell cycle by incubation at low serum concentrations, then restimulated by the addition of 10 per cent serum, were blocked from entering S phase after microinjection with Y13-259 (242). No inhibition of S phase was observed if Y13-259 was injected 22 h after serum stimulation.

Other microinjection experiments have indicated a role for p21*ras* proteins in differentiation. Pheochromocytoma (PC12) cells differentiate into non-replicating, sympathetic neuron-like cells upon exposure to NGF. Microinjection of Y13-259 in PC12 cells before NGF treatment inhibits neurite formation and results in the temporary regression of partially extended neurites. Microinjection of oncogenic H-*ras* proteins into these cells promotes their morphological differentiation into neuron-like cells, with subsequent cessation of cell division (243). Infection of PC12 cells with either the K- or H-murine sarcoma virus also results in a cessation of cell division along with the expression of neuron-specific properties (244).

The tissue distribution of *ras* proteins also suggests a role outside the control of cell growth. Immunohistochemical studies demonstrate p21*ras* to be present in almost all tissues including immature, proliferating cells and mature, non-dividing cells (245). Within tissues, p21 expression is often restricted to cells at specific stages of differentiation. Neural cells of the brain, spinal cord and peripheral nerves all exhibit strong p21*ras* immunoreactivity, as do terminally differentiated muscle and endocrine cells. In normal tissue with separate proliferative and mature compartments, such as the rapidly self-renewing epithelia of the stomach and intestine, no p21*ras* is observed in the proliferative compartments, while the mature parietal cells of the stomach show high levels of p21*ras* (245).

In tumors with varying degrees of cellular differentiation, stronger p21 immunoreactivity is often associated with a more mature cellular phenotype (245). In such tumors, therefore, high levels of p21 may simply reflect the differentiation status of the cell rather than the transformed phenotype *per se*.

3.5 Biochemical properties of *ras* proteins

3.5.1 Guanine nucleotide binding

The known biochemical functions of the mammalian p21*ras* proteins include guanine-nucleotide binding and a GTP hydrolytic activity (246–251). Both GTP and GDP bind to p21*ras* with similar affinities to form a one-to-one stoichiometric complex (250). Scatchard analysis indicated a single class of binding sites with K_d values of 0.83×10^{-8} M for GTP and 1.0×10^{-8} M for GDP for the v-H-*ras* p21 (249). Reported K_d values for nucleotide binding for the human p21 N-*ras* protein are essentially identical: 1.2×10^{-8} M for GTP and 1.3×10^{-8} M for GDP (248). These values, however, were determined under conditions where the purified p21 either contained bound GDP, or else involved the use of partially denatured proteins. Recently, a method has been devised whereby almost all bound nucleotide can be removed rapidly under mild conditions (251). When the nucleotide-free p21 'apoprotein' is isolated, the affinity for guanine nucleotides appears to be about 20-fold greater than that determined previously, with binding constants of 5.7×10^{-10} M for GDP and 6×10^{-10} M for GTP.

The original structural model for p21 was based on the guanine nucleotide binding protein EF-Tu (252,253). Recently, the X-ray structure of the normal human H-*ras* p21 protein has been resolved at 2.7 Å resolution (259), and the mutant H-*ras* p21^{val12} at 2.9 Å (260). The proteins consist of six stranded β-sheets, four α helices, and nine connecting loops, four of which are involved in nucleotide binding. Residues 10–16 lie near the phosphates of GDP, residue 30 near the ribose, while residues 116, 117, 119, 120 and 145–147 form a side of the pocket for the guanine ring. Thus, the overall topology of the nucleotide binding based on the original EF-Tu model (252,253) was generally correct and, as predicted, mutations in the regions of residues 10–16 and 116–120 would be expected to influence nucleotide binding and/or hydrolysis. The major difference in structure between normal and mutant p21 appears in the first loop (L1) in the amino-terminal half of the molecule, corresponding to amino acid residues 9–18, which wraps around the β-phosphate of GDP. This results in L1 being about 2 Å larger in the mutant protein than in normal p21, and the loss of two hydrogen bonds from the protein backbone amino groups of residues 12 and 13 to the β-phosphate of the bound GDP molecule. This loop, as suggested by De Vos *et al.* (259) previously, would normally straddle the phosphodiester bond between the β- and γ-phosphates of GTP, and function as the catalytic site for GTP hydrolysis in normal p21. The loss of these two hydrogen bonds could affect the orientation of the β-phosphate, resulting in a decrease in the rate of GTP hydrolysis. Comparison of the crystal structures of the GDP-bound H-*ras* proteins, therefore, provide a plausible explanation for the decreased GTPase activity of the mutant protein. By maintaining mutant p21*ras* in the GTP-bound state for prolonged periods, propagation of uncontrolled signal for cell growth may result and, therefore, be involved in tumorigenesis (see sections 3.5.2 and 3.7). A more complete understanding of reduced GTP hydrolytic activity by activated *ras* proteins will await structural determination of wild-type and various mutant proteins bound to GTP.

Mutations at position 12 of *ras* appear to have little effect on nucleotide binding (247). In this region, however, within the nucleotide-binding consensus sequence GXGGXGK (residues 10–16) (254), mutations do affect nucleotide binding. Mutagenesis of the H-*ras* gene revealed that substitution of an asparagine residue for lysine-16 reduces the affinity of H-*ras* for both GTP and GDP by a factor of 100 (255), while valine substitutions at positions 10, 13 or 15 of v-H-*ras* result in a loss of nucleotide binding (256). Valine mutations at positions 13 or 15 also lose their ability to transform NIH 3T3 cells. Fifteen of 17 amino acid changes generating mutations at codon 61 of the human H-*ras* gene exhibit an increase in NIH 3T3 transforming activity varying over 1000-fold in potency, while demonstrating no significant change in guanine nucleotide binding (220).

The region spanning amino acids 116–119 has also been examined for its role in guanine nucleotide binding. Within this region is the consensus sequence N–K–X–D, shared between the human *ras* proteins and other GTP-binding proteins (257). The presence of a potential contact site of position 116 with the pyrimidine ring of guanine (252,253), suggests that this residue may be critical in GTP binding. The substitution of isoleucine for asparagine at position 116 of the human H-*ras* protein results in decreased GTP binding (258). Other mutations at positions 116, 117 and 119 of H-*ras* also result in a decreased affinity for GTP; a 10-fold reduction in GTP binding by the mutant 116-His protein, a 100-fold reduction in affinity by the mutant 119-His protein, and a 5000-fold reduction in GTP binding by the mutant 117-Glu protein (257). All four mutant proteins, however, are able to efficiently transform NIH 3T3 cells.

Collectively, these data indicate that for most forms of the *ras* protein, including the position 12 mutants which occur most frequently in human cancers, nucleotide binding appears to be necessary for *ras* transforming activity. Conversely, certain p21 mutants, which apparently lose their GTP-binding activity, may still be able to transform cells (256). It is important to remember, however, that given the high intracellular concentrations of GTP (10^{-3} M) and an affinity constant of 5×10^{-10} M for GTP exhibited by the wild-type *ras* proteins, even mutant *ras* p21 with a 10^{6}-fold reduction in binding would still bind GTP efficiently *in vivo*.

3.5.2 GTPase activity

The second biochemical activity exhibited by the H-, K- and N-*ras* proteins is an intrinsic ability to hydrolyze GTP to GDP and P_i. The reaction rate for GTP hydrolysis by p21*ras* is slow; GTP turnover by purified p21 has been reported at 5–20 mmol/min/mol for the normal *ras* proteins, with a significantly lower GTPase activity observed for most of the mutant forms of p21 (236,237). Biochemical analysis of various p21 mutant proteins has revealed that nearly all codon 12 mutations exhibit a decreased GTPase activity concomitant with acquisition of NIH 3T3 transforming potential (247,248). These findings invited speculation that the reduced GTPase activity played a critical role in malignant transformation by mutated *ras* proteins. A number of subsequent reports, however, indicated a poor correlation between the transforming potential of p21 and its GTPase activity. For example, H-*ras* p21s with a threonine substituted for alanine at position 59 exhibited no impairment of GTPase activity, yet effectively transformed cells (261). When the kinetics of nucleotide release from the thr59 mutant were determined, it was found that the rate of release was increased by 3- to 9-fold relative to normal p21 ($t_{1/2}$ ~100 min for normal H-*ras*). In an environment with a high GTP/GDP ratio, position 59 mutations would act to favor the formation of a GTP-p21 complex. Various amino acid substitutions in codon 61 of the human H-*ras* gene lead to 8- to 10-fold reductions in GTPase activity, yet the mutants exhibit wide variations in transforming efficiency (220). Such observations had suggested initially that a decreased GTPase activity was insufficient for oncogenic activation of p21. As discussed later, however, the discovery of a cytoplasmic protein found to activate the GTPase activity of the normal N-*ras* p21 while being unable to stimulate the GTPase activity of the mutant protein has again raised the possibility that a reduction in GTP hydrolysis plays a key role in oncogenesis.

3.6 Role of *ras* in signal transduction

3.6.1 Yeast adenylate cyclase and p21 ras

The yeast *Saccharomyces cerevisiae* contains two *ras*-related genes, designated *RAS1* and *RAS2*, which encode proteins of 309 and 322 amino acids, respectively (262). These proteins exhibit a structural similarity to the mammalian *ras* proteins, with a high degree of sequence homology in their 172 amino-terminal residues. Like mammalian p21*ras*, the yeast *RAS* proteins bind GTP and GDP and possess a GTP hydrolytic activity (263,264). *RAS* proteins with amino acid substitutions at positions analogous to the sites of oncogenic substitutions in mammalian *ras* proteins demonstrate a reduced GTPase activity. Furthermore, like mammalian p21 proteins, the primary translation products of yeast *RAS* genes are found in the soluble fraction of the cell; the proteins are subsequently modified by fatty acid acylation and then bind to the cell membrane (234).

Genetic studies have shown that at least one functional *RAS* gene is required for growth, since yeast spores lacking both *RAS* genes are incapable of germinating (265). Expression of the human H-*ras* protein in a *ras1⁻ ras2⁻* yeast strain can restore spore viability, albeit with reduced efficiency (266). This suggests that a critical effector domain of *ras* has been at least partially conserved during evolution, and that a functional similarity exists between the yeast and mammalian *ras* proteins.

Yeast mutants deficient in *RAS* function exhibit properties similar to adenylate cyclase-deficient yeast (267). Strains containing a mutant *RAS2* gene coding for valine at position 19 (analogous to the mammalian *ras*^{val12} mutation) appear phenotypically similar to the IAC mutants, which show elevated levels of adenylate cyclase activity. *RAS2*^{val19} strains show a 4-fold increase in intracellular cAMP levels, while in *ras2⁻* strains cAMP levels are significantly depressed (approximately 4-fold). The *bcy*1 mutation suppresses lethality in adenylate cyclase-deficient yeast by causing overexpression of the catalytic subunit of PKA (268). In a *ras1⁻ ras2⁻ bcy*1 strain, intracellular cAMP is virtually undetectable. *In vitro* studies using membranes from *ras1⁻ ras2⁻ bcy*1 cells demonstrated that the human H-*ras* p21 can activate yeast adenylate cyclase in the presence of guanine nucleotides and Mg^{2+} (269). When bound to GTP, both *RAS2* and *RAS2*^{val19} proteins were found to activate cyclase activity equal to that of Mn^{2+}-activated adenylate cyclase, whereas the stimulation of cyclase was reduced approximately four-fold for wild-type *RAS2* and approximately 10-fold for *RAS*^{val19} when the p21 proteins were bound to GTP (270). Since yeast cell membranes contain an active dinucleotide kinase (271), the residual stimulation of cyclase activity observed in the presence of GDP-bound *RAS2* probably results from the conversion of GDP to GTP. A feedback mechanism of regulation may also occur via phosphorylation of *RAS2* by the activated protein kinase A, resulting in inhibition of yeast adenylate cyclase activation (235).

3.6.2 Mammalian adenylate cyclase and p21ras

The similarities in structure and biochemical function between yeast *RAS*, mammalian p21*ras* and the G-protein subunit, Gα (272,273), as well as the ability of mammalian *ras* to regulate adenylate cyclase in yeast membranes (269), suggested that p21 proteins might affect cyclase activity in other species. The first experiments to test this hypothesis were carried out using oocytes from *Xenopus laevis*. Progesterone-induced oocyte maturation, measured by germinal vesicle breakdown (GVBD), is mediated, at least in part, by inhibition of adenylate cyclase (274,275). Microinjection of the monoclonal antibody Y13-238 caused acceleration of steroid-induced GVBD, along with a 50% reduction in cyclase activity (276). The effect on oocyte maturation correlated directly with the inhibition of *ras*-stimulated adenylate cyclase activity by Y13-238 *in vitro*. Whether these data indicate a role for *Xenopus ras* in negatively regulating maturation, or reflect a cross-reactivity of Mab Y13-328 with other oocyte proteins remains to be elucidated.

Microinjection of activated human H-*ras* val12 protein into progesterone-treated *Xenopus* oocytes caused accelerated maturation when compared to oocytes treated with progesterone alone (277). The wild-type H-*ras* protein was 100 times less efficient than the mutant protein in inducing maturation. No significant effects of p21*ras* on the level of cAMP in the oocytes were observed in these experiments. However, the technique used to measure the concentration of cAMP would have detected neither small fluctuations in cAMP levels nor changes of a transient nature.

Incubation of *Xenopus* oocytes in the presence of insulin results in cell maturation by a signalling mechanism distinct from that triggered by progesterone (278). In one study, the injection of *ras* monoclonal antibody

6B7 into oocytes has been claimed to inhibit insulin-induced maturation, while having no effect on progesterone-induced maturation. This result, and the observation that p21*ras* is phosphorylated by the purified insulin receptor *in vitro* (albeit with a very low stoichiometry), led to the proposal that insulin regulates p21 via direct phosphorylation (278). Although attractive, this hypothesis clearly requires further analysis.

In mammalian systems, the role of p21*ras* in the regulation of adenylate cyclase also remains to be clarified. Both the basal and toxin (cholera or pertussis)-stimulated cyclase activities of NIH 3T3 cells transformed by either mutant H-*ras* or high levels of the normal p21 were reported to be similar (279), suggesting that *ras* transformation does not directly affect the interactions of G_s or G_i with the catalytic subunit of adenylate cyclase. However, isoproteronol-treated cells transformed by mutant p21 do exhibit decreased cyclase activity when compared to the similarly-treated parental cell line (280).

Membrane-reconstitution experiments using a mutant lymphoma cell line lacking G_s have been carried out to determine if p21 has any direct effects on the mammalian cyclase system (281). However, even in the presence of excess amounts of p21, no detectable adenylate cyclase activity could be measured in these membranes. Furthermore, when purified p21 was reconstituted into the parental S49 cell membranes, no change in basal adenylate cyclase activity was observed. From the data accumulated thus far, it appears that despite certain structural and biochemical similarities between p21*ras* and G-proteins, the *ras* proteins are not directly involved in regulating mammalian adenylate cyclase activity. However, before conclusively excluding an involvement of p21*ras* in the mammalian cyclase system, further investigation will be required.

3.6.3 The role of p21ras in phosphatidylinositol (PI) turnover

As experiments have failed to establish a direct link between p21 and mammalian adenylate cyclase, attention has focused on the possibility that p21*ras* may act as a guanine nucleotide regulatory protein mediating receptor stimulation of PI metabolism.

Determination of the steady-state levels of PIP2, DAG and inositol phosphates revealed the ratio of DAG to PIP2 was 2.5- to 3-fold higher in the *ras*-transformed cells when compared to the untransformed parental controls (282). The total concentration of PIP2 and IP3 was also reported to be elevated in *ras*-transformed NRK cells (282). When exponentially growing NIH 3T3 fibroblasts were assayed for inositol polyphosphates, however, no significant differences were observed between normal and *ras*-transformed NIH 3T3 cells (282). This suggests that the elevated levels of inositol phosphates reported earlier may simply reflect altered growth rates, rather than be an effect of the transformed phenotype *per se*.

NIH 3T3 cells overexpressing the normal N-*ras* gene product have been demonstrated to exhibit marked increases in inositol phosphate production in response to stimulation by bombesin (38). On the basis of these data, it was suggested that p21 N-*ras* acts as a G-protein which couples bombesin receptors to phospholipase C. While the number of bombesin receptors in this particular cell clone was apparently unaffected by transformation, it is clear that overexpression of p21 can induce increased responsiveness to certain growth factors by effecting total receptor levels rather than by acting directly as a G-protein (283). In addition, direct determination of phospholipase C activity, using membranes isolated from either normal or H-*ras* transformed fibroblasts, has revealed no difference in activity, whether the enzyme is assayed in the presence or absence of non-hydrolyzable GTP analogs (283). In another recent report, NIH 3T3 cells were treated with either phorbol esters or a calcium ionophore, to activate mitogenesis downstream of phospholipase C (284). Microinjection of the monoclonal antibody Y13-259 blocked the stimulation of DNA synthesis by these agents, a result which is inconsistent with p21 acting as a G-protein upstream of phospholipase C-mediated PIP2 breakdown.

While the effect of p21*ras* on PI metabolism remains to be defined, it does appear that DAG levels are elevated in *ras*-transformed cells (285–287). In addition to the hydrolysis of PIP2, DAG can be generated by phospholipase C cleavage of phosphatidylcholine (Ptd.Cho), phosphatidylethanolamine (Ptd.Etn) or phosphatidylserine (Ptd.Ser). *Xenopus* oocytes injected with an oncogenic p21*ras* showed a five-fold increase of DAG levels when compared to oocytes injected with the corresponding normal *ras* protein (286). However, the levels of inositol polyphosphates did not increase in proportion with the increased levels of DAG. Lacal *et al.* have also shown a marked elevation of DAG levels in H-*ras* transformed mammalian cells in the absence of detectable increases in inositol polyphosphates (287). Levels of both phosphorylcholine and phosphorylethanolamine (the products of Ptd.Cho and Ptd.Etd hydrolysis, respectively) were elevated in these H-*ras* transformants. In contrast, the addition of serum to normal NIH 3T3 cells had no effect on these metabolites, while in v-*sis* transformed cells their levels were significantly reduced. Recently, Macara has shown that the increased phosphorylcholine in *ras*-transformed NIH 3T3 fibroblasts arises not from Ptd.Cho hydrolysis by phospholipase C, but rather from the specific induction by *ras* of choline kinase (288). Furthermore, it was suggested that the elevation in DAG is unlikely to be a consequence of either PI or Ptd.Cho turnover.

3.7 A model for transformation by p21ras

The current model for the function of p21*ras* proposes an involvement in signal transduction (199). *ras* proteins are presumed to exist in equilibrium between an inactive (GDP-bound) and active (GTP-bound) state, with most molecules in the inactive form. Upon receiving a stimulus from another, as yet unidentified, protein, exchange of GDP for GTP occurs, followed by a conformational change of the *ras* protein to its active (GTP-bound) state. In this active state, p21*ras* interacts with its presumed effector molecule(s). After interaction of p21*ras* with the effector system has taken place, GTPase activity of the *ras* protein catalyzes the hydrolysis of bound GTP, returning p21 to the inactive GDP-bound form, and leading to the cessation of *ras* – effector interaction. Mutations which either reduce the intrinsic GTPase activity of *ras*, increase the exchange rate between GDP and GTP, or prevent *ras* interaction with any factors which stimulate GTP hydrolysis, would maintain *ras* proteins in the active state. Similarly, mutations causing a reduction in the affinity of *ras* proteins for guanine nucleotides would favor formation of a *ras* – GTP complex, given the greater availability of GTP over GDP in the cell. Any one of these mechanisms could shift the equilibrium to favor the active form of p21*ras*. Those p21*ras* molecules stabilized in the active form would have an extended interaction with the effector system, thereby allowing for continuous signal transduction to the inside of the cell.

Recently, the GTPase activity of the human p21 N-*ras* protein has been shown to be regulated by a novel cytoplasmic protein which was found in extracts of both *Xenopus* oocytes and mammalian cells (289) including a variety of normal tissues, nontransformed cell lines, human tumor cells and SV40, v-H-*ras*, v-*mos* and v-*src* transformed cell lines (290). Two mutant human p21 N-*ras* proteins (asp12 and val12) as well as the normal (gly12) protein, were microinjected into *Xenopus* oocytes. Following immunoprecipitation, both mutant proteins were found to be associated with GTP, whereas the p21^{gly12} was primarily bound to GDP. The high GTP hydrolytic activity of the normal p21 protein *in vivo* appears to result from an interaction with a cytoplasmic protein, termed GAP (for GTPase activating protein), which stimulates GTP hydrolysis of normal p21 N-*ras* by more than 200-fold *in vitro*. GAP does not stimulate the GTPase activity of the mutant N-*ras* p21. The normal function of GAP could be to maintain p21 in an inactive (GDP-bound) state by stimulating the protein's GTPase activity. When position 12 substitutions occur, the modulating protein is unable to stimulate the GTPase activity of the mutant protein, thereby increasing the half-life of the *ras* – GTP active complex. Alternatively, GAP may be the elusive effector target which is regulated by *ras* (290 – 292). Evidence in support of this theory is as follows: (i) the site of interaction of GAP with p21*ras* has been mapped to the previously identified putative effector binding region (amino acids 32 – 40); (ii) regions of p21*ras* which are dispensable for transformation are also dispensable for GAP interaction; and (iii) most mutations in the *ras* putative effector binding site which destroy the biological activity of the oncogenic *ras* protein also destroy the ability of GAP to stimulate the GTPase of normal p21*ras*. In this model, GAP stimulation of p21 GTPase activity would serve as a feedback mechanism to switch off p21 activation of GAP. The inability of GAP to stimulate mutant p21 GTPase would result in an extended activation of GAP by the p21 protein. The existence of such GTPase regulatory proteins may help to clarify the apparent lack of correlation between the transforming potential of certain mutant *ras* proteins and their intrinsic GTPase activities. Thus, potential chemotherapeutic drugs for *ras*-induced malignancies could be identified as agents which increase the GTP hydrolytic rate of the mutant *ras* proteins, block GDP dissociation from p21 or restore an interaction of mutant p21 with GAP.

4. Biochemistry of transformation by oncogene-encoded protein – tyrosine kinases

4.1 Activation of protein – tyrosine kinases in human cancer

At least four oncogene-encoded protein – tyrosine kinases are reported to be involved in the pathogenesis of human cancers (*Table 1*); *abl* in chronic myelogenous leukemia (CML) (9,117,293,294) and in acute lymphoblastic leukemia (ALL) (118,295,296); *src* in colon carcinomas (119); *neu* in breast cancer (5,297,298) and the EGF receptor in squamous cell carcinomas (299). To date, other activated protein – tyrosine kinases have been found only occasionally in human tumors [e.g. *trk* (103,300) and *ret* (104)] The incidence of oncogene-encoded protein – serine/throenine kinases [e.g. *raf* (124,301)] in human malignancies is unknown.

4.1.1 Amplification of the EGF receptor and c-neu

In human cancer, gene amplification is one mechanism whereby an increase in the concentration of a cellular proto-oncogene product can lead to transformation by chronic stimulation of a mitogenic signal transduction pathway (302,303). Amplification of the EGF receptor gene has been observed in breast (304), bladder (305), brain (306) and lung tumors (307,308), while amplification of the closely related proto-oncogene, c-*neu* (309) occurs in approximately 30 per cent of breast carcinomas (5,310). A recent study of 189 breast cancer patients indicates that amplification of the c-*neu* proto-oncogene is a significant predictor of disease prognosis (5,310). Furthermore, studies in model systems have clearly demonstrated that overexpression of either the human EGF receptor (311) or the human c-*neu* (312) is sufficient to induce cellular transformation. In rat neuroblastoma, a single point mutation in the transmembrane domain of the *neu* protein is responsible for induction of its transforming activity (108), but whether an analogous mutation occurs in human cancer remains to be determined.

4.1.2 Activation of c-src

A recent study has implicated activation of the c-*src* protein – tyrosine kinase in human colon cancer. Activation of the c-*src* kinase has been reported in 21 of 21 colon carcinoma cell lines and in 15 of 15 colon carcinoma tissue biopsies (119). Although the mechanism of *src* activation in colon tumors is yet undefined, it appears to involve an increase in the specific activity of the kinase rather than an increase in the amount of protein. In contrast to the EGF receptor and the *neu* protein kinase, overexpression of the normal c-*src* kinase fails to transform cells (313–315). Expression of p60c-*src* to very high levels in NIH 3T3 cells does lead to the appearance of morphologically altered foci, but these cells have a limited capacity for growth in soft agar and are not tumorigenic *in vivo* (316).

The transforming activity of p60v-*src* has been found to be critically dependent not only upon its kinase activity but also on myristylation of the N-terminal glycine residue, a post-translational modification which is required for plasma membrane association (317,318). Additionally, it is now apparent that the viral and cellular forms of *src* are regulated by two distinct sites of tyrosine phosphorylation on the kinase itself. By *in vivo* labeling, the major tyrosine phosphorylation site in p60v-*src* has been mapped to Tyr 416, whereas p60c-*src* is phosphorylated at Tyr 527 (319). *In vitro*, both proteins undergo autophosphorylation at Tyr 416, which appears to correlate with an activation of the enzyme. Phosphorylation at Tyr 527 has been shown to be mediated by a distinct cellular tyrosine kinase which negatively regulates the activity of p60c-*src* (320,321). As a consequence, overexpression of the c-*src* protein does not significantly increase the intracellular c-*src* kinase activity, an observation which accounts for the failure of increased levels of c-*src* protein to transform cells.

Sequences of RSV-derived p60v-*src* proteins diverge from p60c-*src* at the C terminus, starting at residue 515. The C-terminal 19 amino acids of p60c-*scr* are deleted and replaced by a stretch of 12 amino acids, which originate from the 3′ untranslated region of the c-*src* gene. This modification removes the inhibitory Tyr 527 phosphorylation site (322–325). A similar deletion/insertion event may also account for oncogenic activation of the kinase encoded by the *yes* gene (326). The importance of the Tyr 527 site in p60c-*src* has also been demonstrated using site-directed mutagenesis to covert Tyr 527 to Phe. This mutation enables c-*src* to transform NIH 3T3 cells, albeit to a much weaker extent than p60v-*src* (323–325). Mutating Tyr 416 to Phe in p60c-*src* has little effect by itself, but dramatically decreases the transforming activity of the Phe 527 c-*src* mutation (324,325). The importance of Tyr 527 in the activity of p60c-*src* is also illustrated by the association of the transforming protein of polyoma (middle T) with p60c-*src*. This association prevents phosphorylation at Tyr 527 and results in activation of p60c-*src* tyrosine kinase activity (320,323,327).

4.1.3 Translocation of c-abl

The most convincing evidence for the involvement of a proto-oncogene in a human cancer is the activation of c-*abl* in patients with CML (7,328, 329) and ALL (118,295,296). This is extensively discussed in Chapter 3. In leukemic cells of over 95 per cent of patients with CML, approximately 20–25 per cent of adult ALL and 2–5 per cent of pediatriac ALL, the c-*abl* proto-oncogene undergoes a (9;22) chromosomal translocation producing the Philadelphia (Ph') chromosome (6,7,329).

Analogous with c-*src* activation, the c-*abl* protein–tyrosine kinase can be activated by elimination of negative regulatory elements. In the case of viral transduction, viral *gag* sequences have become juxtaposed to the c-*abl* gene, resulting in the loss of c-*abl* sequences at the 5' end (117,330). In humans, a similar mechanism exists in CML and ALL. In both diseases, the Ph' translocation results in the loss of N-terminal c-*abl* sequences, by fusion of c-*abl* DNA to the *phl* gene of chromosome 22. In both cases, this results in the activation of the c-*abl* protein–tyrosine kinase (118,294). In CML, the fusion protein typically appears with $M_r = 210$ kd, while in ALL, the *phl*/c-*abl* protein has $M_r = 190$ kd.

Although the specific activity of p210*phl*/c-*abl* kinase is very similar to viral p160*gag-abl*, expression of the full-length *phl*/c-*abl* cDNA fails to transform NIH 3T3 fibroblasts (331). Thus, an increase in c-*abl* kinase activity *per se* appears to be insufficient to induce fibroblast transformation. In spontaneous recombinants which transform NIH 3T3 cells, NH_2-terminal viral *gag* sequences were found to have become fused to the *phl*/c-*abl* protein (331). The *gag* portion of v-*abl* contains myristic acid linked to a glycine residue. NH_2-terminal mutants of v-*abl* and v-*src* that lack myristylation no longer localize to membranes and are transformation-defective in fibroblasts (317,318,332,333), analogous to the palmitate-deficient *ras* mutants (209). In contrast, crude fractionation studies of CML cells have localized the p210*phl*/c-*abl* protein to the cytosol, and *in vivo* labeling experiments have revealed the protein is not myristylated (S.I.Rayter and J.G.Foulkes, unpublished observation).

4.2 Identification of physiological targets involved in transformation

Despite intensive efforts, no important physiological substrates for the protein–tyrosine kinases have been identified unequivocally (for previous reviews see 1,3). This is due to a number of reasons. First, the promiscuous nature of protein–tyrosine kinases, observed both *in vitro* and *in vivo*, results in the adventitious phosphorylation of a number of highly abundant proteins, therefore obscuring detection of key regulatory targets present at low concentrations. Second, despite the fact that activation of some protein–tyrosine kinases induces a ten-fold increase in total cellular phosphotyrosine, this still accounts for only 0.05 – 2.0 per cent of the total phosphoamino acids (334,335). Third, until recently, the most commonly used techniques for identifying phosphotyrosyl-proteins has been one- and two-dimensional SDS – gel analysis of total proteins from ^{32}P-labeled cells (335). Identifying the function of a protein based on its motility on gel electrophoresis is clearly a major undertaking (334 – 337). Recently, anti-phosphotyrosine antibodies have become available, and the use of these reagents should assist in the identification of phosphotyrosyl-proteins present at low concentrations (338 – 341). Furthermore, it may be fruitful to look for phosphotyrosyl-proteins in cells transformed by oncogenic kinases which do not significantly increase total phosphotyrosine levels (e.g. in cells expressing v-*erb*-B, v-*fms* or v-*ros*), or to analyze proteins that are phosphorylated by growth factors, since in the controlled environment of the normal cell aberrant phosphorylation should be significantly reduced.

Approaches designed to identify substrates can be divided into three categories:

(i)   a search for tissue-specific expression of oncoproteins;
(ii)   examination of the cellular localization of oncoproteins and their substrates;
(iii)   investigation of biochemical changes characteristic of the transformed phenotype, of which we will consider activation of glycolysis, increased PI turnover and S6 phosphorylation in detail.

4.2.1 Tissue-specific expression of protein kinases

The concept that proto-oncogenes play an important role in differentiation is suggested by the observations that oncogenes can arrest the process of cellular differentiation (342,343), and that high levels of proto-oncogene expression occur only in certain tissues (345 – 347) or at particular developmental stages (348).

Protein – tyrosine kinases consitute one group of oncogenes that appear to disrupt the normal linkage between growth control and differentiation (349). The lymphocyte-specific protein – tyrosine kinase gene (*lck*) is transcribed only in cells of lymphoid origin (111,113,350 – 352). When the CML cell line K562 is induced to differentiate through the erythroid pathway, the synthesis of p210*phl*/c-*abl* is reduced, with concomitant reduction in phosphotyrosyl-proteins (353). In contrast, the expression of the human specific protein – tyrosine kinase (*hck*, hemopoietic cell kinase) increases after the granulocytic and monocytic cells have been induced to differentiate (111). In *Drosophila*, both the *abl* and *sevenless* protein – tyrosine kinases are required for normal development of retinal cells (354,355). Thus, in addition to the role of protein – tyrosine kinases in regulating cell growth, the modulation of proteins via tyrosine phosphorylation may be involved in the control of differentiation.

Evidence is emerging that p60c-*src* may also participate in differentiation. The protein has been highly conserved in such widely divergent species as *Homo sapiens, Drosophila melanogaster* (356,357) and the freshwater sponge *Spongilla lacustris* (358). p60v-*src* can potentiate, inhibit or reverse the differentiation program of infected cells, depending upon the developmental system under investigation (359–361). Furthermore, certain cell types appear to be totally refractory to the transforming effects of the v-*src* gene product (362,363). High levels of p60c-*src* are expressed in brain and other neural tissues during embryogenesis of both humans and chickens (364,365). Expression of p60c-*src* in the developing chick neural retina (366) and cerebellum (345) coincides with the onset of neuronal differentiation. In addition, other non-proliferating cells, i.e. platelets (367) and myeloid cells (368,369), also contain high levels of p60c-*src*, suggesting that the *src* kinase may have a role in either the induction or maintenance of the differentiated state.

In cultured neurons from rat embryo striatum, two structurally distinct forms of the *src* kinase have been identified, p60c-*src* (as seen in fibroblasts) and a unique neuron-specific form, p61c-*src* (370). These two forms are the result of differential splicing, resulting in the introduction of six additional amino acids into the N terminus of the neural form (371). Interestingly, both PKC (75,76) and the *abl* protein–tyrosine kinase are also subject to alternative splicing (293). In undifferentiated neurons the predominant species is p60c-*src* whereas upon differentiation p61c-*src* is found (372), concomitant with an increase in *src* protein–tyrosine kinase activity. This elevation in protein kinase activity occurs without a detectable decrease in Tyr 527 phosphorylation or an increase in the phosphorylation of Tyr 416. The p61c-*src* form has two or more additional sites of serine phosphorylation within the NH_2-terminal 18 kd region, one of which is Ser(12). Whether these modifications account for the enzyme's increased activity, however, remains to be determined. Thus, an increasing body of evidence clearly demonstrates that modulation of protein–tyrosine kinase activity can have dramatic effects on the differentiation state of the cell. Bearing in mind that a large proportion of all human cancers develop in tissues involved in self-renewal (373), one model for transformation would have to envisage that activated protein–tyrosine kinases generate malignancies by 'freezing' rapidly dividing stem cell populations in the process of differentiation (363,373).

4.2.2 Subcellular localization of protein kinases

One approach to understanding the mechanism of cell transformation has been to determine the intracellular localization of the transforming protein. Immunofluorescent staining of transformed fibroblasts with antibodies against p60v-*src* of p160v-*abl*, shows these protein–tyrosine kinases to be localized to the inner surface of the plasma membrane, particularly in the region of focal adhesion plaques (374–376).

Temperature-sensitive variants of proteins p60v-*src* (377), p150*gag/fps*

(378) and gp74v-*erb* (379) all dissociate from the plamsa membrane at the non-permissive temperature prior to reversion to the non-transformed phenotype. Similar results have been obtained for the gp140v-*fms* protein (380). Adhesion plaques represent sites where bundles of actin filaments terminate in close association with the plasma membrane, and the disruption of actin-containing microfilament bundles is one of the prominent characteristics of transformed fibroblasts (382). Thus, many of the alterations in cellular morphology may result from phosphorylation of membrane or cytoskeletal components. The importance of cytoskeletal proteins is also implied by sequence analysis of the feline v-*fgr* (383) and the human *trk* oncogenes (300,384). Both oncogenes encode protein–tyrosine kinases which have arisen by recombination with sequences encoding δ-actin (in the case of *fgr*) and tropomyosin (in the case of *trk*). The amino acid sequence 141–269 of p70v-*fgr* is 98 per cent homologous to mammalian cytoplasmic δ-actin. Cytoplasmic actins participate in a variety of functions, such as cell motility, mitosis and maintenance of the cytoskeleton. Thus, p70v-*fgr* may represent an aberrant form of actin which interferes with the formation of proper cytoskeletal structures in cells. The actin moiety of p70v-*fgr* may also serve to direct the tyrosine kinase to a limited set of targets in the cytoskeleton which would not normally be accessible for phosphorylation. Analogous arguments exist for the tropomyosin moiety of p70*trk*. Alternatively, the tropomyosin/actin sequences may simply confer a certain configuration to the tyrosine kinase domain that allows it to function in a deregulated ligand-independent fashion, analogous to the v-*erb*B oncogene product (125).

Cytoskeletal proteins which exhibit elevated levels of phosphotyrosine following transformation by RSV include p36, an abundant membrane-associated protein which displays actin and phospholipid-binding properties *in vitro* (385,386); p81, a submembraneous protein that bears structural homology to the chicken microvillar core protein ezrin (387); vinculin, a 130 kd protein that is a prominent member of cellular adhesion plaques (381); and talin, a 215 kd phosphoprotein which is present in adhesion plaques and in fibrillar arrays paralleling those of external fibronectin (387,388). *In vitro*, talin has been shown to associate with both vinculin (389) and with the cytoplasmic portion of the 140 kd fibronectin receptor (390). Evidence to date, however, would indicate that the increased phosphorylation of vinculin or talin appears to reflect adventitious phosphorylation due to the presence of an active viral protein kinase in adhesion plaques (e.g. p60v-*src*, p90v-*yes*, p160v-*abl*) (381), rather than playing a causal role in altered cell morphology. One recently identified candidate for a physiological substrate is the fibronectin receptor (391). Normal cells contain the fibronectin receptor in focal contacts, whereas transformed cells exhibit a more diffuse distribution. The receptor consists of three membrane glycoproteins of 160 kd, 140 kd and 120 kd (392). Cells transformed by v-*src*, v-*fps*, v-*erb*B or v-*yes* all contain the 140 kd and 120 kd subunits phosphorylated on tyrosine (391). Phosphorylation of the fibronectin receptor in transformed cells may prevent adhesion of the extracellular matrix and thus contribute to the characteristic lack of contact inhibition and altered cell morphology.

p36 and p81 were identified originally by two-dimensional gel analysis of phosphoproteins in src-transformed cells (385,386). p36 and p81 form part of a submembraneous cortical skeleton which is peripherally associated with the inner face of the plasma membrane. In fibroblasts more than half of p36 exists as a heterotetramer in the form $(p36)_2$ $(p10)_2$ (393). p10 is homologous to the brain Ca^{2+}-binding protein S100 (394). The $(p36)_2$ $(p10)_2$ complex associates with phospholipids, actin and spectrin in a Ca^{2+}-dependent fashion (395). p36 is the major phosphotyrosine-containing protein in RSV-transformed cells, and is also phosphorylated on tyrosine in cells transformed by the v-yes, v-fgr, v-fps/fes and v-abl oncogenes (335). At steady-state, up to 25 per cent of p36 molecules contain phosphotyrosine at a single site (Tyr-23). However, the significance of p36 phosphorylation is unknown. It is only weakly phosphorylated in v-$erbB$ transformed fibroblasts (125), and contains no detectable phosphotyrosine when isolated from either v-fms or v-ros-transformed cells.

In addition to a structural role, it has been suggested that p36 may also function in signal transduction (396). Lipocortins are a group of steroid-induced anti-inflammatory proteins that inhibit phospholipase A_2 activity (397). The genes for lipocortins I and II have now been cloned (396,397). Lipocortin I, also termed p35 (398), is 50 per cent homologous to p36 (396) and, like p36, appears to be a substrate for a number of different protein–tyrosine kinases (399). The phosphorylation and subsequent inhibition of lipocortins would provide a mechanism to activate phospholipase A_2 constitutively. This would generate the second messenger arachidonic acid and lead to the activation of other signal transduction pathways. While this is an attractive model, we believe current evidence favors p35 and p36 as having a structural role in the maintenance of the cytoskeleton, rather than participating in the control of signal transduction. The evidence to support this notion is as follows. First, p35 and p36 constitute about 0.4 per cent of total cell protein, levels which are inconsistent with the idea that they represent key regulatory proteins. Second, both p35 and p36 undergo enhanced tyrosine phosphorylation in cells infected by the non-transforming RSV mutants which encode soluble forms of the src kinase (400). Furthermore, kinetic studies have shown that inhibition of phospholipase A_2 by p35 and p36 can be overcome by high concentrations of phospholipid (401). Thus, in $vitro$ inhibition of phospholipase A_2 by these proteins appears to be due to sequestering of the phospholipid substrate rather than a direct interaction with the enzyme (401). Finally, lipocortins must be secreted in order to inhibit the extracellular phospholipase A_2. To date, neither p35 nor p36 has been detected in conditioned media from transformed cell lines (C.M.Isacke and T.Hunter, personal communication).

A major question concerning protein–tyrosine kinases is how their substrates mediate changes in gene expression. The reported internalization of growth factor receptor tyrosine kinases to a juxtanuclear position after hormonal stimulation (402), and a similar localization of a fraction of the src and abl protein kinases (375,376), suggest that signalling pathways for protein–tyrosine kinases could involve the direct phosphorylation of nuclear substrates. The enzyme topoisomerase I was reported to be phosphorylated by p60v-src, in vitro, but it appears that this reaction does not take place in vivo (403). More recently, the phosphorylation of steroid receptors on tyrosine residues in vivo has been reported (404,405). Finally, there is one report of nuclear phosphotyrosyl-proteins in v-abl transformed cells which appear to exhibit sequence specific DNA binding properties (406). Clearly, the binding of phospho-tyrosyl-proteins to specific DNA sequences would represent the most direct mechanism for regulating transcription.

4.2.3 Biochemical analysis of physiological endpoints

The paradigm for protein phosphorylation as a regulatory mechanism is the hormonal control of glycogen metabolism. In the 1940s and early 1950s, it was established that adrenaline causes rapid breakdown of glycogen, while insulin stimulates glycogen synthesis. Analysis of these physiological observations over the next two decades led to the discovery of allosteric enzymes (407), enzyme regulation by reversible protein phosphorylation (408) and the concept of second messengers as intracellular mediators of hormone action (409). By the early 1970s, this pathway was known to contain two protein kinases and involve a simple, one-step protein kinase cascade. Over the next 15 years, work by numerous groups (perhaps most notably in the laboratories of Cohen and Krebs) unravelled a system which has been shown to involve at least eight different protein kinases, four protein phosphatases, four heat-stable regulatory proteins, four allosteric regulatory molecules, and a total of 14 phosphorylation sites. Given the clear success of this approach, a number of laboratories have adopted a similar strategy and have begun to dissect biochemically clearly-defined physiological endpoints which are characteristic of the transformed phenotype. We believe that it is the use of this approach which will ultimately lead to the understanding of the biochemical functions of oncogenes. We will consider three areas in which the transformed phenotype has been studied biochemically.

(i)   *Regulation of glycolysis.* The observations that transformed cells have a 3- to 4-fold increased rate of glycolysis (410,411), and the finding that enolase, phosphoglycerate mutase and lactate dehydrogenase were found to contain phosphotyrosine in RSV-transformed chick cells, suggested that the phosphorylation of these enzymes could play an important physiological role (412). Unfortunately, however, these enzymes are not rate-limiting in the normal glycolytic pathway; there are no detectable changes in their activities in RSV-transformed cells, and their stoichiometry of phosphorylation is low [ranging from 1 to 10 per cent (412)]. Furthermore, neither enolase, phosphoglycerate mutase, nor lactate dehydrogenase contain phosphotyrosine in cells treated with growth factor (334), despite an observable increase in glycolytic flux (413,414). Phospho 2,6-fructokinase, which is the key regulatory enzyme in glycolysis, does not contain phosphotyrosine (415). It seems likely, therefore, that an alternative mechanism must exist for increasing glycolytic flux in transformed cells.

(ii) *Regulation of the phosphatidylinositol (PI) pathway.* The current interest in PI metabolism and cell transformation by protein kinases arose following a report that purified p60v-*src* phosphorylated PI to form mono- and diphosphorylated derivatives (89), but the extremely slow rate of PI phosphorylation by protein – tyrosine kinases makes it very unlikely that this reaction has any physiological role (89,90,418). *In vivo*, however, PI turnover is increased following transformation by the protein – tyrosine kinase family of oncogenes (90,416 – 418), and one mechanism may involve the direct phosphorylation of a PI kinase (179,419). Clearly, confirmation of this observation and the detailed analysis of the phosphorylation state of other enzymes in this pathway represent areas which warrant further investigation.

(iii) *Regulation of ribosomal protein S6 phosphorylation.* Although all protein – tyrosine kinases are specific for the phosphorylation of tyrosine residues *in vitro*, the addition of EGF (420), PDGF (421) or insulin (420) to cells results in the increased phosphorylation of certain proteins on both serine and tyrosine residues. Cell lines transformed by either the v-*abl* or v-*src* protein – tyrosine kinases also show an increase in serine phosphorylation (335,422). Among these phosphoseryl-proteins, ribosomal protein S6 is of particular interest because its phosphorylation is correlated with growth-promoting stimuli in a wide variety of systems, including serum- and growth factor-induced cell proliferation, *Xenopus* oocyte maturation and fertilization (420,423,424), rat liver regeneration (425) and viral transformation (422). In the case of cells transformed with a temperature-sensitive transformation mutant of RSV, enhanced phosphorylation of S6 on serine residues is also temperature sensitive (422).

The widespread correlation between S6 phosphorylation and the growth-promoting actions of such a diverse group of agents suggests that S6 phosphorylation plays an important role in regulating cell growth. Chemical crosslinking studies have mapped S6 to the small head region of the 40s subunit, an area of the ribosome involved in binding of mRNA (426). Phosphorylation of S6 has also been correlated with a conformational change in the ribosome (427,428) and preliminary *in vitro* studies have indicated that phosphorylated ribosomes translate poly(U) more efficiently than non-phosphorylated ribosomes (429). Following serum stimulation of 3T3 cells, ribosomes containing the maximally phosphorylated derivatives of S6 are preferentially incorporated into new polyribosomes, suggesting a possible mechanism for the preferential translation of early growth associated mRNAs (429).

Microinjection of the purified *abl* kinase into *Xenopus* oocytes results in a 7- to 15-fold increase in the phosphorylation of S6 on serine residues (430). As the *abl* kinase cannot phosphorylate serine residues directly, this indicates that the enzyme must either activate an S6 protein kinase and/or inactivate an S6 protein phosphatase. Interestingly, microinjection of the insulin receptor into *Xenopus* oocytes also induces S6 phosphorylation, supporting the hypothesis that the insulin receptor retains biological activity after internalization (424). The activated S6 serine kinase has now been purified to homogeneity and cloned from *Xenopus* oocytes and appears to phosphorylate S6 specifically (431,432; J.L.Maller, personal communication). Antibodies to *Xenopus* oocyte S6 kinase also recognize the S6 kinase activated by either the insulin receptor or p60*src* in fibroblasts (433), suggesting this kinase, or a closely related enzyme, is widely distributed in different cell types. This S6 protein – serine kinase is not affected by incubation with a variety of protein – tyrosine kinases *in vitro*. Furthermore, immuno-precipitates of the activated enzymes labeled *in vivo* do not contain phosphotyrosine. However, recent work from the laboratories of Maller and Erikson has demonstrated that the S6 kinase can be regulated via a phosphorylation reaction catalyzed by a microtubule-associated serine kinase [the MAP-2 kinase (434)]. In turn, the MAP-2 kinase appears to be a substrate for protein – tyrosine kinases (E.Erikson, R.Stugril and J.L.Maller, personal communication). Thus, protein – tyrosine kinases may phosphorylate the MAP-2 serine kinase which then phosphorylates and activates the S6 kinase. Confirmation of this cascade in mammalian cells is now eagerly awaited.

4.3 A model for intracellular protein – tyrosine kinases in signal transduction

Several oncogene-encoded protein – tyrosine kinases which do not span the plasma membrane are myristylated at their N-terminal glycine residue (e.g. p120v-*abl*, p60v-*src*, p90v-*yes*, p85v-*fps*, p56v-*lck* and p110v-*fes*), a post-translational modification which appears to be necessary for both plasma membrane association and fibroblast transformation (350,

435–437). Interestingly, the catalytic subunit of PKA (438) and protein phosphatase-2B (439) also contain myristate, but neither protein is known to interact with membranes. Addition of myristate may, therefore, allow a reversible interaction with other membrane proteins by facilitating hydrophobic protein–protein interactions.

The demonstration that several intracellular protein–tyrosine kinases require an interaction with the plasma membrane in order to mediate their biological effects suggests a number of interesting models. The most direct explanation would be that one or more key physiological targets are localized in this region of the cell. In this regard, a number of membrane proteins involved in signal transduction would make interesting candidate substrates, e.g. the PI kinases, the phospholipases and the guanine nucleotide-binding regulatory proteins. Interestingly, as discussed above [see section 4.2.3(ii)], it has been reported that a PI kinase may be one such target (179,419).

An alternative model to explain the apparent requirement of kinases to be associated with the plasma membrane would be to propose the existence of critical protein–protein interactions which serve to regulate kinase activity. For example, intracellular protein–tyrosine kinases could represent a signal effector system analogous to adenylate cyclase. Extending this scheme, one could envisage a variety of plasma membrane receptors acting via G-proteins to modulate protein–tyrosine kinase activity. Two recent publications demonstrating homology between phospholipase C and protein–tyrosine kinases have provided support for this model (440,441). The sequence of a PI-specific phospholipase C reveals three clustered stretches of homology with the non-catalytic domain of protein–tyrosine kinases (441). One explanation for homology between these two distinct classes of proteins would be to propose these regions correspond to a domain which allows an interaction with putative G-proteins. In this model, mutations in these regions could result in the constitutive activation of a kinase in the absence of an extracellular ligand.

5. Nuclear oncogenes

The localization of several oncogene-encoded proteins to the cell nucleus fuelled hopes that such proteins might constitute a link in the signal transduction pathway from the cell membrane to the nucleus. In the case of *jun, fos, myc, p53* and *erb*A oncogenes, it now appears that their protein products may play a regulatory role in either DNA replication or in the control of gene transcription. In this section, we summarize recent data on the biochemical function of these nuclear oncoproteins.

5.1 *jun*

v-*jun* was discovered as the oncogene of a newly isolated avian sarcoma virus (442). In contrast to other avian retroviral oncogenes, v-*jun* lacks homology to protein – tyrosine kinases; instead, the *jun* protein was found to possess significant homology with the yeast transcriptional activator GCN4 (443). The human c-*jun* gene has a 340 amino acid open reading frame that displays 80 per cent identity to the avian v-*jun* product (27).

Structural and functional similarities between the c-*jun* product and the mammalian transcription factor AP-1 provides strong evidence that AP-1 is the product encoded by c-*jun* (27). Anti-peptide antibodies against the *jun* protein cross-react with AP-1 (28), while expression of cloned c-*jun* in bacteria produces a protein with sequence-specific DNA binding properties identical to those of AP-1 (27). The genes responsive to AP-1 have been shown to be inducible by treatment of cells with the phorbol ester TPA (444) which, as discussed earlier, acts directly on PKC to abrogate the need for activation by DAG. Thus, the activation of PKC by either TPA, or by the chronic stimulation of PI turnover observed in cells transformed by a variety of oncogenes [see sections 3.6.3 and 4.2.3(ii)], may result in the activation of AP-1 (*jun*). In turn, AP-1 activates the transcription of multiple genes involved in cell growth. As discussed below, these include proto-oncogenes such as c-*fos* and c-*myc*, as well as other genes potentially involved in tumor metastasis, such as the proteinases collagenase and stromelysin (444).

5.2 fos

5.2.1 Activation of the fos gene

The *fos* gene was first identified as the transforming sequence in the FGJ and FBR Moloney sarcoma viruses (for general review see 445). Osteogenic sarcomas arise in about three weeks following injection of rodents with either virus, and reach 100 per cent incidence within 40 days. Both the mouse and human c-*fos* genes have been cloned. The human gene consists of four exons which exhibit 90 per cent sequence homology with the mouse c-*fos* gene (446). Conversion of the normal cellular gene to a viral transforming oncogene results from the addition of a transcriptional enhancer sequence, and the disruption of an interaction between the end of the *fos* coding sequences and the 3′ untranslated region of the gene which serves to reduce expression of the normal *fos* protein (447).

In cells stimulated by a variety of agents, the induced c-*fos* product undergoes extensive phosphorylation on both serine and threonine residues (448), concomitant with a shift in the molecular weight from 55 kd to 62 kd (449).

5.2.2 The role of the c-fos protein in normal cells

A role for the normal c-*fos* protein has been suggested in the control of both cell growth and differentiation of embryonal, hematopoietic and neuronal cells (344,454). Expression of the c-*fos* gene is rapidly activated by agents, such as PDGF and TPA, that stimulate cell division (95,

450,451). Other growth stimulatory factors, including EGF and NGF, have similar effects (344,452). Increased levels of c-*fos* mRNA can be detected within 5 min following growth factor treatment, making this one of the earliest nuclear events in response to mitogenic stimuli (95,450). Studies at the cellular level have shown that agents which produce a voltage-dependent calcium influx also lead to an induction of c-*fos* mRNA (453). Cholinergic agonists induce a rapid but transient transcriptional activation of c-*fos* in neuronally differentiated PC12 cells (94). Similarly, administration of the convulsant drug Metrazole to mice induces a specific activation of c-*fos* expression in a neural cell population. Collectively, these data indicate that *fos* expression is tightly coupled to signals elicited by excitation of nerve cells.

5.2.3 Regulation of c-fos expression

The response of c-*fos* to such a broad range of signals has led to a major effort to understand the regulation of *fos* gene expression in both normal and malignant tissue. Several groups, therefore, have analyzed the regulatory sequences around the promoter region of the c-*fos* gene. One strategy has been the identification of sequences upstream of the coding regions which are necessary for c-*fos* induction. One regulatory region, termed the serum response element or SRE, has been shown to mediate transcriptional activation of c-*fos* in response to serum factors (456,457). SRE is a short DNA sequence located about 300 bp upstream of the c-*fos* mRNA cap site and contains a dyad of about 20 bp with a high degree of symmetry (455). Treisman first purified a 62 kd polypeptide from HeLa cells that binds the SRE sequence (455). Roeder's group has also purified a SRE binding protein similar, if not identical, to the protein reported by Treisman (458). This protein has an affinity constant for the enhancer dyad of 3.3×10^{-11}, at least 50 000-fold greater than that of its binding to nonspecific DNA (458). Analysis of insertion and deletion mutants has established a strict correlation between the ability of the dyad symmetry element to promote serum activation of c-*fos* transcription and *in vitro* binding of the protein to the SRE (456). Additional regulatory sequences in the c-*fos* gene may be responsible for *fos* activation via cAMP, PDGF, TPA and Ca^{2+}, although to date these elements have not been as clearly defined.

5.2.4 The fos protein as a trans-acting gene regulator

Binding of the c-*fos* protein to DNA has been demonstrated *in vitro* (449), with release of the c-*fos* protein from nuclei requiring either DNase I or micrococcal nuclease treatment, indicating an association of the *fos* protein with chromatin. Evidence indicating that *fos* acts as a regulator of gene expression came from the analysis of adipocyte differentiation, a phenomenon which is accompanied by the transcriptional activation of multiple genes. Distel *et al.* (459) have obtained immunological evidence that the c-*fos* protein may be part of a regulatory complex that binds the AP2 gene, which encodes a putative lipid-binding protein. The AP2 gene contains a regulatory element, termed FSE2, 124 bases 5′ from its transcription start site. *Trans*-acting factors binding to this region act as negative regulators of AP2 expression. Sequence analysis of FSE2 has revealed a considerable homology to the AP-1 binding site. Recently, Franza *et al.* (460) have obtained data which suggest that the c-*fos* protein binds either directly or indirectly to FSE2 by forming a complex with the c-*jun* protein (AP-1).

5.2.5 *Conclusions on the biochemical role of the fos protein*

To date, *fos* expression has been found to be associated with such diverse processes as proliferation, differentiation, nerve cell adaptation and cellular transformation. Collectively, these observations suggest that the *fos* protein may function as a key mediator in the transduction of cytoplasmic information into the nucleus. Thus, in a manner analogous to the cAMP-dependent protein kinase which serves to transmit information from the extracellular milieu into the cytoplasm, different cytoplasmic signals may activate the *fos* gene expression in different cell types and, in turn, the *fos* protein may interact with different DNA sequences and/or transcription factors to selectively regulate gene expression in a tissue specific manner. Transforming mutations resulting in the constitutive expression of *fos*, could in turn lead to the chronic stimulation of cell growth and/or abnormalities in the differentiation process.

5.3   *myc*

The *myc* oncogene was identified initially as the transforming sequence in the avian retrovirus MC29 (for a general review see 461). Expression of the v-*myc* oncogene can lead to a wide variety of tumors including myelocytomas, carcinomas, sarcomas and lymphomas. Later work demonstrated that v-*myc* belongs to a family of genes which show a high degree of evolutionary conservation (461,462). In humans, this family consists of five genes: c-*myc*, N-*myc*, R-*myc*, L-*myc* and B-*myc* (462).

5.3.1 *Role of myc in control of cell transformation*

In addition to the role of *myc* as a retroviral oncogene, several observations have implicated the c-*myc* gene in malignant transformation. Analysis of chicken B-cell lymphomas induced by the avian luekosis virus (ALV) revealed that the ALV promoter integrates next to the c-*myc* and stimulates its transformation (463) (see Chapter 2). In mouse plasma-cytomas and in human Burkitt's lymphoma, coding sequences from the c-*myc* gene are translocated from their normal chromosomal position into a transformationally active immunoglobulin locus on another chromosome

(464,465, and see Chapter 3). Experiments in which early passage rodent fibroblasts were transfected with different oncogenes demonstrated that the c-*myc*, like a number of other nuclear oncogenes, acts in combination with *ras* to immortalize primary cells (466). Amplification of the c-*myc* gene has been found in several types of human tumors such as lung, breast and colon carcinomas (467–470), while the N-*myc* gene has been found amplified in neuroblastomas, small-cell lung carcinomas and Wilms tumors (471–473). L-*myc* is amplified in small-cell lung cancer (474). Furthermore, elevated levels of *myc* mRNA and protein have been demonstrated in human tumors even in the absence of gene amplification (475). In addition, a correlation between high tissue-specific expression of c-*myc* and cancer has been demonstrated by construction of transgenic mice carrying a c-*myc* gene under the control of a hormonally inducible MMTV promoter. These transgenic mice spontaneously developed mammary adenocarcinomas (476).

5.3.2 Biochemical properties of the c-myc gene product

Using anti-peptide antisera, immunoprecipitation studies have shown that the human c-*myc* gene gives rise to at least two nuclear proteins with molecular weights of 64 and 67 kd (477,478). Both proteins are phosphorylated, exhibit relatively short half-lives *in vivo* (of about 30 min) (479) and exhibit DNA binding properties *in vitro* (477,482), although specific DNA binding sites have not yet been identified. p67 c-*myc* appears to be produced as a result of a novel translational initiation at a CTG codon near the 3' end of exon 1 (480). The two alternative translation initiation sites, which result in the synthesis of two different proteins, may constitute a mechanism for the regulation of *myc* function. For example, the p67 c-*myc* could serve to negatively regulate the activity of p64 c-*myc* protein. The loss of sequences encoding p67 c-*myc*, which occurs following chromosomal translocation in Burkitt's lymphoma, could result in the constitutive activation of the p64 c-*myc* protein (480).

Immunofluorescent and immunoelectron microscopy studies of the c-*myc* protein shows a distribution pattern identical to that of the small nuclear ribonucleoprotein particles (snRNPs) (481). Given the role of snRNPs in mRNA processing, the *myc* protein may also participate in this process. To date, however, no further evidence to support this idea has been reported.

As in the case of *fos*, stimulation of quiescent 3T3 cells with purified growth factors such as EGF, FGF and PDGF, or the pheochromocytoma cell line PC12 with NGF, results in the rapid transient induction of c-*myc* mRNA (450). Such results have led to a number of different experimental approaches in which to analyze the role of the *myc* protein in cellular DNA synthesis. One study has reported that the addition of c-*myc* antibodies to nuclei isolated from a variety of human cell types reversibly inhibits DNA polymerase activity suggesting that the c-*myc* protein is involved in DNA synthesis (483). In another approach, immunoprecipitation techniques using anti-c-*myc* antibodies were used to isolate human DNA fragments containing an autonomous replicative activity (484). The replication of a plasmid containing such an autonomous replicating sequence is inhibited by co-transfection of cells with anti-c-*myc* antibodies. Furthermore, plasmids containing an SV40 origin but lacking the T-antigen gene have been found to replicate in HL60 cells containing amplified c-*myc* (484,485). A third approach (486) has demonstrated that oligonucleotides complementary to codons in the initiation start site of human c-*myc* mRNA inhibit c-*myc* protein expression in human T-lymphocytes and prevent entry of these cells to S phase. In conclusion, although the actual biochemical function of *myc* remains to be fully defined, it is tempting to speculate that the activated *myc* protein may be an example of an oncogene product which induces cell transformation by driving continual rounds of DNA replication.

5.4 p53

The cellular phosphoprotein p53 was first identified as a nuclear protein in SV40-transformed 3T3 cells (487). Formation of a complex with SV40 large T-antigen results in a dramatic increase in the half-life of p53 (487, 488). Experiments in transgenic mice harboring hybrid insulin promoter-SV40 T-antigen genes, demonstrated a complete correlation between expression of large T-antigen and the appearance of p53 (489). In common with other nuclear oncoproteins, transfection of p53 cDNA expression constructs into cells of finite lifespan results in cellular immortalization and an increased susceptibility of these cells to transformation by the *ras* oncogene (490).

The levels of p53 in quiescent cells are normally very low. The synthesis and steady-state levels of p53 then transiently increase prior to the onset of DNA synthesis in late G1, suggesting a role for the normal p53 protein in the progression of cells from a state of growth-arrest to one of active cell division (491). In contrast, constitutively elevated levels of p53 protein are observed in a wide variety of transformed cell lines and in tumor tissue (492,493). Notably, Rotter's laboratory has shown that p53 expression is essential for v-*abl* induced lymphoid tumors in mice (494).

The fact that elevated levels of p53 are found in cells not transformed by SV40 suggested the possibility of a cellular homolog of the large T-antigen which could act to stabilize p53. One such candidate is the heat-shock protein, HSP70, which appears to form a complex with p53 in transformed cells which overexpress this oncoprotein (495). Current models for the biochemical function of p53 are discussed in Chapter 5.

5.5 erbA

5.5.1 Biological properties of v-erbA

v-*erb*A was isolated originally from the avian erythroblastosis virus

AEV (496). AEV transforms both chicken fibroblasts and erythrocytic progenitor cells *in vitro* and induces both erythro-leukemia and sarcomas *in vivo* (497). AEV, one of several avian leukemia viruses which carry more than one oncogene, also contains the v-*erb*B oncogene, which encodes the protein – tyrosine kinase domain of the EGF receptor.

The functional implications of AEV encoding two distinct oncogenes have been investigated by comparing the oncogenic properties of mutant forms of AEV (498). Wild-type AEV erythrocytic transformants are blocked at an early stage of differentiation (499). Expression of *erb*B by an AEV variant which lacks *erb*A is sufficient to induce fibroblast and erythrocyte cell transformation. Although the growth of transformed erythroid progenitor cells is independent of exogenous erythropoietin, these erythrocytic transformants have a high rate of spontaneous differentiation into apparently normal mature cells which then cease to divide (500). In the case of normal chick embryo fibroblasts, cells that express p75*gag/erb*A multiply in growth factor-depleted culture medium at low cell density, while expression of the *erb*A protein in fibroblasts transformed by v-*erb*B enhances their tumorigenicity *in vivo* (498).

5.5.2 The superfamily of hormone-receptor DNA binding proteins

The sequence similarities between v-*erb*A and the human glucocorticoid receptor first led to the suggestion that the v-*erb*A product may be a protein that binds to DNA enhancer elements (501). Recently, DNA sequence determination of a number of steroid and thyroid hormone receptors has revealed they constitute a superfamily of related genes. To date, this family includes the human estrogen (hER) (49,502), glucocorticoid (hGR) (502,503), thyroid (52), aldosterone (504) and retinoic acid receptors (hRR) (505). Similarities between the predicted amino acid sequences of c-*erb*A with chicken progesterone (cPR) and human aldosterone receptors (hAR) provided further evidence that the c-*erb*A protein might function in an analogous manner.

Steroids modulate a wide range of physiological actions including metabolism, differentiation and cell growth. Thyroid hormones also induce a wide variety of effects, including stimulation of the rate of basal metabolism, activation of growth hormone production in the pituitary, and brain development in neonates. According to current models, these hormones enter the cell by facilitated diffusion and bind specific receptors which act as signal transducers. Binding of the ligand to the receptor presumably effects allosteric changes that enable the ligand – receptor complex to bind specific DNA sequences in the chromatin, thereby regulating the transcription of specific genes.

5.5.3 Biochemical properties of the erbA protein

The highest degree of homology between v-*erb*A, human c-*erb*A and the hGR is found in a cysteine-rich sequence of 65 amino acids, beginning at c-*erb*A amino acid residue 102 (506). This region, which contains several Cys – Lys – Arg sequences, represents the DNA-binding domain of the hGR. Within this domain, the c-*erb*A protein shows 47 per cent homology to hGR and 52 per cent homology to hER. The more distant homology exhibited by the carboxy terminus of the c-*erb*A protein in comparison to the immediate members of the steroid receptor family indicated that the 52 kd c-*erb*A protein was not a steroid receptor, but was more likely to correspond to a receptor for another class of ligands. *In vitro* translation of the cloned human c-*erb*A gene led to the identification of its ligand as triiodothyronine (T3), a member of the thyroid hormone family (52). Although the p75*gag*/v-*erb*A is defective in T3 binding, the protein is localized to the nucleus and binds DNA *in vitro* (51). Thus, this oncogene may correspond to a constitutively activated form of the receptor. Since the presumed DNA-binding region of p75*gag*/v-*erb*A has two amino acid substitutions, it is unclear whether p75*gag*/v-*erb*A affects the expression of the same genes as the normal receptor. Interestingly, a second c-*erb*A-related gene has been found in the human genome. The human homolog of the avian gene (hc-*erb*Aα) is located on human chromosome 17, whereas a distinct but closely related gene (hc-*erb*Aβ) has been mapped to chromosome 3 (52). Thus, the identification of *erb*A as the T3 receptor, and c-*jun* as AP-1, firmly establishes the existence of a class of oncogenes whose products act by regulating transcription. Clearly, the major task ahead is to identify which genes they regulate.

5.5.4 The role of the estrogen receptor in human breast cancer

Involvement of steroid hormones in modulating cellular transformation is probably best documented for estrogen in human breast cancer. Increased levels of the hER have been found in about 50 per cent of human breast cancers. Approximately 70 per cent of hER-positive tumors respond to anti-estrogen therapy, compared to about 5 per cent of hER-negative tumors, suggesting a strong correlation between the growth of breast tumors and the presence of the hER (49). Estrogens have been shown to be mitogenic for the MCF-7 breast cancer cell line, which contains elevated levels of hER (49).

6. Anti-oncogenes and tumor suppressors

6.1 Definition

Anti-oncogenes can be defined as genes whose repression, inactivation, dysfunction or loss results in cell transformation. Tumor suppressor genes have been defined experimentally as genes whose introduction, activation or expression results in the inhibition or suppression of the tumorigenic phenotype. Clearly, these two definitions may represent alternate descriptions of a single class of genes.

6.2 Discovery

Hereditary predisposition to cancer is a well-documented phenomenon (507,508). Loss or mutational inactivation of anti-oncogenes is strongly implicated by the characteristic chromosomal deletions observed in a number of tumors, such as pediatric retinoblastoma (chromosome 13) and Wilms tumor (chromosome 11), as well as colorectal cancer in adults (chromosome 5) (507–509). These anti-oncogenes appear to act recessively, requiring the loss or inactivation of both gene copies for tumor development. Individuals who possess one copy of an anti-oncogene would have an increased propensity to a particular cancer.

Non-tumorigenic hybrid cells generated from fusion between tumorigenic and normal cells first suggested the existence of tumor suppressor genes (510). Normal human fibroblasts fused with HeLa cells were found to be non-tumorigenic (510). Morphologically transformed revertants of these hybrids were found to have lost a single copy of fibroblast chromosomes 11 and 14. The subsequent introduction of a single copy of fibroblast chromosome 11 into either tumorigenic-revertant hybrids or the parental HeLa cells result in a non-tumorigenic phenotype (511). In a separate study, the introduction of a normal human chromosome 11 into a Wilms tumor cell line also suppressed the ability of these cells to form tumors in mice (512). Virally transformed cells can also be rendered non-tumorigenic by tumor suppressor genes. Hybrids formed between *src*-transformed rat cells and their normal counterparts result in cells of normal morphology (513). Flat revertants of Kirsten sarcoma virus-transformed cells have also been isolated which still express the v-K-*ras* p21 gene product but can no longer grow in soft agar (514). Furthermore, these revertants are resistant to transformation by certain other oncogenes, such as *src* or *fes*.

Hybrids of related tumorigenic cells apparently share common genetic lesion(s) and maintain tumorigenicity, whereas hybrids of unrelated tumorigenic cell types can complement one another in terms of their tumor-suppressor activity. For example, human carcinoma cells fused with other carcinoma cells remain highly tumorigenic, whereas hybrids formed between carcinomas and sarcomas, or carcinomas and melanomas, result in non-tumorigenic cells (515). Thus, it is clear that normal cells can express gene products that act to maintain the normal cell phenotype. Loss of such genes could lead to tumor development.

6.3 Biochemical functions of anti-oncogenes

While the biochemical functions of anti-oncogene products remain to be defined, known biochemical properties of oncoproteins allow one to propose potential activities. From this perspective, the following section briefly reviews how certain gene products could function as anti-oncogenes.

6.3.1 Growth inhibitors as anti-oncogenes

Among the increasing number of polypeptides with growth inhibitory properties (516), the TGF-β gene family has been the most extensively characterized. TGF-β1 is the prototypic member of a group of homologous polypeptides that control tissue growth and development in such diverse organisms as *Drosophila, Xenopus* and man (517). Depending upon the type and physiological status of the target cell, TGF-βs can either inhibit or stimulate both cell growth and differentiation. Recently, we discovered the existence of a third form of TGF-β, which we termed TGF-β3 (159). TGF-βs are potent growth inhibitors for a variety of cell types. These include various tumor cell lines (518), endothelial cells (519), thymocytes (158), normal human prokeratinocytes (520), hematopoietic progenitor cells (521), B-cells (522) and T-cells (523).

The effect of TGF-β1 and TGF-β2 on cell growth is determined by the cell type, the physiological state of the cell, the presence or absence of other growth factors, and the surrounding microenvironment (e.g. extracellular matrix and type of adjacent cells). These different factors apparently account for the TGF-β1's bifunctional effects on cell growth *in vitro*. For example, in combination with either EGF or TGF-α, TGF-β1 stimulates the anchorage-independent growth of NRK cells in soft agar (524). Conversely, TGF-β1 even at pM concentrations is a potent inhibitor of NRK cells grown in monolayer culture (518). Fibroblasts transfected with a *myc* gene are stimulated to grow in soft agar by TGF-β1 in the presence of PDGF, but inhibited by TGF-β1 in the presence of EGF (518).

Although TGF-β1 and TGF-β2 are equally potent at inhibiting the growth of certain cell types (158), dramatic differences in the biological response to these factors have been reported in other systems. For example, TGF-β1 is 100-fold more potent than TGF-β2 in inhibiting hematopoietic stem cell proliferation (521). These differential biological effects are mediated by a class of at least three glycoprotein receptors with varying affinities for the different TGF-βs (158). As with other growth inhibitory peptides, however, the biochemical mechanisms whereby these receptors convert the signal of extracellular ligand binding into a potent growth-inhibitory response remain to be defined. To date, no effects of the TGF-βs on cyclic nucleotide metabolism, phosphoinositide turnover or protein – tyrosine kinase activity have been reported. TGF-β1 does not stimulate the autophosphorylation of its own receptors, nor has it been observed to induce significant changes in membrane phosphoproteins (525). Furthermore, TGF-β1 can antagonize the mitogenic effects of EGF and insulin without affecting their receptor tyrosine kinase activity (526).

Thus, TGF-β, through its interaction with specific receptors, is a potent growth inhibitor and apparently plays a critical role in maintenance of growth homeostasis. The loss of TGF-β function, therefore, could lead to a constant stimulation of cell growth. Of interest in this context is the recent report that seven of eight retinoblastoma cell lines had no detectable TGF-β receptors, with the eighth cell line having an anomalous type III receptor (527). None of the eight cell lines tested were inhibited in their growth by TGF-β1. The human gene (RB) that determines susceptibility to hereditary retinoblastoma encodes a phosphoprotein with DNA-binding activity (528). Since it appears that the loss of the RB gene product results in the loss of the ability of cells to respond to TGF-β, the RB gene product may normally regulate expression of the TGF-β receptor or it may interact with the TGF-β receptor signal transduction pathway to mediate inhibition of cell growth.

6.3.2 Phosphotyrosyl-protein phosphatases

As discussed previously in this chapter, elevated levels of protein–tyrosine kinase activity can result in cell transformation. For protein phosphorylation to be reversible, specific protein phosphatases must exist. An important role for phosphotyrosyl-protein phosphatases in transformation is suggested by the observations involving tumor viruses which encode temperature-sensitive (ts) protein–tyrosine kinases. Cells transformed by ts p60v-src (377), ts p140v-fps (378) or ts gp74v-erbB (379) undergo reversion to the normal phenotype in a temperature-dependent manner. Preceding reversion, however, there is a rapid loss of protein-bound phosphorylation demonstrating the presence of one or more potent phosphatase activities. Thus, transformation by oncogenic protein–tyrosine kinases should be envisaged as a dynamic process in which the normal equilibrium is shifted to favor the protein kinase. In theory, a decrease in phosphotyrosyl-protein phosphatase activity would elevate phosphotyrosine levels and could lead to cell transformation in an analogous manner. The DNA sequences encoding phosphotyrosyl-protein phosphatases, therefore, have the potential to function as tumor suppressors or anti-oncogenes. To date, little is known about phospho-tyrosyl-protein phosphatases, but there appear to be multiple forms, all of which are distinct from the better characterized phosphoseryl-phosphothreonyl protein phosphatases (reviewed in 529).

6.3.3 G-proteins as anti-oncogenes

Another potential control for tumor-suppressor activity is represented by the G-protein family. As has been discussed, mutated forms of p21ras are associated with a wide variety of human cancers. Fusions of human fibroblasts expressing an activated c-H-ras oncogene with normal cells result in cell hybrids which are no longer tumorigenic, even though these cells continue to express the mutated p21ras protein (530).

'Classical' G-proteins consist of three distinct subunits: α, β and γ (238).

The α subunits represent the catalytic moiety which, when bound to GTP, activate the second-messenger generating pathway. The β/γ subunits function to inhibit α. G-proteins have been shown to regulate cyclic nucleotide metabolism, PI turnover and possibly certain protein – tyrosine kinases. Loss or dysfunction of either the β or γ subunits would have dramatic effects on the major signal transduction pathways. In the case of p21*ras*, no associated β/γ type of subunits has yet been identified. However, a p21 binding protein, termed GAP, has recently been described which induces a 200-fold activation of the GTPase of normal p21 (292; see section 3.7). Clearly, the functional loss of such a GAP-like protein would lead to an increase in the steady-state levels of the active p21*ras* – GTP complex, analogous to oncogenic mutations which inhibit the GTPase activity of p21.

6.3.4 Transcription factors as anti-oncogenes

One final class of anti-oncogenes is represented by proteins which serve to regulate proto-oncogene expression at the transcriptional level, through recognition of *cis*-regulatory elements, or through interaction with DNA binding proteins. One example of a *cis*-regulatory sequence has been found near c-*fos*, which acts to prevent the inappropriate transcriptional activation of this gene (531). Loss or inactivation of these sequences could lead to constitutive expression of the c-*fos* protein.

An intriguing aspect of the role of intracellular hormone receptors in human cancer has been uncovered recently by Evans' group (505). Screening a cDNA library with an oligonucleotide homologous to the DNA binding domain of steroid receptors resulted in the isolation of a new member of this gene family. In order to identify the ligand of this putative receptor, sequences corresponding to the DNA-binding domain of the gene product were replaced with those from the DNA-binding domain of the hGR, with the idea of producing a hybrid receptor capable of activating a glucocorticoid inducible promoter in the presence of the appropriate ligand (505). Analysis of a variety of candidate ligands in this system led to the discovery that the gene encoded the receptor for retinoic acid.

Retinoids comprise a group of compounds including retinoic acid and retinol (vitamin A) that exert profound effects on development, differentiation and morphogenesis in a wide variety of systems (reviewed in 532). One interesting aspect of the discovery of the retinoic acid receptor is the fact that retinoic acid has been shown to inhibit tumor progression in animals and to block the action of tumor promoters *in vitro*. Thus, the different members of this family of transcriptional modulators may play both positive and negative roles in the generation of human malignancies.

6.4 Conclusions on anti-oncogenes

Both anti-oncogenes and tumor-suppressor genes encode proteins with biochemically unrelated functions which have the common feature of regulating the processes of normal cell growth and differentiation. Only now can we begin to speculate on the possible functions of these gene products and what roles they may play in both the generation of human malignancies and in tumor suppression. It is apparent, however, that the analysis of their biochemical activities will provide valuable insights towards our understanding of the transformation process.

7. References

1. Hunter,T. and Cooper,J.A. (1985) Protein–tyrosine kinases. *Annu. Rev. Biochem.,* **54**, 897.
2. Bishop,J.M. (1985) Viral oncogenes. *Cell,* **42**, 23.
3. Foulkes,J.G. and Rich-Rosner,M. (1985) Tyrosine-specific protein kinases as mediators of growth control. In *Molecular Aspects of Cellular Regulation,* Cohen,P. and Housley, M.D. (eds), Vol. 4, Elsevier, Amsterdam.
4. Rayter,S.I., Bell,J.C., Fry,M.J. and Foulkes,J.G. (1987) Oncogene function and mechanism of action. In *New Avenues in Development Cancer Chemotherapy,* Harrup,K.R. and Conners,T.A. (eds), Academic Press, New York.
5. Slamon,D.J., Clark,G.M., Wong,S.G., Levin,W.J., Ullrich,S.A. and McGuire,W.L. (1987) Human breast cancer: correlation of relapse and survival with amplification of the HER-2/*neu* oncogene. *Science,* **235**, 177.
6. Heisterkamp,N., Groffen,J., Stephenson,J.R., Spurr,N.K., Goodfellow,P.N., Solomon,E., Carrit,B. and Bodmer,W.F. (1982) Chromosomal translocation of human cellular homologues of two viral oncogenes. *Nature,* **299**, 747.
7. de Klein,A., van Kessel,A.G., Grosveld,G., Bartram,C.R., Hageneijer,A., Bootsma,D., Spur,N.K., Heisterkamp,N., Groffen,J. and Stephenson,J.R. (1982) A cellular oncogene is translocated to the Philadelphia chromosome in chronic myelocytic leukemia. *Nature,* **303**, 765.
8. Konopka,J.B. and Witte,O.N. (1985) Detection of c-*abl* tyrosine kinase activity in vitro permits direct comparison of normal and altered c-*abl* products. *Mol. Cell. Biol.,* **5**, 3116.
9. Ben-Neriah,Y., Daley,G.Q., Mes-Masson,A.M., Witte,O.N. and Baltimore,D. (1986) The chronic myelogenous leukemia specific P210 protein is the product of the *bcr/c-abl* hybrid gene. *Science,* **233**, 212.
10. Reddy,E.P., Reynolds,R.K., Santos,E. and Barbacid,M. (1982) A single point mutation is responsible for the acquisition of transforming properties of the T24 bladder carcinoma oncogene. *Nature,* **300**, 149.
11. Tabin,C., Bradley,S., Bargmann,C., Weinberg,R., Papageorge,A., Scolnick,E., Dhar,R., Lowy,D. and Chang,E. (1982) Mechanism of activation of a human oncogene. *Nature,* **300**, 143.
12. Balmain,A. (1986) Molecular events associated with tumor initiation, promotion and progression in mouse skin. In *Oncogenes and Growth Control,* Kah,P. and Graf,T. (eds), Springer-Verlag, Berlin.
13. Balmain,A., Ramsden,N., Bowden,G.T. and Smith,J. (1984) Activation of the mouse cellular H-*ras* gene in chemically induced skin papillomas. *Nature,* **307**, 658.
14. Quintanilla,M., Brown,K., Ramsden,M. and Balmain,A. (1986) Carcinogen-specific mutation and amplification of Ha-*ras* during mouse skin carcinogenesis. *Nature,* **322**, 78.
15. Sukumar,S., Notario,V., Martin-Zanca,D. and Barbacid,M. (1983) Induction of mammary carcinomas in rat by nitroso-methylurea involves malignant activation of H-*ras*-1 locus by single point mutations. *Nature,* **306**, 658.
16. Zarbl,H., Sukomar,S., Arthur,A.V., Martin-Zanca,D. and Barbacid,M. (1985) Direct mutagenesis of H-*ras*-1 oncogenes by N-nitroso-N-methylurea during initiation of mammary carcinogenesis in rats. *Nature,* **315**, 382.
17. Roussel,M.F., Downing,J.R., Rettenmier,C.W. and Sherr,C.J. (1988) A point mutation in the extracellular domain of the human CSF-1 receptor (c-*fms* proto-oncogene product) activates its transforming potential. *Cell,* **55**, 979.

18. Waterfield,M.D., Scrace,G.T., Whittle,N., Stroobant,B.P., Johnson,A., Wasteson,A., Westermark,B., Heldin,C.H., Huang,J.S. and Deuel,T.F. (1983) PDGF is structurally related to the putative transforming protein P28^{sis} of simian sarcoma virus. *Nature*, **304**, 35.

19. Sakamoto,H., Mori,M., Taira,M., Yoshida,T., Matsukawa,S., Shimizu,K., Sekiguchi, M., Terada,M. and Sugimura,T. (1986) Transforming gene from human stomach cancers and a noncancerous portion of stomach mucosa. *Proc. Natl. Acad. Sci. USA*, **83**, 3997.

20. Bovi,P.D., Curatola,A.M., Kern,F.G., Greco,A., Ittmann,M. and Basilico,C. (1987) An oncogene isolated by transfection of Kaposi's sarcoma DNA encodes a growth factor that is a member of the FGF family. *Cell*, **50**, 729.

21. Moore,R., Casey,G., Brookes,S., Dixon,M., Peters,G. and Dickson,C. (1986) Sequence, topography and protein coding potential of mouse int-2: a putative oncogene activated by mouse mammary tumour virus. *EMBO J.*, **5**, 919.

22. Dickson,C. and Gordon,P. (1987) Potential oncogene product related to growth factors. *Nature*, **326**, 833.

23. Sporn,M.B. and Todaro,G.J. (1980) Autocrine secretion and malignant transformation of cells. *New. Engl. J. Med.*, **303**, 878.

24. Downward,J., Yarden,Y., Mayes,E., Scrace,G., Totty,N., Stockwell,P., Ullrich,A., Schlessinger,J. and Waterfield,M.D. (1984) Close similarity of EGF receptor and v-*erb*-B oncogene protein sequences. *Nature*, **307**, 521.

25. Sherr,C.J., Rettenmier,C.W., Sacca,R., Roussel,M.F., Look,A.T. and Stanley,E.R. (1985) The c-*fms* proto-oncogene product is related to the receptor for the mononuclear phagocyte growth factor, CSF-1. *Cell*, **4**, 665.

26. Gilman,A.G. (1987) G-proteins: transducers of receptor-generated signals. *Annu. Rev. Biochem.*, **56**, 615.

27. Bohmann,D., Bos,T.J., Admon,A., Nishimura,T., Vogt,P.K. and Tjian,R. (1987) Human proto-oncogene c-*jun* encodes a DNA binding protein with structural and functional properties of transcription factor AP-1. *Science*, **238**, 1386.

28. Lee,W., Mitchell,P. and Tjian,R. (1987) Purified transcription factor AP-1 interacts with TPA-inducible enhancer elements. *Cell*, **49**, 741.

29. Dolhman,H.G., Caron,M.G. and Lefkowitz,R.J. (1987) A family of receptors coupled to guanine nucleotide regulatory proteins. *Biochem.: Persp. in Biochem.*, **26**, 2657.

30. Cohen,P. (1982) The role of protein phosphorylation in neural and hormonal control of cellular activity. *Nature*, **296**, 613.

31. Lefkowitz,R.J., Stadel,J.M. and Caron,M.G. (1983) Adenylate cyclase coupled beta adrenergic receptors structure and mechanisms of activation and desensitization. *Annu. Rev. Biochem.*, **52**, 159.

32. Nishizuka,Y. (1984) Turnover of inositol phospholipids and signal transduction. *Science*, **225**, 1365.

33. Houslay,M.D. (1985) The insulin receptor and signal generation at the plasma membrane. In *Molecular Aspects of Cellular Regulation*, Cohen,P. and Houslay,M.D. (eds), Vol. 4, Elsevier, Amsterdam.

34. Berridge,M.J. and Irvine,R.F. (1984) Inositol trisphosphate, a novel second messenger in cellular signal transduction. *Nature*, **312**, 315–321.

35. Parker,P.J. and Ullrich,A. (1986) Protein kinase C. In *Ocogenes and Growth Control*, Kahns,P. and Graf,T. (eds), Springer-Verlag, Berlin.

36. Irvine,R.F., Letcher,A.J., Lander,D.J. and Berridge,M.J. (1986) Specificity of inositol phosphate-stimulated calcium mobilization from Swiss mouse 3T3 cells. *Biochem. J.*, **240**, 301.

37. Irvine,R.F. and Moor,R.M. (1986) Micro-injection of inositol 1,3,4,5-tetrakisphosphate activates sea urchin eggs by a mechanism dependent on external calcium. *Biochem. J.*, **240**, 917.

38. Wakelam,M.J., Davies,S.A., Houslay,M.D., McKay,I., Marshall,C.J. and Hall,A. (1986) Normal p21 N-*ras* couples bombesin and other growth factor receptors to inositol phosphate production. *Nature*, **323**, 173.

39. Ushiro,H. and Cohen,S. (1980) Identification of phosphotyrosine as a product of the EGF-activated protein kinase in A431 cell membranes. *J. Biol. Chem.*, **255**, 8363.

40. Nishimura,J., Huang,J.S. and Deuel,T.F. (1982) PDGF stimulates tyrosine-specific protein kinase activity in Swiss mouse 3T3 cell membranes. *Proc. Natl. Acad. Sci. USA*, **79**, 4303.

41. Jacobs,S., Kull,F.C., Earp,H.S., Svoboda,M.E., Van Wyk,J.J. and Cuatrecasas,P. (1983) Somatomedin C stimulates the phosphorylation of the subunit of its own receptor. *J. Biol. Chem.*, **258**, 9581.

42. Kasuga,M., Zick,U., Blithe,D.L., Crettaz,M. and Kahn,C.R. (1982) Insulin stimulates tyrosine phosphorylation of the insulin receptor in a cell-free system. *Nature*, **298**, 667.

43. Rettenmier,C.W., Chen,J.M., Roussel,M.F. and Sherr,C.J. (1985) The product of the c-*fms* proto-oncogene: a glycoprotein with associated tyrosine kinase activity. *Science*, **228**, 320.

44. Moolenaar,W.H. (1986) Cytoplasmic pH and free Ca^{2+} in the action of growth factors. In *Oncogenes and Growth Control*, Kahn,P. and Graf,T. (eds), Springer-Verlag, Berlin.

45. Jeltsch,J.M., Krozowski,Z., Stricker,C.Q., Gronemeyer,H., Simpson,J.M., Garnier,J.M., Krust,A., Jacob,F. and Chambon,P. (1986) Cloning of the chicken progesterone receptor. *Proc. Natl. Acad. Sci. USA*, **83**, 5424.

46. Connelly,O.M., Sullivan,W.P., Toft,D.O., Birnbaumer,M., Cook,R.G., Maxwell,B.L., Zarucki-Shulz,T., Greene,G.L., Schrader,W.T. and O'Malley,B.W. (1986) Molecular cloning of the chicken progesterone receptor. *Science*, **233**, 767.

47. King,W.J. and Greene,G.L. (1984) Monoclonal antibodies localize oestrogen receptor in the nuclei of target cells. *Nature*, **307**, 745.

48. Welshons,W.V., Lieberman,M.E. and Garski,J. (1984) Nuclear localization of unoccupied oestrogen receptors. *Nature*, **307**, 747.

49. Green,S., Walter,P., Kumar,V., Krust,A., Bornert,J.M., Argos,P. and Chambon,P. (1986) Human estrogen receptor complementary DNA sequence expression and homology to v-*erb*-A. *Nature*, **320**, 134.

50. Klein-Hitpass,L., Schorpp,M., Wagner,U. and Ryffel,G.U. (1986) An estrogen-responsive element derived from the 5' flanking region of the *Xenopus* vitellogenin A2 gene functions in transfected human cells. *Cell*, **46**, 1053.

51. Sap,J., Munoz,A., Damm,K., Goldberg,Y., Ghysdael,J., Leutz,A., Beug,H. and Vennstrom,B. (1986) The c-*erb*-A protein is a high-affinity receptor for thyroid hormone. *Nature*, **324**, 635.

52. Weinberger,C., Thompson,C.C., Ong,E.S., Lebo,R., Gruol,D.J. and Evans,R.M. (1986) The c-*erb*-A gene encodes a thyroid hormone receptor. *Nature*, **324**, 641.

53. Walsh,D.A., Perkins,J.P. and Krebs,E.G. (1968) An adenosine $3^1,5^1$-monophosphate-dependent protein kinase from rabbit skeletal muscle. *J. Biol. Chem.*, **243**, 3763.

54. Edelman,A.M., Blumenthal,D.K. and Krebs,E.G. (1987) Protein serine/threonine kinases. *Annu. Rev. Biochem.*, **56**, 567.

55. Kuo,J.F. and Greengard,P. (1969) Cyclic nucleotide-dependent protein kinases in widespread occurrence of adenosine 3',5'-monophosphate-dependent protein kinase in various tissues and phyla of the animal kingdom. *Proc. Natl. Acad. Sci. USA*, **64**, 1349.

56. Krebs,E.G. and Beavo,J.A. (1979) Phosphorylation dephosphorylation of enzymes. *Annu. Rev. Biochem.*, **48**, 923.

57. Lohmann,S.M. and Walter,U. (1984) Regulation of the cellular and subcellular concentrations and distribution of cyclic nucleotide-dependent protein kinases. *Protein Phos. Res.*, **18**, 63.

58. Weldon,S.L., Munby,M.C. and Taylor,S.S. (1985) The regulatory subunit of neural cAMP-dependent protein kinase II represents a unique gene product. *J. Biol. Chem.*, **260**, 6440.

59. Uhler,M.D., Carmichael,D.F., Lee,D.C., Chrivia,J.C., Krebs,E.G. and McKnight,G.S. (1986) Isolation of cDNA clones coding for the catalytic subunit of mouse cAMP-dependent protein kinase. *Proc. Natl. Acad. Sci. USA*, **83**, 1300.

60. Jungmann,R.A. and Russell,D.H. (1977) Cyclic AMP, cyclic AMP-dependent protein kinase, and the regulation of gene expression. *Life Sci.*, **20**, 1787.

61. Lohmann,S.M., DeCamilli,P., Einig,I. and Walter,U. (1984) High-affinity binding of the regulatory subunit (RII) of cAMP-dependent protein kinase to microtubule-associated and other cellular proteins. *Proc. Natl. Acad. Sci. USA*, **81**, 6723.

62. Constantinou,A.I., Squinto,S.P. and Jungmann,R.A. (1985) The phosphoform of the regulatory subunit RII of cyclic AMP-dependent protein kinase possesses intrinsic topoisomerase activity. *Cell*, **42**, 429.

63. Weber,I.T., Takio,K., Titani,K. and Steitz,T.A. (1982) The cAMP-binding domains

of the regulatory subunit of cAMP-dependent protein kinase and the catabolite gene activator protein are homologous. *Proc. Natl. Acad. Sci. USA*, **79**, 7679.

64. Peterkofsy,A. (1976) Cyclic nucleotides in bacteria. *Adv. Cyclic Nucl. Res.*, **7**, 1.
65. Grove,J.R., Price,D.J., Goodman,H.M. and Avruch,J. (1987) Recombinant fragment of protein kinase inhibitor blocks cyclic AMP-dependent gene transcription. *Science*, **238**, 530.
66. Pastan,I. and Willingham,M. (1978) Cell transformation and the 'morphologic phenotype' of transformed cells. *Nature*, **274**, 645.
67. Cho-Chung,Y.S., Clair,T., Yi,P.N. and Parkson,C. (1977) Comparative studies on cyclic AMP binding and protein kinase in cyclic AMP-responsive and unresponsive Walker 256 mammary carcinomas. *J. Biol. Chem.*, **252**, 6335.
68. Roth,C.W., Richert,N.D., Pastan,I. and Gottesman,M.M. (1983) Cyclic AMP treatment of Rous sarcoma virus—transformed Chinese hamster ovary cells increases phosphorylation of pp60*src* and increases pp60*src* kinase activity. *J. Biol. Chem.*, **258**, 10768.
69. Klee,C.B., Crouch,T.H. and Richman,P.C. (1980) Calmodulin. *Annu. Rev. Biochem.*, **49**, 489.
70. Nairn,A.C. and Palfrey,H.C. (1987) Identification of the major Mr 100,000 substrate for calmodulin-dependent protein kinase III in mammalian cells as elongation factor-2. *J. Biol. Chem.*, **262**, 17299.
71. Palfrey,H.C., Nairn,A.C., Muldoon,L.L. and Villereal,M.L. (1987) Rapid activation of calmodulin-dependent protein kinase II in mitogen-stimulated human fibroblasts. Correlation with intracellular Ca^{2+} transients. *J. Biol. Chem.*, **262**, 9785.
72. Nishizuka,Y. (1986) Studies and perspectives of protein kinase C. *Science*, **233**, 305.
73. Inoue,M., Kishimoto,A., Takai,Y. and Nishizuka,Y. (1977) Studies on a cyclic nucleotide independent protein kinase and its pro-enzyme in mammalian tissues: part 2, pro-enzyme and its activation by calcium dependent protease from rat brain. *J. Biol. Chem.*, **252**, 7610.
74. Coussens,L., Parker,P.J., Rhee,L., Yang-Feng,T.L., Chen,E., Waterfield,M.D., Francke,U. and Ullrich,A. (1986) Multiple, distinct forms of bovine and human protein kinase C suggest diversity in cellular signalling pathways. *Science*, **233**, 859.
75. Knopf,J.L., Lee,M., Sutlzman,L.A., Kriz,R.W., Loomis,C.R., Hewick,R.M. and Bell,R.M. (1986) Cloning and expression of multiple protein kinase C cDNAs. *Cell*, **46**, 491.
76. Parker,P.J., Coussens,L., Totty,N., Rhee,L., Young,S., Chen,E., Stabel,S., Waterfield,M.D. and Ullrich,A. (1986) The complete primary structure of protein kinase C—the major phorbol ester receptor. *Science*, **233**, 854.
77. Ohno,S., Kawasaki,H., Imajoh,S., Suzuki,K., Inagaki,M., Yokokwa,H., Sakoh,T. and Hikada,H. (1987) Tissue-specific expression of three distinct types of rabbit protein kinase C. *Nature*, **325**, 161.
78. Scott,D., Fischer,E.H., Demaille,J.G. and Krebs,E.G. (1985) Identification of an inhibitory region of the heat-stable protein inhibitor of the cAMP-dependent protein kinase. *Proc. Natl. Acad. Sci. USA*, **82**, 4379.
79. Cheng,H.C., Kemp,B.E., Pearson,R.B., Smith,A.J., Misconi,L., Van Patten,S.M. and Walsh,D.A. (1986) A potent synthetic peptide inhibitor of the cAMP-dependent protein kinase. *J. Biol. Chem.*, **261**, 989.
80. House,C. and Kemp,B.E. (1987) Protein kinase C contains a pseudosubstrate prototype in its regulatory domain. *Science*, **238**, 1726.
81. Downes,C.P. and Michell,R.M. (1985) Inositol phospholipid breakdown as a receptor-controlled generator of second messengers. In *Molecular Mechanisms of Transmembrane Signalling*. Cohen,P. and Houslay,M.D. (eds), Elsevier, Amsterdam.
82. Klausner,R.D., Harford,J. and van Rorswoude,J. (1984) Rapid internalization of the transferrin receptor in K562 cells is triggered by ligand binding or treatment with a phorbol ester. *Proc. Natl. Acad. Sci. USA*, **81**, 3005.
83. Takayama,S., White,M.F., Lewis,V. and Kahn,C.R. (1984) Phorbol esters modulate insulin receptor phosphorylation and insulin action in cultured hepatoma cells. *Proc. Natl. Acad. Sci. USA*, **81**, 7797.
84. Minani,Y., Samelson,L.E. and Klausner,R.D. (1987) Internalization and cycling of the T cell antigen receptor. Role of protein kinase C. *J. Biol. Chem.*, **262**, 13342.
85. Garcia-Sainz,J.A., Mendlovic,F. and Martinez-Olmedo,M.A. (1985) Effects of phorbol

esters on alpha-1-adrenergic-mediated glucagon-mediated actions in isolated rat hepatocytes. *Biochem. J.*, **228**, 277.

86. MacIntyre,D.E., McNicol,A. and Drummond,A.H. (1985) Tumor-promoting phorbol esters inhibit agonist-induced phosphatidate formation and calcium flux in human platelets. *FEBS Lett.*, **180**, 160.

87. Cochet,C., Gill,G.N., Meisenhelder,J., Cooper,J.A. and Hunter,T. (1984) C kinase phosphorylates the epidermal growth factor receptor and reduces its epidermal growth factor stimulated tyrosine protein kinase activity. *J. Biol. Chem.*, **259**, 2553.

88. Diringer,H. and Friis,R.R. (1977) Changes in phosphatidylinositol metabolism correlated to growth state of normal and RSV-transformed Japanese quail cells. *Cancer Res.*, **37**, 2979.

89. Sugimoto,Y., Whitman,M., Cantley,L.C. and Erikson,R. (1984) Evidence that the Rous sarcoma virus transforming gene product phosphorylates phosphatidylinositol and diacylglycerol. *Proc. Natl. Acad. Sci. USA*, **81**, 2117.

90. Fry,M.J., Gebhardt,A., Parker,P.J. and Foulkes,G. (1985) Phosphatidylinositol turnover and transformation of cells by Abelson murine leukemia virus. *EMBO J.*, **4**, 3173.

91. Wolfman,A. and Macara,I.G. (1987) Elevated levels of diacylglycerol and decreased phorbol ester sensitivity in *ras*-transformed fibroblasts. *Nature*, **325**, 359.

92. Parker,P.J., Stabel,S. and Waterfield,M.D. (1984) Purification to homogeneity of protein kinase C from bovine brain identity with the phorbol ester receptor. *EMBO J.*, **3**, 953.

93. Castagna,M., Takai,Y., Kaibuchi,K., Sano,K., Kikkawa,U. and Nishizuka,Y. (1982) Direct activation of calcium activated phospholipid dependent protein kinase by tumor promoting phorbol esters. *J. Biol. Chem.*, **257**, 7847.

94. Greenberg,M.E. and Ziff,E.B. (1984) Stimulation of 3T3 cells induces transcription of the c-*fos* proto-oncogene. *Nature*, **311**, 433.

95. Kruijer,W., Cooper,J.A., Hunter,T. and Verma,I.M. (1984) Platelet-derived growth factor induces rapid but transient expression of the c-*fos* gene and protein. *Nature*, **312**, 711.

96. Kelly,K., Cochran,B.H., Stiles,C.D. and Leder,P. (1983) Cell-specific regulation of the c-*myc* gene by lymphocyte mitogens and platelet-derived growth factor. *Cell*, **35**, 603.

97. Colamonici,O.R., Trepel,J.B., Vidal,C.A. and Neckers,L.M. (1986) Phorbol ester induces c-*sis* gene transcription in stem cell line K562. *Mol. Cell. Biol.*, **6**, 1874.

98. Berg,J.M. (1986) Potential metal-binding domains in nucleic acid binding proteins. *Science*, **232**, 485.

99. Cambier,J.C., Newell,M.K., Justement,L.B., McGuire,J.C., Leach,K.L. and Chen,Z.Z. (1987) Ia binding ligands and cAMP stimulate nuclear translocation of PKC in B lymphocytes. *Nature*, **327**, 629.

100. Courtneidge,S.A. and Smith,A.E. (1983) Polyoma virus transforming protein associates with the product of the cellular *src* gene. *Nature*, **303**, 435.

101. Erikson,R.L., Collett,M.S., Erikson,E. and Purchio,A.F. (1979) Evidence that the avian sarcoma virus transforming gene product is a cyclic AMP-independent protein kinase. *Proc. Natl. Acad. Sci. USA*, **76**, 6260.

102. Hunter,T. and Sefton,B.M. (1980) Transforming gene product of Rous sarcoma virus phosphorylates tyrosine. *Proc. Natl. Acad. Sci. USA*, **77**, 1311.

103. Martin-Zanca,D., Mitra,G., Long,L.K. and Barbacid,M. (1986) Molecular characterization of the human *trk* oncogene. In *Cold Spring Harbor Symp. Quant. Biol.*, Vol. LI.

104. Takahashi,M., Ritz,J. and Cooper,G.M. (1985) Activation of a novel human transforming gene, *ret*, by DNA rearrangement. *Cell*, **42**, 581.

105. Gonzatti-Haces,M., Seth,A., Park,M., Copeland,T., Orozlan,S. and Vande Woude,G.F. (1988) Characterization of the TPR-MET oncogene p65 and the MET protooncogene p140 protein–tyrosine kinases. *Proc. Natl. Acad. Sci. USA*, **85**, 21.

106. Birchmeier,C., Young,D. and Wigler,M. (1986) Characterization of two new human oncogenes. *Cold Spring Harbor Symp. Quant. Biol.*, Vol. LI.

107. Birchmeier,C., Birnbaum,D., Waitches,G., Fasano,O. and Wigler,M. (1986) Characterization of an activated human *ros* gene. *Mol. Cell. Biol.*, **6**, 3109.

108. Bargmann,C.I., Hurg,M.-C. and Weinberg,R.A. (1986) The *neu* oncogene encodes an epidermal growth factor receptor-related protein. *Nature*, **319**, 2226.

109. Semba,K., Nishizawa,M., Miyajma,N., Yoshida,M.C., Sukegawa,J., Yamanashi,Y., Sasaki,M., Yamamoto,T. and Toyoshima,K. (1986) *yes*-related proto-oncogene, *syn*, belongs to the protein–tyrosine kinase family. *Proc. Natl. Acad. Sci. USA*, **83**, 5459.

110. Kawakami,T., Pennington,C.Y. and Robbins,K.C. (1986) Isolation and oncogene potential of a novel human *src*-like gene. *Mol. Cell. Biol.*, **6**, 4195.

111. Quintrell,N., Lebo,R., Varmus,H.E., Bishop,J.M., Peltenati,M.J., Le Beau,M.M., Diaz,M.O. and Rowley,J.D. (1987) Identification of a human gene (*HCK*) that encodes a protein–tyrosine kinase and is expressed in hemopoietic cells. *Mol. Cell. Biol.*, **7**, 2267.

112. Yamanashi,Y., Fukushige,S.I., Semba,K., Sukegawa,J., Miyajima,N., Matsubase,K.I., Yamoto,T. and Toyoshima,K. (1987) The *yes*-related cellular gene *lyn* encodes a possible tyrosine kinase similar to p56 *lck*. *Mol. Cell. Biol.*, **7**, 237.

113. Ziegler,S.F., Marth,J.D., Lewis,B. and Perlmutter,R.M. (1986) Novel protein–tyrosine kinase gene (*hck*) preferentially expressed in cells of hematopoietic origin. *Mol. Cell. Biol.*, **1**, 2276.

114. Kruh,G.D., King,C.R., Kraus,M.H., Popescu,N.C., Amsbaugh,S.C., McBride,W.O. and Aaronson,S.A. (1986) A novel human gene closely related to the *abl* oncogene. *Science*, **234**, 1545.

115. Foster,D.A., Levy,J.B., Daley,G.Q., Simon,M.C. and Hanafusa,H. (1986) Isolation of chicken cellular DNA sequences with homology to the region of viral oncogenes that encodes the tyrosine kinase domain. *Mol. Cell. Biol.*, **6**, 325.

116. Hirai,H., Maru,Y., Hagiwara,K., Nishida,J. and Takaku,F. (1987) A novel putative tyrosine kinase receptor encoded by the *eph* gene. *Science*, **238**, 1717.

117. Konopka,J.B., Watanabe,S.M., Singer,J.W., Collins,S.J. and Witte,O.N. (1985) Cell lines and clinical isolates derived from Ph'-positive chronic myelogenous leukemia patients express c-*abl* proteins with a common structural alteration. *Proc. Natl. Acad. Sci. USA*, **82**, 1810.

118. Chan,L.C., Karhi,K.K., Rayter,S.I., Heisterkamp,N., Eridari,S., Powles,R., Lawler, S.D., Groffen,J., Foulkes,J.G., Greaves,M.F. and Wiedemann,L.M. (1987) A novel c-*abl* protein is expressed in Philadelphia chromosome positive acute lymphoblastic leukaemia which does not involve the breakpoint cluster region. *Nature*, **325**, 635.

119. Bolen,J.B., Veillette,A., Schwartz,A.M., DeSeau,V. and Rosen,N. (1987) Activation of pp60^{c-src} protein kinase activity in human colon carcinoma. *Proc. Natl. Acad. Sci. USA*, **84**, 2251.

120. Moelling,K., Heimann,B., Beimling,P., Rapp,U.R. and Sander,T. (1984) Serine- and threonine-specific protein kinase activities of purified *gag-mil* and *gag-raf* proteins. *Nature*, **312**, 558.

121. Kloetzer,W.S., Maxwell,S.A. and Arlinghaus,R.B. (1984) Further characterization of the p85*gag-mos* associated protein kinase activity. *Virology*, **138**, 143.

122. Sutrave,P., George,M. and Bonner,T. (1984) The human cellular homologues of the *raf-mil* oncogene. In *Symposium on Genes and Cancer*, Los Angeles, CA.

123. Koenen,M., Sippel,A.E., Trachmann,C. and Bister,K. (1988) Primary structure of the chicken c-*mil* protein: identification of domains shared with or absent from the retroviral v-*mil* protein. *Oncogene*, **2**, 179.

124. Schultz,A.M. Copeland,T., Oroszlan,S. and Rapp,U.R. (1988) Identification and characterization of c-*raf* phosphoproteins in transformed murine cells. *Oncogene*, **2**, 187.

125. Gilmore,T., DeClue,J.E. and Martin,G.S. (1985) Tyrosine kinase activity associated with the v-*erb*-B gene product. In *Cancer Cells 3; Growth Factors and Transformation*, Cold Spring Harbor Laboratory Press.

126. Kaplan,D.R., Whitman,M., Schaffhausen,B., Raptis,L., Garcea,R.L., Pallas,D., Robert,T.M. and Cantley,L. (1986) Phosphatidylinositol metabolism and polyoma-mediated transformation. *Proc. Natl. Acad. Sci. USA*, **83**, 3624.

127. De Larco,J.E. and Todaro,G.J. (1978) Growth factors from murine sarcoma virus-transformed cells. *Proc. Natl. Acad. Sci. USA*, **75**, 4001.

128. Anzano,M.A., Roberts,A.B., Smith,J.M., Sporn,M.B. and De Larco,J.E. (1983) Sarcoma growth factor from conditioned medium is composed of both type alpha and type beta transforming growth factors. *Proc. Natl. Acad. Sci. USA*, **80**, 6264.

129. Roberts,A.B. and Sporn,M.B. (1985) Transforming growth factors. *Cancer Surveys*, **4**, 783.

130. Derynck,R. (1986) Transforming growth factor-α: structure and biological activities. *J. Cell. Biochem.*, **32**, 293.
131. Marquardt,H., Hunkapiller,M.W., Hood,L. and Todaro,G.J. (1984) Rat transforming growth factor type 1: structure and relation to epidermal growth factor. *Science*, **223**, 1079.
132. Marquardt,H. and Todaro,G.J. (1982) Human transforming growth factor. *J. Biol. Chem.*, **257**, 5220.
133. Derynck,R., Roberts,A.B., Winkler,M.E., Chen,E.Y. and Goeddel,D.V. (1984) Human transforming growth factor-alpha: precursor structure and expression in *E.coli. Cell*, **38**, 297.
134. Lee,D.C., Rose,T.M., Webb,N.R. and Todaro,G.J. (1985) Cloning and sequence analysis of a cDNA for rat growth transforming factor-alpha. *Nature*, **313**, 489.
135. Bringman,T.S., Lindquist,P.B. and Derynck,R. (1987) Different transforming growth factor-α species are derived from a glycosylated and palmitoylated transmembrane precursor. *Cell*, **48**, 429.
136. Linsley,P.S., Hargreaves,W.R., Twardzik,D.R. and Todaro,G.J. (1985) Detection of larger polypeptides structurally and functionally related to type I transforming growth factor. *Proc. Natl. Acad. Sci. USA*, **82**, 356.
137. Derynck,R., Goeddel,D.V., Ullrich,A., Gutterman,J.V., Williams,R.D., Bringman,T.S. and Berger,W.H. (1987) Synthesis of messenger RNA for transforming growth factors α and β and the epidermal growth factor receptor by human tumors. *Cancer Res.*, **47**, 707.
138. Sherwin,S.A., Twardzik,D.R., Bohn,W.H., Cockley,K.D. and Todaro,G.J. (1983) High-molecular-weight transforming growth factor activity in the urine of patients with disseminated cancer. *Cancer Res.*, **43**, 403.
139. Lee,D.C., Rochford,R., Todaro,G.J. and Villareal,L. (1985) Developmental expression of rat transforming growth factor-α mRNA. *Mol. Cell. Biol.*, **5**, 3644.
140. Coffey,R.J., Jr, Derynck,R., Wilcox,J.N., Bringham,T.S., Goustin,A.S., Moses,H.L., Pillelkow,M. and Pihelkow,M. (1987) Production and auto-induction of transforming growth factor-α in human keratinocytes. *Nature*, **328**, 817.
141. Kudlow,J.E., Kobrin,M.E., Purchio,A.F., Twardzik,D.R., Hernandez,E.R., Asa,S.L. and Adashi,E.Y. (1987) Ovarian transforming growth factor-α gene expression; immunohistochemical localization to the theca-interstitial cells. *Endocrinology*, **121**, 1577.
142. Twardzik,D.R., Sherwin,S.A., Ranchalis,J. and Todaro,G. (1982) Transforming growth factors in the urine of normal, pregnant and tumor bearing humans. *J. Natl. Cancer Inst.*, **69**, 793.
143. Tam,J. (1985) Physiological effects of transforming growth factor in the newborn mouse. *Science*, **229**, 673.
144. Rhodes,J.A., Tam,J.P., Finke,V., Sounders,M., Bernanke,J., Silen,W. and Murphy, R.A. (1986) Transforming growth factor-α inhibits secretion of gastric acid. *Proc. Natl. Acad. Sci. USA*, **83**, 3844.
145. Schultz,G.S., White,M., Michell,R., Brown,G., Lynch,J., Twardzik,D.R. and Todaro,G.J. (1987) Epithelial wound healing enhanced by transforming growth factor-α and vaccinia growth factor. *Science*, **235**, 350.
146. Stern,P.H., Krieger,N.S., Nissenson,R.A., Williams,R.D., Winkler,M.E., Derynck,R. and Strewler,G.J. (1985) Human transforming growth factor-α stimulates bone resorption in vitro. *J. Clin. Invest.*, **76**, 2016.
147. Korc,M., Haussler,C.A. and Trookman,N.S. (1987) Divergent effects of epidermal growth factor and transforming growth factor on a human endometrial carcinoma cell line. *Cancer Res.*, **47**, 4909.
148. Shreiber,A.B., Winkler,M.E. and Derynck,R. (1986) Transforming growth factor-α: a more potent angiogenic mediator than epidermal growth factor. *Science*, **232**, 1250.
149. Reynolds,F.H., Jr, Todaro,G., Fryling,C. and Stephenson,J.R. (1981) Human transforming growth factors induce tyrosine phosphorylation of EGF receptors. *Nature*, **292**, 259.
150. Pike,L.J. and Eakes,A.T. (1987) Epidermal growth factor stimulates the production of phosphatidylinositol monophosphate and the breakdown of polyphosphoinositides in A431 cells. *J. Biol. Chem.*, **262**, 1644.
151. Helper,J.R., Nakahata,N., Lorenber,T.W., DiGuiseppi,J., Herman,B., Earp,H.S. and

Harden,T.K. (1987) Epidermal growth factor stimulates the rapid accumulation of inositol (1,4,5)-trisphosphate and a rise in cytosolic calcium mobilized from intracellular stores in A431 cells. *J. Biol. Chem.*, **262**, 2952.

152. Johnson,R.M. and Garrison,J.C. (1987) Epidermal growth factor and angiotensin II stimulate formation of inositol 1,4,5 and inositol 1,3,4-triphosphate in hepatocytes differential inhibition by pertussis toxin and phorbol 12-myristate, 13-acetate. *J. Biol. Chem.*, **262**, 17285.

153. Besterman,J.M., Watson,S.P. and Cuatrecasas,P. (1986) Lack of association of epidermal growth factor-, insulin- and serum-induced mitogenesis with stimulation of phosphoinositide degradation in BALB/c 3T3 fibroblasts. *J. Biol. Chem.*, **261**, 723.

154. Moses,H.L., Branum,E.L., Proper,J.A. and Robinson,R.A. (1981) Transforming growth factor production by chemically transformed cells. *Cancer Res.*, **41**, 2842.

155. Assoian,R.K., Komoriya,A., Meyers,C.A., Miller,D.M. and Sporn,M.B. (1983) Transforming growth factor-β in human platelets: identification of a major storage site, purification and characterization. *J. Biol. Chem.*, **258**, 7155.

156. Derynck,R., Jarret,J.A., Chen,E.Y., Eaton,D.H., Bell,J.R., Assoian,R.K., Roberts,A.B., Sporn,M.B. and Goeddel,D.V. (1985) Human transforming growth factor-β complementary DNA sequence and expression in normal and transformed cells. *Nature*, **316**, 701.

157. Derynck,R., Jarret,J.A., Chen,E.Y. and Goeddel,D.V. (1986) The murine transforming growth factor-β precursor. *J. Biol. Chem.*, **261**, 4377.

158. Cheifetz,S., Weatherbee,J.A., Tsang,M.L.-S., Anderson,J.K., Mole,J.E., Lucas,R. and Massague,J. (1987) The transforming growth factor-β system, a complex pattern of cross-reactive ligand and receptors. *Cell*, **48**, 409.

159. ten Dijke,P., Hansen,P., Iwata,K.K., Pieler,C. and Foulkes,J.G. (1988) Identification of a new member of the transforming growth factor-β gene family. *Proc. Natl. Acad. Sci. USA*, **85**, 4715.

160. Anastasi,A., Erspamer,V. and Bucci,M. (1971) Isolation and structure of bombesin and alytesin, two analogous active peptides from the skin of European amphibians Bombina and Alytes. *Experientia*, **27**, 166.

161. Rozengurt,E. and Sinnett-Smith,J. (1983) Bombesin stimulation of DNA synthesis and cell division in cultures of Swiss 3T3 cells. *Proc. Natl. Acad. Sci. USA*, **80**, 2936.

162. Moody,T.W., Pert,C.B., Gazdar,A.F., Carney,D.N. and Minna,J.D. (1981) High levels of intracellular bombesin characterize human small-cell lung carcinoma. *Science*, **214**, 1246.

163. Cutitta,F., Carney,D.N., Mulshine,J., Moody,T.W., Fedorko,J., Fishler,A. and Minna,J.D. (1985) Autocrine growth factor in human small lung cancer. *Cancer Surveys*, **4**, 707.

164. Cirillo,D.M., Gaudino,G., Naldini,L. and Comoglio,P.M. (1986) Receptor for bombesin with associated tyrosine kinase activity. *Mol. Cell. Biochem.*, **6**, 4641.

165. Isacke,C.M., Meisenhelder,J., Brown,K.D., Gould,K.L., Gould,S.J. and Hunter,T. (1986) Early phosphorylation events following the treatment of Swiss 3T3 cells with bombesin and the mammalian bombesin-related peptide, gastrin-releasing peptide. *EMBO J.*, **5**, 2889.

166. Takuwa,N., Takuwa,Y., Bollag,W.E. and Rusmussen,H. (1987) The effects of bombesin on phosphoinositide and calcium metabolism in Swiss 3T3 cells. *J. Biol. Chem.*, **262**, 182.

167. Zachary,I., Sinnett-Smith,J.W. and Rozengurt,E. (1986) Early events elicited by bombesin and structurally related peptides in quiescent Swiss 3T3 cells: activation of protein kinase C and inhibition of epidermal growth factor binding. *J. Cell. Biol.*, **102**, 2211.

168. Matuoka,K., Fukamik,K., Nakanishi,O., Kawai,S. and Takenawa,T. (1988) Mitogenesis in response to PDGF and bombesin abolished by microinjection of antibody to PIP_2. *Science*, **239**, 640.

169. Gospodarowicz,D., Neufeld,G. and Schwiegerer,L. (1987) Fibroblast growth factor: structural and biological properties. *J. Cell. Phys. Suppl.*, **5**, 15.

170. Pelech,S.L., Olwin,B.B. and Krebs,E.G. (1986) Fibroblast growth factor treatment of Swiss 3T3 cells activates a subunit S6 kinase that phosphorylates a synthetic peptide substrate. *Proc. Natl. Acad. Sci. USA*, **83**, 5968.

171. Magnaldo,I., L'Allemain,G., Chambard,J.C., Moenner,M., Barritault,D. and

Pouyssegur,J. (1986) The mitogenic signalling pathway of fibroblast growth factor is not mediated through phosphoinositide hydrolysis and protein kinase C activation in hamster fibroblasts. *J. Biol. Chem.*, **261**, 16916.

172. Heldin,C.H., Wasteson,A. and Westermark,B. (1982) Interaction of platelet-derived growth factor with its fibroblast receptor. *J. Biol. Chem.*, **257**, 4216.

173. Heldin,C.H. and Westermark,B. (1984) Growth factors: mechanism of action and relation to oncogenes. *Cell*, **37**, 9.

174. Betsholtz,C., Johnson,A., Heldin,C.H., Westermark,B., Lind,P., Undea,M.S., Eddy,R., Shows,T.B., Philpott,K., Mellor,A.L., Knott,T.J. and Scott,J. (1986) cDNA sequence and chromosomal location of human platelet-derived growth factor A-chain and its expression in tumor cell lines. *Nature*, **320**, 695.

175. Doolittle,R.F., Hunkapiller,N.M.W., Hood,L.E., Devare,S.G., Robbins,K.C., Aaronson,S.A. and Antoniades,H.N. (1983) Simian sarcoma virus onc gene, v-*sis* is derived from the gene (or genes) encoding a platelet-derived growth factor. *Science*, **221**, 275.

176. Robbins,K.C., Antoniades,H.N., Devare,S.G., Hunkapiller,M.W. and Aaronson,S.A. (1983) Structural and immunological similarities between simian sarcoma virus gene product(s) and human platelet-derived growth factor. *Nature*, **305**, 605.

177. Yarden,Y., Escobedo,J.A., Kuang,W.-J., Yang Feng,T.L., Daniel,T.O., Tremble,P.M., Chen,E.Y., Ando,M.E., Harkins,R.N., Francke,V., Fried,V.A., Ullrich,A. and Williams,L.T. (1986) Structure of the receptor for platelet-derived growth factor helps define a family of closed related growth factors. *Nature*, **323**, 226.

178. Williams,L.T., Tremble,P.M., Daniel,T.O., Coughlin,S.R., Giels,G. and Wang,J.Y.J. (1985) Platelet-derived growth factor receptors in normal and transformed cells. In *Cancer Cells 3: Growth Factors and Transformation.* Cold Spring Harbor Laboratory Press.

179. Kaplan,D.R., Whitman,M., Schaffhausen,B., Pallas,D.C., White,M., Cantley,L. and Roberts,T.M. (1987) Common elements in growth stimulation and oncogenic transformation: 85 kd phosphoprotein and phosphatidylinositol kinase activity. *Cell*, **50**, 1021.

180. Huang,J.S., Haung,S.S., Kennedy,B. and Deuel,T.F. (1982) Platelet-derived growth factor specific to target cells. *J. Biol. Chem.*, **257**, 8130.

181. Keating,M.T. and Williams,L.T. (1988) Autocrine stimulation of intracellular PDGF receptor in v-*sis* transformed cells. *Science*, **239**, 914.

182. Fry,M.J., Rayter,S.I., Heldin,C. and Foulkes,J.G. (1989) Secretion of a PDGF-like growth factor from transformed cell lines: a possible autocrine effect on phosphoinositide turnover. Manuscript in preparation.

183. Rosenthal,A., Lindquist,P.B., Bringman,T.S., Goedell,D.V. and Derynck,R. (1986) Expression in rat fibroblasts of a human transforming growth factor-α cDNA results in transformation. *Cell*, **46**, 301.

184. Smith,J.J., Derynck,R. and Korc,M. (1987) Production of transforming growth factor-α in human pancreatic cancer cells: evidence for a superagonist autocrine cycle. *Proc. Natl. Acad. Sci. USA*, **84**, 7567.

185. Gebhardt,A., Bell,J.C. and Foulkes,J.G. (1986) Abelson transformed fibroblasts lacking the EGF receptor are not tumorigenic in nude mice. *EMBO J.*, **5**, 2191.

186. Stern,D.F., Hare,D.L., Cecchini,M.A. and Weinberg,P.A. (1987) Construction of a novel oncogene based on synthetic sequences encoding epidermal growth factor. *Science*, **235**, 321.

187. Shimizu,K., Birnbaum,D., Ruley,M.A., Fasano,O., Suard,Y., Edlund,L., Taparowsky, E., Goldfarb,M. and Wigler,M. (1983) Structure of the Ki-*ras* gene of the human lung carcinoma cell line Calu-1. *Nature*, **304**, 497.

188. Capon,D.J., Chen,E.Y., Levinson,A.D., Seeburg,P.H. and Goeddel,D.V. (1983) Complete nucleotide sequence of the T-24 human bladder carcinoma oncogene and its normal homologue. *Nature*, **301**, 33.

189. Taparowsky,E., Shimizu,K., Goldfarb,M. and Wigler,M. (1983) Structure and activation of the human N-*ras* gene. *Cell*, **34**, 581.

190. Capon,D.J., Seeburg,P.H., McGrath,J.P., Hayflick,J.S., Edman,U., Levinson,A.D. and Goeddel,D.V. (1983) Activation of Ki-*ras* 2 gene in human colon and lung carcinomas by two different point mutations. *Nature*, **304**, 507.

191. Madaule,P. and Axel,R. (1985) A novel *ras*-related gene family. *Cell*, **41**, 31.

192. Lowe,D.G., Capon,D.J., Delwart,E., Sakaguchi,A.Y., Naylor,S.L. and Goeddel,D.V.

(1987) Structure of the human and murine R-*ras* genes, novel genes closely related to *ras* proto-oncogenes. *Cell,* **48**, 137.

193. Chardin,P. and Tavitian,A. (1986) The ra1 gene: a new *ras* related gene isolated by the use of a synthetic probe. *EMBO J.,* **5**, 2203.

194. Schmitt,H.D., Wagner,P., Pfaff,E. and Gallwitz,D. (1986) The *ras*-related YPT1 gene product in yeast: a GTP-binding protein that might be involved in microtubule organization. *Cell,* **47**, 401.

195. Touchot,N., Chardin,P. and Tavitian,A. (1987) Four additional members of the *ras* gene superfamily isolated by an oligonucleotide strategy: molecular cloning of YPT-related cDNAs from a rat brain library. *Proc. Natl. Acad. Sci. USA,* **84**, 8210.

196. Weinberg,R.A. (1984) *ras* oncogenes and the molecular mechanisms of carcinogenesis. *Blood,* **64**, 1143.

197. Bos,J.L., Fearon,E.R., Hamilton,S.R., Verlaan-de Vries,M. van Boom,J.H., van der Eb,A.J. and Vogelstein,B. (1987) Prevalence of *ras* gene mutations in human colorectal cancers. *Nature,* **327**, 293.

198. Forrester,K., Almoguera,C., Han,K., Grizzle,W.E. and Perucho,M. (1987) Detection of high incidence of K-*ras* oncogenes during human colon tumorigenesis. *Nature,* **327**, 298.

199. Barbacid,M. (1987) *ras* genes. *Annu. Rev. Biochem.,* **56**, 779.

200. Ellis,R.W., DeFeo,D., Shih,T.Y., Gonda,M.A., Tsuchida,N., Lowy,D.R. and Scolnick,E.M. (1981) The p21 ras genes of Harvey and Kirsten sarcoma viruses originate from divergent members of a family of normal vertebrate genes. *Nature,* **292**, 506.

201. Souyri,M., Koehne,C.F., O'Donnelly,P.V., Aldrich,T.H., Furth,M.E. and Fleissner,E. (1987) Biological effect of a murine retrovirus carrying an activated N-*ras* gene of human origin. *Virology,* **158**, 69.

202. Westaway,Y.D., Papkoff,J., Moscovici,C. and Varmus,H.E. (1986) Identification of a provirally activated c-Ha-*ras* oncogene in an avian nephroblastoma via a novel procedure: cDNA cloning of a chimeric viral-host transcript. *EMBO J.,* **5**, 301.

203. Silberberg,F.S., Schejter,E., Hoffman,F.M. and Shilo,B.-Z. (1984) The Drosophila *ras* oncogenes: structure and nucleotide sequence. *Cell,* **37**, 1027.

204. DeFeo-Jones,D., Scolnick,E.M., Koller,R. and Dhar,R. (1983) *ras*-related gene sequences identified and isolated from *Saccharomyces cerevisiae. Nature,* **306**, 707.

205. Fukui,Y. and Kaziro,Y. (1985) Molecular cloning and sequence analysis of a *ras* gene from *Schizosaccharomyces pombe. EMBO J.,* **4**, 687.

206. McGrath,J.P., Capon,D.J., Smith,D.H., Chen,E.Y., Seeburg,P.H., Goeddel,D.V. and Levinson,A.D. (1983) Structure and organization of the human Ki-*ras* proto-oncogene and a related processed pseudogene. *Nature,* **304**, 501.

207. Chen,Z.-Q., Ulsh,L.S., DuBois,G. and Shih,T.Y. (1985) Post-translational processing of p21 *ras* proteins involves palmitylation of the C-terminal tetrapeptide containing cysteine-186. *J. Virol.,* **56**, 607.

208. Magee,A.I., Gutierrez,L., McKay,I.A., Marshall,C.J. and Hall,A. (1987) Dynamic fatty acylation of p21 N-*ras. EMBO J.,* **6**, 3353.

209. Willumsen,B.M., Norris,K., Papageorge,A.G., Hubbert,N.L. and Lowy,D.R. (1984) Harvey murine sarcoma virus p21 ras protein: biological and biochemical significance of the cysteine nearest the carboxyl terminus. *EMBO J.,* **3**, 2581.

210. Ballester,R., Furth,M.E. and Rosen,O.M. (1987) Phorbol ester- and protein kinase C-mediated phosphorylation of the cellular Kirsten *ras* gene product. *J. Biol. Chem.,* **262**, 2688.

211. Fujita,J., Oshida,O., Yuasa,Y., Rhim,J.S., Hatanaka,M. and Aaronson,S.A. (1984) Ha-*ras* oncogenes are activated by somatic alterations in human urinary tract tumors. *Nature,* **309**, 464.

212. McCoy,M., Toole,J.J., Cunningham,J.M., Chang,E.H., Lowy,D.R. and Weinberg,R.A. (1983) Characterization of a human colon/lung carcinoma oncogene. *Nature,* **302**, 79.

213. Santos,E., Martin-Zanca,D., Reddy,E.P., Pierotti,M.A., Della Porta,G. and Barbacid, M. (1984) Malignant activation of a K-*ras* oncogene in lung carcinoma but not in normal tissue of the same patient. *Science,* **223**, 661.

214. Albino,A.P., LeStrange,R., Oliff,A.I., Old,L.J. and Furth,M.E. (1984) Transforming ras genes from human melanoma: a manifestation of tumor heterogeneity. *Nature,* **308**, 69.

215. Eva,A., Tronick,S.R., Gol,R.A., Pierce,J.H. and Aaronson,S.A. (1983) Transforming genes of human hematopoietic tumors: frequent detection of ras-related oncogenes whose activation appears to be independent of tumor. *Proc. Natl. Acad. Sci. USA,* **80**, 4926.
216. Bos,J.L., Toksoz,D., Marshall,C.J., Verlaan-de Vries,M., Veeneman,G.H., Van der Eb,A.J., van Boom,J.H., Hanssen,J.W.G. and Steenvoorden,A.C.M. (1985) Amino-acid substitutions at codon 13 of the N-*ras* oncogene in human acute myeloid leukemia. *Nature,* **315**, 726.
217. Hirai,H., Kobayashi,Y., Mano,H., Hagiwara,K., Maru,Y., Omine,M., Mizoguchi,H., Nishida,J. and Takaku,F. (1987) A point mutation at codon 13 of the N-*ras* oncogene in myelodysplastic syndrome. *Nature,* **327**, 430.
218. Fasano,O., Aldrich,T., Tamanoi,F., Taparowski,E., Furth,M. and Wigler,M. (1984) Analysis of the transforming potential of the human H-*ras* gene by random mutagenesis. *Proc. Natl. Acad. Sci. USA,* **81**, 4008.
219. Yuasa,Y., Srivastava,S.K., Dunn,C.Y., Rhim,J.S., Reddy,E.P. and Aaronson,S.A. (1983) Acquisition of transforming properties by alternative point mutations within c-*ras* human proto-oncogene. *Nature,* **303**, 775.
220. Der,C.J., Finkel,T. and Cooper,G.M. (1986) Biological and biochemical properties of human *ras* H genes mutated at codon 61. *Cell,* **44**, 167.
221. Bizob,D., Wood,A.W. and Skalka,A.M. (1986) Mutagenesis of the Ha-*ras* oncogene in mouse skin tumors induced by polycyclic aromatic hydrocarbons. *Proc. Natl. Acad. Sci. USA,* **83**, 6048.
222. Almoguera,C., Shibata,D., Forrester,K., Martin,J., Arnheim,N. and Perucho,M. (1988) Most human carcinomas of the exocrine pancreas contain mutant c-K-*ras* genes. *Cell,* **53**, 549–554.
223. Viola,M.V., Fromowitz,F., Oravez,S., Deb,S., Finkel,G., Lundy,J., Handi,P., Thor,A. and Schlom,J. (1986) Expression of *ras* oncogene p21 in prostate cancer. *New Engl. J. Med.,* **314**, 133.
224. Viola,M.V., Fromowitz,F., Oravez,S., Deb,S. and Schlom,J. (1985) *ras* oncogene p21 expression is increased in premalignant lesions and high grade bladder carcinoma. *J. Exp. Med.,* **161**, 1213.
225. Gerosa,M.A., Tolarico,D., Fagnani,C., Raimondi,E., Calombatti,N., Tridente,G., DeCarli,L. and Della Valle,G. (1989) Overexpression of N-*ras* oncogene and epidermal growth factor receptor gene in human glioblastoma. *J. Natl. Cancer Inst.,* **81**, 63.
226. Gallick,G.E., Kurzrock,R. Kloetzer,W.S., Arlinghaus,R.B. and Gotterman,J.V. (1985) Expression of p21*ras* in fresh primary and metastatic human colorectal tumors. *Proc. Natl. Acad. Sci. USA,* **82**, 1795.
227. Fujita,K., Ohuchi,N., Yau,T., Okumura,M., Fukushima,Y., Kanakura,Y., Kitamura,Y. and Fujita,J. (1987) Frequent overexpression, but not activation by point mutation, of *ras* genes in primary human gastric cancers. *Gastroenterology,* **93**, 1339.
228. Stewart,T.A., Pattengale,P.K. and Leder,P. (1984) Spontaneous mammary adeno-carcinomas in transgenic mice that carry and express MmTV/myc fusion genes. *Cell,* **38**, 627.
229. Sinn,E., Moller,W., Pattengale,P., Tepler,I., Wallace,R. and Leder,P. (1987) Coexpression of MMTV/v-Ha-*ras* and MMTV/c-*myc* genes in transgenic mice: synergistic action of oncogenes *in vivo. Cell,* **49**, 465.
230. Quaife,C.J., Pinkert,C.A., Ornitz,D.M., Palmiter,R.D. and Brinster,R.L. (1987) Pancreatic neoplasia induced by *ras* expression in acinar cells of transgenic mice. *Cell,* **48**, 1023.
231. Kleinberg,D.L. (1987) Prolactin and breast cancer. *New Engl. J. Med.,* **316**, 269.
232. Willingham,M.C., Pastan,I., Shih,T.Y. and Scolnick,E.M. (1980) Localization of the *ras* gene product of the Harvey strain of MSV to plasma membrane of transformed cells by electron microscopic immunocytochemistry. *Cell,* **19**, 1005.
233. Willingham,M.C., Banks-Schlegel,S.P. and Pastan,I.H. (1983) Immunocytochemical localization in normal and transformed human cells in tissue culture using a monoclonal antibody to the *ras* protein of the Harvey strain of murine sarcoma virus. *Exp. Cell Res.,* **149**, 141.
234. Fujiyama,A. and Tamanoi,F. (1986) Processing and fatty acid acylation of RAS1 and RAS2 proteins in *Saccharomyces cerevisiae. Proc. Natl. Acad. Sci. USA,* **83**, 1266.
235. Resnick,R.J. and Racker,E. (1988) Phosphorylation of the RAS2 gene product by

protein kinase A inhibits the activation of yeast adenylate cyclase. *Proc. Natl. Acad. Sci. USA*, **85**, 2474.

236. Gibbs,J.B., Sigal,I.S., Peo,M. and Scolnick,E.M. (1984) Intrinsic GTPase activity distinguishes normal and oncogenic *ras* p21 molecules. *Proc. Natl. Acad. Sci. USA*, **81**, 5704.

237. Manne,V., Bekesi,E. and Kung,H.-F. (1985) Ha-*ras* proteins exhibit GTPase activity: point mutations that activate Ha-*ras* gene products result in decreased GTPase activity. *Proc. Natl. Acad. Sci. USA*, **82**, 376.

238. Gilman,A.G. (1984) G-proteins and dual control of adenylate cyclase. *Cell*, **36**, 577.

239. Majerus,P.W., Connolly,T.M., Deckmyn,H., Ross,T.S., Bross,T.E., Ishii,H., Bansal,V.S. and Wilson,D.B. (1986) The metabolism of phosphatidylinositide-derived messenger molecules. *Science*, **234**, 1519.

240. Stacey,D.W. and Kung,H.-F. (1984) Transformation of NIH3T3 cells by microinjection of Ha-ras p21 protein. *Nature*, **310**, 508.

241. Kung,H.-F., Smith,M.R., Bekesi,E., Manne,V. and Stacey,D.W. (1986) Reversal of transformed phenotype by monoclonal antibodies against H-*ras*. *Exp. Cell. Res.*, **162**, 363.

242. Mulcahy,L.S., Smith,M.R. and Stacey,D.W. (1985) Requirement for *ras* proto-oncogene function during serum-stimulated growth of NIH3T3 cells. *Nature*, **313**, 241.

243. Bar-Sagi,D. and Feramisco,J.R. (1985) Microinjection of the *ras* oncogene protein into PC12 cells induces morphological differentiation. *Cell*, **42**, 841.

244. Noda,M., Ko,M., Ogura,A., Lio,D.-G., Amano,T., Takano,T. and Ikawa,Y. (1985) Sarcoma viruses carrying *ras* oncogenes induce differentiation-associated properties in a neuronal cell line. *Nature*, **318**, 73.

245. Chesa,P.G., Rettig,W.J., Melamed,M.R., Old,L.J. and Niman,H.L. (1987) Expression of p21*ras* in normal and malignant human tissues: lack of association with proliferation and malignancy. *Proc. Natl. Acad. Sci. USA*, **84**, 3234.

246. Scolnick,E.M., Papageorge,A.G. and Shih,T.Y. (1979) Guanine nucleotide-binding activity as an assay for *src* protein of rat-derived murine sarcoma viruses. *Proc. Natl. Acad. Sci. USA*, **76**, 5355.

247. Colby,W.W., Hayflick,J.S., Clark,S.G. and Levinson,A.D. (1986) Biochemical characterization of polypeptides encoded by mutated human Ha-*ras* 1 genes. *Mol. Cell. Biol.*, **6**, 730.

248. Trahey,M., Milley,R.J., Cole,G.E., Innis,M., Paterson,H., Marshall,C.J., Hall,A. and McCormick,F. (1987) Biochemical and biological properties of the human N-*ras* p21 protein. *Mol. Cell. Biol.*, **7**, 541.

249. Hattori,S., Ulsh,L.S., Halliday,K. and Shih,T.Y. (1985) Biochemical properties of a highly purified v-*ras* H protein overproduced in *Escherichia coli* and inhibition of its activities by a monoclonal antibody. *Mol. Cell. Biol.*, **5**, 1449.

250. Poe,M., Scolnick,E.M. and Stein,R.B. (1985) Harvey *ras* p21 expressed in *Escherichia coli* purifies as a binary one-to-one complex with GDP. *J. Biol. Chem.*, **260**, 3906.

251. Feuerstein,J., Goody,R.S. and Wittinghofer,A. (1987) Preparation and characterization of nucleotide-free and metal ion-free p21 Apoprotein. *J. Biol. Chem.*, **262**, 8455.

252. McCormick,F., Clark,B.F.C., LaCour,T.F.M., Kjeldgaard,M., Norskov-Lauritsen,L. and Nyborg,J. (1985) A model for the tertiary structure of p21, the product of the *ras* oncogene. *Science*, **230**, 78.

253. Jurnak,F. (1985) Structure of the GDP domain of EF-Tu and location of the amino acids homologous to *ras* oncogene products. *Science*, **230**, 32.

254. Gay,N.J. and Walker,J.E. (1983) Homology between human bladder carcinoma oncogene product and mitochondrial ATP-synthase. *Nature*, **301**, 262.

255. Sigal,I.S., Gibbs,J.B., D'Alonzo,J.S., Temeles,G.L., Wolanski,B.S., Socher,S.H. and Scolnick,E.M. (1986) Mutant *ras*-encoded proteins with altered nucleotide binding exert dominant biological effects. *Proc. Natl. Acad. Sci. USA*, **83**, 952.

256. Clanton,D.J., Lu,Y., Blair,D.G. and Shih,T.Y. (1987) Structural significance of the GTP-binding domain of *ras* p21 studied by site-directed mutagenesis. *Mol. Cell. Biol.*, **7**, 3092.

257. Der,C.J., Pan,B.-T. and Cooper,G.M. (1986) *ras* H mutants deficient in GTP binding. *Mol. Cell. Biol.*, **6**, 3291.

258. Walter,M., Clark,S.G. and Levinson,A.D. (1986) The oncogenic activation of human p21*ras* by a novel mechanism. *Science*, **233**, 649.

259. De Vos,A.M., Tong,L., Milburn,M.V., Matias,P.M., Jancarik,J., Noguchi,S., Nishimora,S., Niora,K., Ohtsuka,E. and Kim,S.-H. (1988) Three dimensional structure of an oncogene protein: catalytic domain of human c-H-*ras* p21. *Science*, **239**, 888.

260. Tong,L., deVos,A.M., Milburn,M.V., Jancarik,J., Naguchi,S., Nishimura,S., Miura,K., Ohtsuka,E. and Kim,S.H. (1989) Structural differences between a *ras* oncogene protein and the normal proteins. *Nature*, **337**, 90.

261. Lacal,J.C. and Aaronson,S.A. (1986) Activation of *ras* p21 transforming properties associated with an increase in the release of bound guanine nucleotide. *Mol. Cell. Biol.*, **6**, 4214.

262. Powers,S., Kataoka,T., Fasano,O., Goldfarb,M., Strathern,J., Broach,J. and Wigler,M. (1984) Genes in *S.cerevisiae* encoding proteins with domains homologous to the mammalian *ras* proteins. *Cell*, **36**, 607.

263. Temeles,G.L., Gibbs,J.B., D'Alonzo,J.S., Sigal,I.S. and Scolnick,E.M. (1985) Yeast and mammalian *ras* proteins have conserved biochemical properties. *Nature*, **313**, 700.

264. Tamanoi,F., Walsh,M., Kataoka,T. and Wigler,M. (1984) A product of yeast RAS2 gene is a guanine nucleotide binding protein. *Proc. Natl. Acad. Sci. USA*, **81**, 6924.

265. Kataoka,T., Powers,S., McGill,C., Fasano,O., Strathern,J., Broach,J. and Wigler,M. (1984) Genetic analysis of yeast RAS1 and RAS2 genes. *Cell*, **37**, 437.

266. Kataoka,T., Powers,S., Cameron,S., Fasano,O., Goldfarb,M., Broach,J. and Wigler,M. (1985) Functional homology of mammalian and yeast *ras* genes. *Cell*, **40**, 19.

267. Toda,T., Uno,I., Ishikawa,T., Powers,S., Kataoka,T., Broek,D., Cameron,S., Broach,J., Matsumoto,K. and Wigler,M. (1985) In yeast, RAS proteins are controlling elements of adenylate cyclase. *Cell*, **40**, 27.

268. Matsumoto,K., Uno,I., Oshima,Y. and Ishikawa,T. (1982) Isolation and characterization of yeast mutants deficient in adenylate cyclase and cyclic AMP dependent protein kinase. *Proc. Natl. Acad. Sci. USA*, **79**, 2355.

269. Broek,D., Samiy,N., Fasano,O., Fujiyama,A., Tamanoi,F., Northup,J. and Wigler,M. (1985) Differential activation of yeast adenylate cyclase by wild-type and mutant RAS oncogenes. *Cell*, **41**, 763.

270. Field,J., Broek,D., Kataoka,T. and Wigler,M. (1987) Guanine nucleotide activation, and competition between, *ras* proteins from *Saccharomyces cerevisiae*. *Mol. Cell. Biol.*, **7**, 2128.

271. Kimura,N. and Shimada,N. (1983) GDP does not mediate but rather inhibits hormonal signal to adenylate cyclase. *J. Biol. Chem.*, **258**, 2278.

272. Hurley,J.B., Simon,M.I., Teplow,D.B., Robishalu,J.D. and Gilman,A.G. (1984) Homologies between signal transducing G-proteins and *ras* gene products. *Science*, **226**, 860.

273. Tanabe,T., Nukada,T., Nishikawa,Y., Sugimoto,K., Suzuki,H., Takahashi,H., Noda,M., Haga,T., Ichiyama,A., Kangawa,K., Minamino,N., Matsuo,H. and Numa,S. (1985) Primary structure of the alpha-subunit of transducin and its relation to *ras* proteins. *Nature*, **315**, 242.

274. Finidori-Lepicard,J., Schorderet-Slatkine,S., Hanoune,J. and Baulieu,E.-E. (1981) Progesterone inhibits membrane-bound adenylate cyclase in *Xenopus laevis* oocytes. *Nature*, **292**, 255.

275. Sadler,S.E. and Maller,J.L. (1983) Inhibition of *Xenopus* oocyte adenylate cyclase by progesterone and 2',5'-dideoxyadenosine is associated with slowing of guanine nucleotide exchange. *J. Biol. Chem.*, **258**, 7935.

276. Sadler,S.E., Schechter,A.L., Tabin,C.J. and Maller,J.L. (1986) Antibodies to the *ras* gene product inhibit adenylate cyclase and accelerate progesterone-induced cell division in *Xenopus laevis* oocytes. *Mol. Cell. Biol.*, **6**, 719.

277. Birchmeier,C., Broek,D. and Wigler,M. (1985) *ras* proteins can induce meiosis in *Xenopus* oocytes. *Cell*, **43**, 615.

278. Korn,L.J., Siebel,C.W., McCormick,F. and Roth,R.A. (1987) *ras* p21 as a potential mediator of insulin action in Xenopus oocytes. *Science*, **236**, 840.

279. Levitzki,A., Rudick,J., Pastan,I., Vass,W.C. and Lowy,D.R. (1986) Adenylate cyclase activity of NIH3T3 cells morphologically transformed by *ras* genes. *FEBS Lett.*, **197**, 134.

280. Tarpley,W.G., Hopkins,N.K. and Gorman,R.P. (1986) Reduced hormone-stimulated adenylate cyclase activity in NIH3T3 cells expressing the EJ human bladder *ras* oncogene. *Proc. Natl. Acad. Sci. USA*, **83**, 3703.

281. Beckner,S.K., Hattori,S. and Shih,T.Y. (1985) The *ras* oncogene is not a regulatory component of adenylate cyclase. *Nature, 317*, 71.
282. Fleischman,L.F., Chahwala,S.B. and Cantley,L. (1986) *ras*-transformed cells: altered levels of phosphatidylinositol-4,5-bisphosphate and catabolites. *Science, 231*, 407.
283. Downward,J., DeGunzburg,J. and Weinberg,R.A. (1988) p21 *ras* induced responsiveness of phosphatidylinositol turnover to bradykinin is a receptor number effect. *Proc. Natl. Acad. Sci. USA, 85*, 5774.
284. Yu,C.-L., Tsai,M.-H. and Stacey,D.W. (1988) Cellular *ras* activity and phospholipid metabolism. *Cell, 52*, 63.
285. Preiss,J., Loomis,C.R., Bishop,W.R., Stein,R., Neidel,J.E. and Bell,R.M. (1986) Quantitative measurement of sn-1,2-diacylglycerols present in platelets, hepatocytes, and *ras*- and *sis*-transformed normal rat kidney cells. *J. Biol. Chem., 261*, 8597.
286. Lacal,J.C., DeLaPena,P., Moscat,J., Carcia-Barreno,P., Anderson,P.S. and Aaronson, S.A. (1987) Rapid stimulation of diacylglycerol production in *Xenopus* oocytes by microinjection of Ha-*ras* p21. *Science, 238*, 533.
287. Lacal,J.C., Moscat,J. and Aaronson,S.A. (1987) Novel source of 1,2-diacylglycerol elevated in cells transformed by Ha-*ras* oncogene. *Nature, 330*, 269.
288. Macara,I.G. (1989) Elevated phosphorylcholine concentration in *ras*-transformed NIH 3T3 cells arises from increased choline kinase activity, not from phosphatidylcholine breakdown. *Mol. Cell. Biol., 9*, 325.
289. Trahey,M. and McCormick,F. (1987) A cytoplasmic protein stimulates normal N-*ras* p21 GTPase, but does not affect oncogenic mutants. *Science, 238*, 542.
290. Adari,H., Lowy,D.R., Williamson,B.M., Der,C. and McCormick,F. (1988) Guanosine triphosphate activity protein (GAP) interacts with the p21 *ras* effector binding domain. *Science, 240*, 518.
291. Colen,C., Hancock,J.F., Marshall,C.J. and Hall,A. (1988) The cytoplasmic protein GAP is implicated as the target for regulation by the *ras* gene product. *Nature, 332*, 548.
292. McCormick,F. (1989) *ras* GTPase activity protein: signal transmitter and signal terminator. *Cell, 56*, 5.
293. Ben-Neriah,Y., Bernards,A., Pasking,M., Daley,G.Q. and Baltimore,D. (1986) Alternative 5′ exons in c-*abl* mRNA. *Cell, 44*, 577.
294. Konopka,J.B., Watanabe,S.M. and Witte,O.N. (1984) An alteration of the human c-*abl* protein in K562 leukemia cells unmasks associated tyrosine kinase activity. *Cell, 37*, 1035.
295. Kurzrock,R., Shtalrid,M., Romero,P., Klowtzer,W.S., Talpas,M., Trujillo,J.M., Blick,M., Beran,M. and Gutterman,J.U. (1987) A novel c-*abl* protein product in Philadelphia positive acute lymphoblastic leukemia. *Nature, 325*, 631.
296. Clark,S.S., McLaughlin,J., Crist,W.M., Champlin,R. and Witte,O.N. (1987) Unique forms of the c-*abl* tyrosine kinase distinguish Ph′-positive CML from Ph′-positive ALL. *Science, 234*, 85.
297. Venter,D.J., Tuzi,N.L., Kumar,S. and Gullick,W.J. (1987) Overexpression of the c-*erb* B-2 oncoprotein in human breast carcinomas: immunohistological assessment correlates with gene amplification. *Lancet, ii*, 69.
298. Kraus,M.H., Popescu,N.C., Amsbaugh,S.C. and King,C.R. (1987) Overexpression of the EGF receptor-related proto-oncogene *erb*-B-2 in human mammary tumor cell lines by different molecular mechanisms. *EMBO J., 6*, 605.
299. Yamamoto,T., Kamata,N., Kawano,H., Shimizu,S., Kuroki,T., Toyoshima,K., Rikimaru,K., Nomura,N., Ishizaki,R., Pastan,I., Gamon,S. and Shimizu,N. (1986) High incidence of amplification of the epidermal growth factor receptor gene in human squamous carcinoma cell lines. *Cancer Res., 46*, 414.
300. Martin-Zanca,D., Hughes,S.H. and Barbacid,M. (1986) A human oncogene formed by the fusion of truncated tropomyosin and protein tyrosine kinase sequences. *Nature, 319*, 743.
301. Fukai,M., Yamamota,T., Kawai,S., Masuo,K. and Toyoshima,K. (1985) Detection of a *raf*-related and two other transforming DNA sequences in human tumors maintained in nude mice. *Proc. Natl. Acad. Sci. USA, 82*, 5954.
302. Blick,M., Westin,E., Gutterman,J., Wong-Staal,F., Gallo,R., McGedie,K., Keating,M. and Murphy,E. (1984) Oncogene expression in human leukaemia. *Blood, 64*, 1234.
303. Schwab,M., Varmus,H.E. and Bishop,J.M. (1984) Amplification of cellular oncogenes

in tumor cells. *Cancer Cells,* **2**, 215.

304. Ro,J., North,S.M., Gallick,G.E., Hortobagyi,G.N., Gutterman,J.U. and Blick,M. (1988) Amplified and overexpression of epidermal growth factor receptor gene in uncultured primary human breast carcinoma. *Cancer Res.,* **48**, 161.

305. Berger,M.S., Greenfield,C., Gullick,W.J., Haley,J., Downward,J., Neal,D.E., Harris,A.L. and Waterfield,M.D. (1987) Evaluation of epidermal growth factor receptors in bladder tumors. *Br. J. Cancer,* **56**, 533.

306. Wong,A.J., Bijner,S.H., Bigner,D.D., Kinzler,K.W., Hamilton,S.R. and Vogelstein,B. (1987) Increased expression of the epidermal growth factor receptor gene in malignant gliomas is invariably associated with gene amplification. *Proc. Natl. Acad. Sci. USA,* **84**, 6899.

307. Berger,M.S., Gullick,W.J., Greenfield,C., Evans,S., Addis,B.J. and Waterfield,M.D. (1987) Epidermal growth factor receptors in lung tumors. *J. Path.,* **152**, 297.

308. Kamata,N., Chida,K., Rikimaru,K., Horikoshi,M., Enomoto,S. and Kuroki,T. (1986) Growth-inhibitory effects of epidermal growth factor and overexpression of its receptors on human squamous cell carcinomas in culture. *Cancer Res.,* **46**, 1648.

309. Yamamoto,T., Ikawa,S., Akiyama,T., Semba,K., Nomura,N., Miyajima,N., Saito,T. and Toyoshima,K. (1986) Similarity of protein encoded by the human c-*erb*-B-2 gene to epidermal growth factor receptor. *Nature,* **319**, 23.

310. Varley,J.M., Swallow,J.E., Brammar,W.J., Whittaker,J.L. and Walder,R.A. (1987) Alterations to either c-*erb* B-2 (*neu*) or c-*myc* proto-oncogenes in breast carcinomas correlate with poor short-term prognosis. *Oncogene,* **1**, 423.

311. Velu,T.J., Beguinot,L., Vass,W.C., Willingham,M.C., Mertino,G.M., Pastan,I. and Lowy,D.R. (1987) Epidermal growth factor-dependent transformation by a human EGF receptor proto-oncogene. *Science,* **238**, 1408.

312. DiFiore,P.P., Pierce,J.H., Kraus,M.H., Segalto,O., King,R. and Aaronson,S.A. (1987) *erb* B-2 is a potent oncogene when overexpressed in NIH/3T3 cells. *Science,* **237**, 178.

313. Parker,R.C., Varmus,H.E. and Bishop,J.M. (1984) Expression of v-*src* and chicken c-*src* in rat cells demonstrates qualitative differences between pp60 v-*src* and pp60 c-*src*. *Cell,* **37**, 131.

314. Shalloway,D., Coussens,P.M. and Yaciuk,P. (1984) Overexpression of the c-*src* protein does not induce transformation of NIH3T3 cells. *Proc. Natl. Acad. Sci. USA,* **81**, 7071.

315. Iba,H., Takeya,T., Cross,F.R., Hanafusa,T. and Hanafusa,H. (1984) Rous sarcoma virus variants that carry the cellular *src* gene instead of the viral *src* gene cannot transform chicken embryo fibroblasts. *Proc. Natl. Acad. Sci. USA,* **81**, 4424.

316. Johnson,P.J., Coussens,P.M., Darko,A.V. and Shalloway,D. (1985) Overexpressed pp60 c-*src* can induce focus formation without complete transformation of NIH3T3 cells. *Mol. Cell. Biol.,* **5**, 1073.

317. Cross,F.R., Garber,E.A., Pellman,D. and Hanafusa,H. (1984) A short sequence in p60*src* N-terminus is required for p60*src* myristylation and membrane association and for cell transformation. *Mol. Cell. Biol.,* **4**, 1834.

318. Kamps,M.P., Buss,J.E. and Sefton,B.M. (1985) Mutation of NH_2-terminal glycine of p60 *src* prevents myristylation and morphological transformation. *Proc. Natl. Acad. Sci. USA,* **82**, 4625.

319. Smart,J.E., Oppermann,H., Czernilofsky,A.P., Purchio,A.F., Erikson,R.L. and Bishop,J.M. (1981) Characterization of sites of tyrosine phosphorylation in the transforming protein of Rous sarcoma virus (pp60 v-*src*) and its normal cellular homologue (pp60 c-*src*). *Proc. Natl. Acad. Sci. USA,* **78**, 6013.

320. Courtneidge,S.A. (1985) Activation of the pp60 c-*src* kinase by middle T antigen binding or by dephosphorylation. *EMBO J.,* **4**, 1471.

321. Cooper,J.A., Gould,K.L., Cartwright,C.A. and Hunter,T. (1986) Tyr^{527} is phosphorylated in pp60 *src*: implications for regulation. *Science,* **231**, 1431.

322. Takeya,T. and Hanafusa,H. (1983) Structure and sequence of the cellular gene homologous to the RSV *src* gene and the mechanism for generating the transforming virus. *Cell,* **32**, 881.

323. Cartwright,C.A., Eckhart,W., Simon,S. and Kaplan,P.L. (1987) Cell transformation by pp60 c-*src* mutated in the carboxy-terminal regulatory domain. *Cell,* **49**, 75.

324. Piwnica-Worms,H., Sanders,K.B., Roberts,T.M., Smith,A.E. and Cheng,S.M. (1987) Tyrosine phosphorylation regulates the biochemical and biological properties of pp60 c-*src*. *Cell,* **49**, 75.

325. Kmiecik,T.E. and Shalloway,D. (1987) Activation and suppression of pp60 c-*src* transforming ability by mutation of its primary sites of tyrosine phosphorylation. *Cell*, **49**, 65.

326. Sukegawa,A.J., Semba,K., Yamanashi,Y., Nishizawa,M., Miyajima,N., Yamamoto,T. and Toyoshima,K. (1987) Characterization of cDNA clones for the human c-*yes* gene. *Mol. Cell. Biol.*, **7**, 41.

327. Bolen,J.B., Thiele,C.J., Israel,M.A., Yonemoto,W., Lipsich,L.A. and Brugge,J.S. (1984) Enhancement of cellular *src* gene product associated tyrosyl kinase activity following polyoma virus infection and transformation. *Cell*, **38**, 767.

328. Wiedemann,L.M., Karhi,K.K., Shivji,M.K.K., Rayter,S.I., Pegran,S.M., Dowden,G., Bevan,D., Will,A., Galton,D.A.G. and Chan,L.C. (1988) The correlation of bcr rearrangement of p210 *phl*/c-*abl* expression with morphological analysis of Ph negative CML and other myeloproliferative diseases. *Blood*, **71**, 349.

329. Bartram,C.R., de Klein,A., Hagemeijer,A., van Agthoven,T., van Kessel,A.G., Bootsma,D., Grosveld,G., Ferguson-Smith,M.A., Davies,T., Stone,M., Heisterkamp, N., Stephenson,J.R. and Groffen,J. (1983) Translocation of the human c-*abl* oncogene occurs in variant Ph'-positive and not Ph-negative chronic myelocytic leukemia. *Nature*, **306**, 277.

330. Shtivelman,E., Lifshitz,B., Gale,R.P., Roe,B.A. and Garaani,E. (1986) Alternative splicing of RNAs transcribed from the human *abl* gene and from the *bcr-abl* fused gene. *Cell*, **47**, 277.

331. Daley,G.Q., McLaughlin,J., Witte,O.N. and Baltimore,D. (1987) The CML-specific P210 *bcr/abl* protein, unlike v-*abl*, does not transform NIH/3T3 fibroblast. *Science*, **237**, 532.

332. Pellman,D., Garber,E.A., Cross,F.R. and Hanafusa,H. (1985) Fine structural mapping of a critical NH_2-terminal region of p60[src]. *Proc. Natl. Acad. Sci. USA*, **82**, 1623.

333. Prywes,R., Foulkes,J.G., Rosenberg,N. and Baltimore,D. (1983) Sequences of the A-MuLV protein needed for fibroblast and lymphoid cell transformation. *Cell*, **34**, 569.

334. Cooper,J.A., Bowen-Pope,D.F., Raines,E., Ross,R. and Hunter,T. (1982) Similar effects of PDGF and EGF on the phosphorylation of tyrosine in cellular proteins. *Cell*, **31**, 263.

335. Cooper,J.A. and Hunter,T. (1983) Identification and characterization of cellular targets for tyrosine protein kinases. *J. Biol. Chem.*, **258**, 1108.

336. Nakamura,K.D., Martinez,R. and Weber,M.J. (1983) Tyrosine phosphorylation of specific proteins after mitogen stimulation of chicken embryo fibroblasts. *Mol. Cell. Biol.*, **3**, 380.

337. Cooper,J., Nakamura,K.D., Hunter,T. and Wever,M.J. (1983) Phospho-tyrosine containing proteins and expression of transformation parameters in cells infected with partial transformation mutants of Rous sarcoma virus. *J. Virol.*, **46**, 15.

338. Frackelton,A.R., Tremble,P.M. and Williams,L.T. (1986) Evidence for the PDGF-stimulated tyrosine phosphorylation of the PDGF receptor *in vivo*: immunopurification using a monoclonal antibody to phosphotyrosine. *J. Biol. Chem.*, **25**, 7909.

339. Comoglio,P.M., DiRenzo,M.F., Tarone,G., Giancotti,F.G., Naldini,L. and Marchisio, P.C. (1984) Detection of phospho-tyrosine containing proteins in the detergent insoluble fraction of Rous sarcoma virus transformed fibroblasts by azo-benzenephosphonate antibodies. *EMBO J.*, **3**, 483.

340. Foulkes,J.G., Chow,M., Gorka,C., Frackelton,R. and Baltimore,D. (1985) Purification and characterization of a protein–tyrosine kinase encoded by the Abelson murine leukaemia virus. *J. Biol. Chem.*, **260**, 8070.

341. Huhn,R.D., Posner,M.R., Rayter,S.I., Foulkes,J.G. and Frackelton,A.R., Jr (1987) Cell lines and peripheral blood leucocytes derived from individuals with chronic myelogenous leukemia display virtually identical phosphotyrosyl-proteins. *Proc. Natl. Acad. Sci. USA*, **84**, 4408.

342. Beug,H., Leutz,A., Kahn,P. and Graf,T. (1984) Ts mutants of E26 leukaemia virus allow transformed myeloblasts but not erythroblasts or fibroblasts to differentiate at the non-permissive temperature. *Cell*, **39**, 579.

343. Falcone,G., Tato,F. and Alemo,S. (1985) Distinctive effects of the viral oncogenes *myc*, *erb*, *fps* and *src* on the differentiation program of quail myogenic cells. *Proc. Natl. Acad. Sci. USA*, **82**, 426.

344. Muller,R. and Wagner,E.F. (1984) Differentiation of F9 teratocarcinoma stem cells after transfer of c-*fos* proto-oncogenes. *Nature*, **311**, 438.

345. Fults,D.W., Towle,A.C., Lauder,J.M. and Maness,P.F. (1985) pp60 c-*src* in the developing cerebellum. *Mol. Cell. Biol.*, **5**, 27.
346. Strebhardt,K., Mullins,J.I., Bruck,D. and Ruebsamen-Waigmann,H. (1987) Additional member of the protein – tyrosine kinase family: the *src*- and *lck*-related proto-oncogene c-*tkl*. *Proc. Natl. Acad. Sci. USA*, **84**, 8778.
347. Muller,R., Slamon,D.J., Tremblay,J.M., Cline,M.J. and Verma,I.M. (1982) Differential expression of cellular oncogenes during pre- and post-natal development of the mouse. *Nature*, **299**, 640.
348. Gonda,T.J. and Metcalf,D. (1984) Expression of *myb*, *myc* and *fos* proto-oncogenes during the differentiation of murine myeloid leukemia. *Nature*, **310**, 249.
349. Mitchells,R.L., Zokas,L., Schreiber,R.D. and Verma,I.M. (1985) Rapid induction of the expression of proto-oncogenes *fos* during human monocytic differentiation. *Cell*, **40**, 208.
350. Marchildon,G.A., Casnellie,J.E., Walsh,K.A. and Krebs,E.G. (1984) Covalently bound myristate in a lymphoma tyrosine protein kinase. *Proc. Natl. Acad. Sci. USA*, **81**, 7679.
351. Marth,J.D., Disteche,C., Pravtcheva,D., Ruddle,F., Krebs,E.G. and Potmultes,R.M. (1986) Localization of a lymphocyte-specific protein tyrosine kinase gene (*lck*) at a site of frequent chromosomal abnormalities in human lymphomas. *Proc. Natl. Acad. Sci. USA*, **83**, 7400.
352. Marth,J.D., Peet,R., Krebs,E.G. and Perlmutter,R.M. (1985) A lymphocyte-specific protein – tyrosine kinase gene is rearranged and overexpressed in the murine T cell lymphoma LSTRA. *Cell*, **43**, 393.
353. Richardson,J.M., Marla,A.O. and Wang,J.Y.J. (1987) Reduction of protein – tyrosine phosphorylation during differentiation of human leukemia cell line K562. *Cancer Res.*, **47**, 4066.
354. Henkemeyer,M.J., Gertler,F.B., Goodman,W. and Hoffmann,F.M. (1987) The *Drosophila* Abelson proto-oncogene homolog; identification of mutant alleles that have pleiotropic effects late in development. *Cell*, **51**, 821.
355. Hafen,E., Basler,K., Edstroen,J.-E. and Rubin,G.M. (1987) *Sevenless*, a cell-specific homoeotic gene of Drosophila, encodes a putative transmembrane receptor with a tyrosine kinase domain. *Science*, **236**, 55.
356. Shilo,B.Z. and Weinberg,R.A. (1981) DNA sequences homologous to vertebrate oncogenes are conserved in *Drosophila melanogaster*. *Proc. Natl. Acad. Sci. USA*, **78**, 6789.
357. Simon,M.A., Kornberg,T.B. and Bishop,J.M. (1983) Three loci related to the *src* oncogene and tyrosine-specific protein kinase activity in *Drosophila*. *Nature*, **302**, 837.
358. Barnekow,A. and Schartl,M. (1984) Cellular *src* gene product detected in the freshwater sponge *Spongilla lacustris*. *Mol. Cell. Biol.*, **4**, 1179.
359. Alema,S., Casalbore,P., Agostini,E. and Tato,F. (1985) Differentiation of PC 12 pheochromocytoma cells induced by v-*src* oncogene. *Nature*, **316**, 557.
360. Boettiger,D. and Durban,E.M. (1979) Progenitor-cell populations can be infected by RNA tumor viruses, but transformation is dependent on the expression of specific differentiated functions. *Cold Spring Harbor Symp. Quant. Biol.*, **44**, 1249.
361. Boettiger,D., Anderson,S.A. and Dexter,T.M. (1984) Effect of *src* infection on long term marrow cultures: increased self renewal of hemopoietic progenitor cells without leukemia. *Cell*, **36**, 763.
362. Boettiger,D. (1985) Effect of oncogenes on stem cells. *BioEssays*, **2**, 106.
363. Durban,E.M. and Boettiger,D. (1981) Differential effects of transforming avian RNA tumor virus on avian macrophages. *Proc. Natl. Acad. Sci. USA*, **78**, 3600.
364. Brugge,J.S., Colton,P.C., Queral,A.E., Barnett,J.N., Nonner,D. and Keane,R.W. (1985) Neurons express high levels of a structurally modified activated form of pp60 c-*src*. *Nature*, **316**, 554.
365. Levy,B.T., Sorge,L.K., Meymandi,A. and Maness,P.F. (1984) pp60 c-*src* kinase is in chick and human embryonic tissues. *Dev. Biol.*, **104**, 9.
366. Sorge,L.K., Levy,B.T. and Maness,P.F. (1984) pp60 c-*src* is developmentally regulated in the neural retina. *Cell*, **36**, 249.
367. Gleden,A., Nemeth,S.P. and Brugge,J.S. (1986) Blood platelets express high levels of the pp60 c-*src* specific tyrosine kinase activity. *Proc. Natl. Acad. Sci. USA*, **83**, 852.
368. Barnekow,A. and Gessler,M. (1986) Activation of the pp60 c-*src* kinase during differentiation of monomyelocytic cells *in vitro*. *EMBO J.*, **5**, 701.

369. Gee,C.E., Griffin,J., Sastre,L., Miller,L.J., Springer,T.A., Piwnica-Worms,H. and Roberts,T.M. (1986) Differentiation of myeloid cells is accompanied by increased levels of pp60c-*src* protein and kinase activity. *Proc. Natl. Acad. Sci. USA,* **83**, 5131.

370. Cartwright,C.A., Simantov,R., Kaplan,P.L., Hunter,T. and Eckhart,W. (1987) Alterations in pp60 c-*src* accompany differentiation of neurons from rat embryo striatum. *Mol. Cell. Biol.,* **7**, 1830.

371. Martinez,R., Mathey-Prevot,B., Bernards,A. and Baltimore,D. (1987) Neuronal pp60 c-*src* contains a six-amino acid insertion relative to its non-neuronal counterpart. *Science,* **237**, 411.

372. Lynch,S.A., Brugge,J.S. and Levine,J.M. (1986) Induction of altered c-*src* product during neural differentiation of embryonal carcinoma cells. *Science,* **234**, 873.

373. Buick,R.N. and Pollack,M.A. (1984) Perspectives on clonogenic tumor cells, stem cells and oncogenes. *Cancer Res.,* **44**, 4909.

374. Shriver,K. and Rohrschneider,L. (1981) Organization of pp60*src* and selected cytoskeletal proteins within adhesion plaques and junctions of Rous sarcoma virus transformed rat cells. *J. Cell. Biol.,* **89**, 525.

375. Rohrschneider,L.R. and Najita,L.M. (1984) Detection of the v-*abl* gene product at cell-substratum contact sites in Abelson murine leukemia virus transformed fibroblasts. *J. Virol.,* **51**, 547.

376. Resh,M.D. and Erikson,R.L. (1985) Highly specific antibody to Rous sarcoma virus *src* gene product recognizes a novel population of pp60 v-*src* and pp60 c-*src* molecules. *J. Cell. Biol.,* **100**, 409.

377. Collett,M.S., Erikson,E. and Erikson,R.L. (1979) Structural analysis of the avian sarcoma virus transforming protein: sites of phosphorylation. *J. Virol.,* **29**, 770.

378. Pawson,T., Guyden,J., Kung,T.H., Radke,K., Gilmore,T. and Martin,G.S. (1980) A strain of Fujinami sarcoma virus which is temperature sensitive in protein phosphorylation and cellular transformation. *Cell,* **22**, 767.

379. Beug,H. and Hayman,M.J. (1984) Temperature sensitive mutants of avian erythroblastosis virus surface: expression of the *erb*B product correlates with transformation. *Cell,* **36**, 963.

380. Rousell,M.F., Rettenmier,C.W., Look,A.T. and Sherr,C.J. (1984) Cell surface expression of v-*fms*-coded glycoproteins is required for transformation. *Mol. Cell. Biol.,* **4**, 1999.

381. Nigg,E.A., Sefton,B.M., Singer,S.J. and Vogt,P.K. (1986) Cytoskeleton organization, vinculin-phosphorylation and fibronectin expression in transformed fibroblasts with different cell morphologies. *Virology,* **151**, 50.

382. Wang,E. and Goldberg,A.R. (1976) Changes in microfilament organization and surface topography upon transformation of chick embryo fibroblasts with Rous sarcoma virus. *Proc. Natl. Acad. Sci. USA,* **73**, 4065.

383. Naharro,G., Robbins,K.C. and Reddy,E.P. (1984) Gene product of v-*fgr* onc hybrid protein containing a portion of actin and a tyrosine specific protein kinase. *Science,* **223**, 63.

384. Mitra,G., Martin-Zanca,D. and Barbacid,M. (1987) Identification and biochemical characterization of p70 *trk*, the gene product of the human *TRK* oncogene. *Proc. Natl. Acad. Sci. USA,* **84**, 6707.

385. Radke,K., Carter,V.C., Moss,P., Dehayza,P., Schliwa,M. and Martin,G.S. (1983) Membrane association of a 36,000 dalton substrate for tyrosine phosphorylation in chicken embryo fibroblasts transformed by avian sarcoma virus. *J. Cell. Biol.,* **97**, 1601.

386. Gould,K.L., Woodgett,J.R., Isacke,C.M. and Hunter,T. (1986) The protein–tyrosine kinase substrate p36 is also a substrate for protein kinase C *in vitro* and *in vivo*. *Mol. Cell. Biol.,* **6**, 2738.

387. Burridge,K. and Connell,L. (1983) A new protein of adhesion plaques and ruffling membranes. *J. Cell. Biol.,* **97**, 359.

388. de Clue,J.E. and Martin,G.S. (1987) Phosphorylation of talin at tyrosine in Rous sarcoma virus-transformed cells. *Mol. Cell. Biol.,* **7**, 371.

389. Burridge,K. and Margeat,P. (1984) An interaction between vinculin and talin. *Nature,* **308**, 744.

390. Horwitz,A., Duggon,K., Budz,C., Beckerle,M.C. and Burridge,K. (1986) Interaction of plasma membrane fibronectin receptor with talin: a transmembrane linkage. *Nature,* **320**, 531.

391. Hirst,R., Horwitz,A., Buck,C. and Rohrschneider,L. (1986) Phosphorylation of the fibronectin receptor complex in cells transformed by oncogenes that encode tyrosine kinases. *Proc. Natl. Acad. Sci. USA*, **83**, 6470.
392. Horwitz,A., Duggan,K., Greggs,R., Decker,C. and Buck,C. (1985) The cell substrate attachment antigen has properties of a receptor for laminin and fibronectin. *J. Cell. Biol.*, **101**, 2134.
393. Gerke,V. and Weber,K. (1985) The regulatory chain in the 36-kD protein substrate complex of viral tyrosine-specific protein kinases is related in sequence to the S-100 protein of glial cells. *EMBO J.*, **4**, 2917.
394. Glenney,J.R. and Tack,B.F. (1985) Amino-terminal sequences of p36 and associated p10: identification of the site of tyrosine phosphorylation and homology with S-100. *Proc. Natl. Acad. Sci. USA*, **82**, 7884.
395. Huang,K.-S., Wallner,B.P., Mattaliano,R.J., Tizard,R., Burne,C., Frey,A., Hession,C., McGray,P., Sinclair,L.K., Chow,E.P., Browning,J.L., Ramachandran,K.L., Tang,J., Smart,J.E. and Pepinsky,R.B. (1986) Two human 35 kD inhibitors of phospholipase A-2 are related to substrates of pp60v-*src* and of the epidermal growth factor receptor-kinase. *Cell*, **46**, 191.
396. Saris,C.J.M., Tack,B.F., Kristensen,T., Glenney,J.R., Jr and Hunter,T. (1986) The cDNA sequence for the protein–tyrosine kinase substrate p36 (calpactin I heavy chain) reveals a multidomain protein with internal repeats. *Cell*, **46**, 201.
397. Wallner,B.P., Mattaliano,R.J., Hession,C., Cate,R.L., Tizard,R., Sinclair,L.K., Foeller,C., Chow,E.P., Browning,J.L., Ramachandran,K.L. and Pepinsky,R.B. (1986) Cloning and expression of human lipocortin: a phospholipase A-2 inhibitor with potential anti-inflammatory activity. *Nature*, **320**, 77.
398. De,B.K., Misono,K.S., Lukas,T.J., Mroczkowski,B. and Cohen,S. (1986) A calcium-dependent 35-kilodalton substrate for epidermal growth factor receptor/kinase isolated from normal tissue. *J. Biol. Chem.*, **261**, 13787.
399. Varticovski,L., Chahwala,S.B., Whitman,M., Cantley,L., Chow,E.P., Sinclair,L.K. and Pepinksy,R.B. (1988) Location of sites in human lipocortin I that are phosphorylated by protein tyrosine kinases and protein kinases A and C. In press.
400. Kamps,M.P., Buss,J.E. and Sefton,B.M. (1986) Rous sarcoma virus transforming protein lacking myristic acid phosphorylates known polypeptide substrates without inducing transformation. *Cell*, **45**, 105.
401. Davidson,F.F., Dennis,E.A., Powell,M. and Glenney,J.R., Jr (1987) Inhibition of phospholipase A_2 by lipocortins and calpactins: an effect of binding to substrate phospholipids. *J. Biol. Chem.*, **262**, 1698.
402. Miller,K., Beardmore,J., Karety,H., Schlessinger,J. and Hopkins,C.R. (1986) Localization of the epidermal growth factor (EGF) receptor within the endosome of EGF-stimulated epidermoid carcinoma (A431) cells. *J. Cell. Biol.*, **102**, 500.
403. Colledge,W.H., Edge,M. and Foulkes,J.D. (1986) A comparison of topoisomerase activity in normal and transformed cells. *Biosci. Rep.*, **6**, 301–307.
404. Migliaccio,A., Rotondi,A. and Auricchio,F. (1984) Calmodulin-stimulated phosphorylation of 17 β-estradiol receptor on tyrosine. *Proc. Natl. Acad. Sci. USA*, **81**, 5921.
405. Mendel,D.B., Bodwell,J.E. and Munch,A. (1987) Activation of cytosolic glucocorticoid-receptor complexes in intact WEHI-7 cells does not dephosphorylate the steroid-binding protein. *J. Biol. Chem.*, **262**, 5644.
406. Bell,J.C., Mahadevan,L.C., Colledge,W.H., Frackelton,A.R., Jr, Sargent,M.G. and Foulkes,J.G. (1987) Abelson-transformed fibroblasts contain nuclear phosphotyrosyl-proteins which preferentially bind to murine DNA. *Nature*, **325**, 552.
407. Cori,G.T., Colowick,S.P. and Cori,C.F. (1938) *J. Biol. Chem.*, **123**, 381.
408. Krebs,E.G. and Fischer,E.M. (1956). *Biochem. Biophys. Acta*, **20**, 150.
409. Sutherland,E.W. and Rall,T.W. (1958) Fractionation and characterization of a cyclic adenine ribonucleotide formed by tissue particles. *J. Biol. Chem.*, **232**, 1077.
410. Warburg,O. (1930) *The Metabolism of Tumours*. Constable, London.
411. Hatanaka,M., Huebner,R.J. and Gilden,R.V. (1969) Alterations in characteristics of sugar uptake by mouse cells transformed by murine sarcoma viruses. *J. Natl. Cancer Inst.*, **43**, 1091.
412. Cooper,J.A., Reiss,N.A., Schwartz,R.J. and Hunter,T. (1983) Three glycolylic enzymes are phosphorylated at tyrosine in cells transformed by RSV. *Nature*, **302**, 218.
413. Diamond,I., Legg,A., Schneider,J.A. and Rozengurt,E. (1978) Glycolysis in quiescent

cultures of 3T3 cells stimulated by serum, epidermal growth factor and insulin in intact cells and persistence of the stimulation after cell homogenization. *J. Biol. Chem.*, **253**, 866.

414. Schneider,J.A., Diamond,I. and Rozengurt,E. (1978) Glycolysis in quiescent cultures of 3T3 cells: addition of serum, epidermal growth factor and insulin increases the activity of phospho-fructo-kinase in a protein synthesis independent manner. *J. Biol. Chem.*, **253**, 872.

415. Hunter,T. and Cooper,J.A. (1986) Viral oncogenes and tyrosine phosphorylation. In *The Enzymes: Control by Phosphorylation, Part A.* Boyer,P.D. and Krebs,E.G. (eds), Vol. XVII.

416. Jackowski,S., Rettermeier,C.W., Scherr,C.J. and Rock,C.O. (1986) A guanine nucleotide-dependent phosphatidylinositol 4,5-diphosphate phospholipase C in cells transformed by the v-*fms* and v-*fes* oncogenes. *J. Biol. Chem.*, **261**, 4978.

417. Kato,M., Kawai,S. and Takenawa,T. (1987) Altered signal transduction in *erb*B-transformed cells. *J. Biol. Chem.*, **262**, 5696.

418. Macara,I.G., Marinetti,G.V. and Balduzzi,P.C. (1984) Transforming protein of avian sarcoma virus UR2 is associated with phosphatidylinositol kinase activity; possible role in tumorigenesis. *Proc. Natl. Acad. Sci. USA*, **81**, 2728.

419. Courtneidge,S.A. and Heber,A. (1987) An 81 kd protein complexed with middle T antigen and pp60 c-*src*: a possible phosphatidylinositol kinase. *Cell*, **50**, 1031.

420. Thomas,G., Martin-Perez,J., Seigmann,M. and Otto,A.M. (1982) The effect of serum, EGF, PGR2 and insulin on S6 phosphorylation and initiation of protein and DNA synthesis. *Cell*, **30**, 235.

421. Nishimura,J. and Deuel,T.F. (1983) PDGF stimulates the phosphorylation of ribosomal protein S6. *FEBS Lett.*, **156**, 130.

422. Decker,S. (1981) Phosphorylation of ribosomal protein S6 in avian sarcoma virus-transformed chicken embryo fibroblasts. *Proc. Natl. Acad. Sci. USA*, **78**, 4112.

423. Nielsen,P.J., Thomas,G. and Maller,J.L. (1982) Increased phosphorylation of ribosomal protein S-6 during meiotic maturation of *Xenopus laevis* oocytes. *Proc. Natl. Acad. Sci. USA*, **79**, 2937.

424. Stith,B.J. and Maller,J.L. (1984) The effect of insulin on intracellular pH and ribosomal protein S-6 phosphorylation in oocytes of *Xenopus laevis*. *Dev. Biol.*, **102**, 79.

425. Gressner,A.M. and Wool,I.G. (1974) The phosphorylation of liver ribosomal proteins; in-vivo evidence that only a single small subunit protein S-6 is phosphorylated. *J. Biol. Chem.*, **249**, 6917.

426. Collatz,E., Wool,I.G., Lin,A. and Stoffler,G. (1976) The isolation of eukaryotic ribosomal proteins. *J. Biol. Chem.*, **251**, 4666.

427. Martin-Perez,J. and Thomas,G. (1983) Ordered phosphorylation of 40S ribosomal protein S6 after serum stimulation of quiescent 3T3 cells. *Proc. Natl. Acad. Sci. USA*, **80**, 926.

428. Hallberg,R., Wilson,P. and Sulton,C. (1981) Regulation of ribosome phosphorylation and antibiotic sensitivity in tetrahymena therimophilia: a correlation. *Cell*, **26**, 47.

429. Burkhardt,S.J. and Traugh,J.A. (1983) Changes in ribosome function by cAMP-dependent and cAMP-independent phosphorylation of ribosomal protein S6. *J. Biol. Chem.*, **258**, 14003.

430. Maller,J.L., Foulkes,J.G., Erikson,E. and Baltimore,P. (1985) Phosphorylation of ribosomal protein S6 on serine after microinjection of the Abelson murine leukemia virus tyrosine-specific protein kinase into *Xenopus* oocyte. *Proc. Natl. Acad. Sci. USA*, **82**, 272.

431. Erikson,E. and Maller,J.L. (1985) A protein kinase from *Xenopus* eggs for ribosomal protein S6. *Proc. Natl. Acad. Sci. USA*, **82**, 742.

432. Erikson,E. and Maller,J.L. (1986) Purification and characterization of a protein kinase from *Xenopus* eggs highly specific for ribosomal protein S6. *J. Biol. Chem.*, **261**, 350.

433. Erikson,E., Stefanovic,D., Blenis,J., Erikson,R.L. and Maller,J.L. (1988) Antibodies to *Xenopus* egg S6 kinase II recognize S6 kinase from progesterone and insulin stimulated *Xenopus* oocytes and from proliferating chicken embryo fibroblasts. *Mol. Cell. Biol.*, **7**, 3147.

434. Maller,J.L. (1987) Mitogenic signalling and protein phosphorylation in *Xenopus* oocytes. *J. Cycl. Nuc. Prot. Phos. Res.*, **11**, 543.

435. Sefton,B.M., Trowbridge,T.S., Cooper,J.A. and Scolnick,E.M. (1982) The

transforming proteins of Rous sarcoma virus, Harvey sarcoma virus and Abelson virus contain highly bound lipid. *Cell,* **31**, 465.

436. Schultz,A.M., Henderson,L.E., Oroszlan,S., Garber,E.A. and Hanafusa,H. (1985) Amino-terminal myristylation of the protein kinase p60*src*: a retroviral transforming protein. *Science,* **227**, 427.

437. Schultz,A. and Oroszlan,S. (1974) Myristylation of gag onc fusion proteins in mammalian transforming retroviruses. *Virology,* **133**, 431.

438. Carr,S.A., Biemann,K., Shoji,S., Parmelee,D.C. and Titari,K. (1982) *n*-Tetradecanoyl is the NH_2-terminal blocking group of the catalytic subunit of cyclic AMP-dependent protein kinase from bovine cardiac muscle. *Proc. Natl. Acad. Sci. USA,* **79**, 6128.

439. Aitken,A. and Cohen,P. (1982) Isolation and characterization of active fragments of protein phosphatase. *FEBS Lett.,* **1**, 54.

440. Mayer,B.J., Hamaguchi,M. and Hanafusa,H. (1988) A novel viral oncogene with structural similarity to phospholipase C. *Nature,* **332**, 272.

441. Stahl,M.L., Ferenz,R., Kelleher,K.L., Kriz,R.W. and Knopf,J.L. (1988) Sequence similarity of phospholipase C with the non-catalytic region of *src*. *Nature,* **332**, 269.

442. Maki,Y., Bos,T.J., Davis,C., Starbuck,M. and Vogt,P.K. (1987) Avian sarcoma virus 17 carries the *jun* oncogene. *Proc. Natl. Acad. Sci. USA,* **84**, 2848.

443. Vogt,P.K., Bos,T.J. and Doolittle,R.F. (1987) Homology between the DNA-binding domain of the GCN4 regulatory protein of yeast and the carboxyl-terminal region of a protein coded for by the oncogene *jun*. *Proc. Natl. Acad. Sci. USA,* **84**, 3316.

444. Angel,P., Imagawa,M., Chiu,R., Stein,B., Imbra,R.J., Rahmsdorf,H.J., Jonat,C., Herrlich,P. and Karin,M. (1987) Phorbol ester-inducible genes contain a common cis element recognized by a TPA-modulated trans-acting factor. *Cell,* **49**, 729.

445. Curran,T. (1988) In *The Oncogene Handbook*, Reddy,E.P., Skalka,A.M. and Curran,T. (eds), Elsevier, Amsterdam.

446. Van Straaten,F., Muller,R., Curran,T., Van Beveren,C. and Verma,I.M. (1983) Complete nucleotide sequence of a human c-*onc* gene: deduced amino acid sequence of the human c-*fos* gene protein. *Proc. Natl. Acad. Sci. USA,* **80**, 3183.

447. Miller,A.D., Curran,T. and Verma,I.M. (1984) c-*fos* protein can induce cellular transformation: a novel mechanism of activation of a cellular oncogene. *Cell,* **36**, 51.

448. Curran,T. and Teich,N.M. (1982) Candidate product of the FBJ murine osteosarcoma virus oncogene: characterization of a 55,000 dalton phosphoprotein. *J. Virol.,* **42**, 114.

449. Sambucetti,L.C. and Curran,T. (1986) The *fos* protein complex is associated with DNA in isolated nuclei and binds to DNA cellulose. *Science,* **234**, 1417.

450. Muller,R., Bravo,R., Burckhardt,J. and Curran,T. (1984) Induction of c-*fos* gene and protein by growth factors precedes activation of c-*myc*. *Nature,* **312**, 716.

451. Fisch,T.M., Prywes,R. and Roeder,R. (1987) c-*fos* sequences necessary for basal expression and induction by epidermal growth factor, 12-O-tetradecanoyl phorbol-13-acetate, and the calcium ionophore. *Mol. Cell. Biol.,* **7**, 3490.

452. Muller,R., Muller,D., Verrier,B., Bravo,R. and Herbst,H. (1986) Evidence that expression of c-*fos* protein in amnion cells is regulated by external signals. *EMBO J.,* **5**, 311.

453. Morgan,J.I. and Curran,T. (1986) Role of ion flux in the control of c-*fos* expression. *Nature,* **322**, 552.

454. Ruther,U., Wagner,E.F. and Muller,R. (1985) Analysis of the differentiation-promoting potential of inducible c-*fos* genes introduced into embryonal carcinoma cells. *EMBO J.,* **4**, 1775.

455. Treisman,R. (1987) Identification and purification of a polypeptide that binds to the c-*fos* serum response element. *EMBO J.,* **6**, 2711.

456. Greenberg,M.E., Siegfried,Z. and Ziff,E.B. (1987) Mutation of the c-*fos* gene dyad symmetry element inhibits serum inducibility of transcription *in vivo* and the nuclear regulatory factor binding *in vitro*. *Mol. Cell. Biol.,* **7**, 1217.

457. Prywes,R. and Roeder,R.G. (1986) Inducible binding of a factor to the c-*fos* enhancer. *Cell,* **47**, 777.

458. Prywes,R. and Roeder,R.G. (1987) Purification of the c-*fos* enhancer-binding protein. *Mol. Cell. Biol.,* **7**, 3482.

459. Distel,R.J., Ro,H., Rosen,B.S., Groves,D.L. and Spiegelman,B.M. (1987) Nucleoprotein complexes that regulate gene expression in adipocyte differentiation: direct participating of c-*fos*. *Cell,* **49**, 835.

460. Franza,B.R., Jr, Rauccher,F.J., Josephs,S.F. and Curran,T. (1988) The fos complex and *fos*-related antigens recognize sequence elements that contain AP-1 binding sites. *Science*, **239**, 1150.

461. Cole,M.D. (1986) The *myc* oncogene: its role in transformation and differentiation. *Annu. Rev. Genet.*, **20**, 361.

462. LeGouy,E., DePinho,R., Zimmerman,D., Ferrier,P., Collum,R. and Alt,F.W. (1987) Structure and expression of *myc*-family genes. In *Nuclear Oncogenes*, Cold Spring Harbor Laboratory Press.

463. Hayward,W.S., Nell,B.G. and Astrim,S.M. (1981) Activation of a cellular onc gene by promoter insertion in ALV-induced lymphoid leukemias. *Nature*, **290**, 475.

464. Adams,J., Gerondakis,S., Webb,E., Corcoran,L.M. and Cory,S. (1983) Cellular *myc* oncogene is altered by chromosome translocation to the immunoglobulin locus in murine plasmacytomas and is rearranged similarly in human Burkitt lymphomas. *Proc. Natl. Acad. Sci. USA*, **80**, 1982.

465. Crews,S., Barth,R., Hood,L., Prehn,J. and Calame,K. (1982) Mouse c-*myc* oncogene is located on chromosome 15 and translocated to chromosome 12 in plasmacytomas. *Science*, **218**, 1319.

466. Land,H., Parada,L.F. and Weinberg,R.A. (1983) Tumorigenic conversion of primary embryo fibroblasts requires at least two cooperating oncogenes. *Nature*, **304**, 596.

467. Little,C.D., Nau,M.M., Carney,D.N., Gazdar,A.F. and Minna,J.D. (1983) Amplification and expression of the c-*myc* oncogene in human lung cancer cell lines. *Nature*, **306**, 194.

468. Alitalo,K., Schwab,M., Lin,C.C., Varmus,H.E. and Bishop,M. (1983) Homogeneously staining chromosomal regions contain amplified copies of an abundantly expressed cellular oncogene (c-*myc*) in malignant neuroendocrine cells from a human colon carcinoma. *Proc. Natl. Acad. Sci. USA*, **80**, 1707.

469. Schwab,M., Klempnauer,K.-H., Alitalo,K., Varmus,H. and Bishop,M. (1986) Rearrangement at the 5' end of an amplified c-*myc* in human COLO 320 cells is associated with abnormal transcription. *Mol. Cell. Biol.*, **6**, 2752.

470. Escot,C., Thiellet,C., Lidereau,R., Spyratos,F., Champeme,M., Gest,J. and Callahan,R. (1986) Genetic alteration of the c-*myc* proto-oncogene (MYC) in human primary breast carcinomas. *Proc. Natl. Acad. Sci. USA*, **83**, 4834.

471. Seeger,R.C., Brodeur,G.M., Sather,H., Dalton,A., Seigel,S.E., Wong,K.Y. and Hammond,D. (1985) Association of multiple copies of the N-*myc* oncogene with rapid progression of neuroblastomas. *New Engl. J. Med.*, **313**, 1111.

472. Nau,M.M., Brooks,B.J., Jr, Carney,D.N., Gazdar,A.F., Battey,J.F., Sausville,E.A. and Minna,J.D. (1986) Human small-cell lung cancers show amplification and expression of the N-*myc* gene. *Proc. Natl. Acad. Sci. USA*, **83**, 1092.

473. Nisen,P.D., Zimmerman,K.A., Cotter,S.V., Gilbert,F. and Alt,F.W. (1986) Enhanced expression of the N-*myc* gene in Wilms' tumors. *Cancer Res.*, **46**, 6217.

474. Nau,M.N., Burke,B.J., Battey,J., Sausville,E., Gazdar,A.F., Kirsch,I.R., McBride, O.W., Bertness,V., Hollis,G.F. and Minna,J.D. (1985) L-*myc*, a new *myc*-related gene amplified and expressed in human small-cell lung cancer. *Nature*, **318**, 69.

475. Erisman,M.D., Rothberg,P.G., Diehl,R.E., Morse,C.C., Spandorfer,J.M. and Astrin, S.M. (1985) Deregulation of c-*myc* gene expression in human colon carcinoma is not accompanied by amplification or rearrangement of the gene. *Mol. Cell. Biol.*, **5**, 1969.

476. Stewart,T.A., Pattengale,P.K. and Leder,P. (1984) Spontaneous mammary adeno-carcinomas in transgenic mice that carry and express MMTV/*myc* fusion genes. *Cell*, **38**, 627.

477. Persson,H., Hennighausen,L., Taub,R., DeGrado,W. and Leder,P. (1984) Antibodies to human c-*myc* oncogene product: evidence of an evolutionarily conserved protein induced during cell proliferation. *Science*, **225**, 687.

478. Hann,S.R., Abrams,H.D., Rohrschneider,L.R. and Eisenman,R.N. (1983) Proteins encoded by v-*myc* and c-*myc* oncogenes. Identification and localization in acute leukemia virus transformants and bursal lymphoma cell lines. *Cell*, **34**, 789.

479. Ramsay,G., Evan,G.I. and Bishop,M. (1984) The protein encoded by the human proto-oncogene c-*myc*. *Proc. Natl. Acad. Sci. USA*, **81**, 7742.

480. Hann,S.R., King,M.W., Bentley,D.L., Anderson,C.W. and Eisenman,R.N. (1988) A non-AUG translational initiation in c-*myc* exon 1 generates an N-terminally distinct protein whose synthesis is disrupted in Burkitt's lymphomas. *Cell*, **52**, 185.

481. Spector,D.L., Watt,R.A. and Sullivan,N.F. (1987) The v- and c-*myc* oncogene proteins

colocalize in situ with small nuclear ribonucleoprotein particles. *Oncogene*, **1**, 5.

482. Watt,R.A., Shatzman,A.R. and Rosenberg,M. (1985) Expression and characterization of the human c-*myc* DNA-binding protein. *Mol. Cell. Biol.*, **5**, 448.

483. Studzinski,G.P., Brelvi,Z.S., Feldman,S.C. and Watt,R.A. (1986) Participation of c-*myc* protein in DNA synthesis of human cells. *Nature*, **234**, 467.

484. Iguchi-Ariga,S.M.M., Itani,T., Yamaguchi,M. and Ariga,H. (1987) c-*myc* protein can be substituted for SV40 T antigen DNA replication. *Nucleic Acids Res.*, **15**, 4888.

485. Iguchi-Ariga,S.M.M., Itgani,T., Kiji,Y. and Ariga,H. (1987) Possible function of the c-*myc* product: promotion of cellular DNA replication. *EMBO J.*, **6**, 2365.

486. Heikkila,R., Schwab,G., Wickstrom,E., Loke,S.L., Pluznik,D.H., Watt,R. and Neckers,L.M. (1987) A c-*myc* antisense oligodeoxynucleotide inhibits entry into S phase but not progress from G_0 to G_1. *Nature*, **328**, 455.

487. Lane,D.P. and Crawford,L.V. (1979) T antigen is bound to a host protein in SV40-transformed cells. *Nature*, **278**, 261.

488. McCormick,F. and Harlow,E. (1985) Association of a murine 53,000-dalton phosphoprotein with simian virus 40 large-T antigen in transformed cell. *J. Virol.*, **34**, 213.

489. Efrat,S., Baekkeskov,S., Lane,D. and Hanahan,D. (1987) coordinate expression of the endogenous p53 gene in β cells of transgenic mice expressing hybrid insulin-SV40 T antigen genes. *EMBO J.*, **6**, 2699.

490. Parada,L.F., Land,H., Weinberg,R.A., Wolf,D. and Rotter,V. (1984) Cooperation between gene encoding p53 tumour antigen and ras in cellular transformation. *Nature*, **312**, 649.

491. Reich,N.C. and Levine,A.J. (1984) Growth regulation of a cellular tumor antigen, p53, in nontransformed cells. *Nature*, **308**, 199.

492. Koeffler,H.P., Miller,C., Nicolson,M.A., Ranyard,J. and Bosselman,R.A. (1986) Increased expression of p53 protein in human leukemia cells. *Proc. Natl. Acad. Sci. USA*, **83**, 4035.

493. Rogel,A., Popliker,M., Webb,C.G. and Oren,M. (1985) p53 cellular tumor antigen: analysis of mRNA levels in normal adult tissues, embryos and tumors. *Mol. Cell. Biol.*, **5**, 2851.

494. Wolf,D., Harris,N. and Rotter,V. (1984) Reconstitution of p53 expression in a nonproducer Ab-MuLV-transformed cell line by transfection of a functional p53 gene. *Cell*, **38**, 119.

495. Pinhasi-Kimhi,O., Michalovitz,D., Ben-Zeev,A. and Oren,M. (1986) Specific interaction between the p53 cellular tumour antigen and major heat shock proteins. *Nature*, **320**, 182.

496. Debuire,B., Henry,C., Benaissa,M., Biserte,G., Claverie,J.M., Saule,S., Martin,P. and Stehelin,D. (1984) Sequencing the erbA gene of avian erythroblastosis virus reveals a new type of oncogene. *Science*, **224**, 1456.

497. Gazzolo,L., Samarut,J., Bouabdelli,M. and Blanchet,J.P. (1980) Early precursors in the erythroid lineages are the specific target cells of avian erythroblastosis virus *in vitro*. *Cell*, **22**, 683.

498. Gandrillon,O., Jurdic,P., Benchaibi,M., Xiao,J.-H., Ghysdael,J. and Samarut,J. (1987) Expression of the v-*erbA* oncogene in chicken embryo fibroblasts stimulates their proliferation *in vitro* and enhances tumor growth *in vivo*. *Cell*, **49**, 687.

499. Samarut,J. and Gazzolo,L. (1982) Target cells infected by avian erythroblastosis virus differentiate and become transformed. *Cell*, **28**, 921.

500. Frykberg,L., Palmieri,S., Beug,H., Graf,T., Hayman,M.J. and Vennstrom,B. (1983) Transforming capacities of avian erythroblastosis virus mutants deleted in the *erb*B or *erb*B oncogenes. *Cell*, **32**, 227.

501. Weinberger,C., Hollenberg,S.M., Rosenfeld,M.G. and Evans,R.M. (1985) Domain structure of human glucocorticoid receptor and its relationship to the v-*erb*-A oncogene product. *Nature*, **318**, 670.

502. Green,G.L., Gilna,P., Waterfield,M., Baker,A., Hort,Y. and Shine,J. (1986) Sequence and expression of human estrogen receptor complementary DNA. *Science*, **231**, 1150.

503. Hollenberg,S.M., Weinberger,C., Ong,E.S., Cerelli,G., Oro,A., Lebo,R., Thompson, E.B., Rosenfeld,M.G. and Evans,R.M. (1985) Primary structure and expression of a functional human glucocorticoid receptor cDNA. *Nature*, **318**, 635.

504. Arriza,J.L., Weinberger,C., Cerelli,G., Glaser,T.M., Handelin,B.L., Houseman,D.E.

and Evans,R.M. (1987) Cloning of human mineralocorticoid receptor complementary DNA: structural and functional kinship with the glucocorticoid receptor. *Science*, **237**, 268.

505. Giguere,V., Ong,E.S., Prudimar,S. and Evans,R.M. (1987) Identification of a receptor for the morphogen retinoic acid. *Nature*, **330**, 624.

506. Weinberger,C., Hollenberg,S.M., Rosenfeld,M.G. and Evans,R.M. (1985) Domain structure of human glucocorticoid receptor and its relationship to the v-*erb*-A oncogene product. *Nature*, **318**, 635.

507. Knudson,A.G. (1985) Hereditary cancer, oncogenes and antioncogenes. *Cancer Res.*, **45**, 1437.

508. Hansen,M.F. and Cavenee,W.K. (1987) Genetics of cancer predisposition. *Cancer Res.*, **47**, 5518.

509. Solomon,E., Voss,R., Hall,V., Bodmer,W.F., Jass,J.R., Jeffreys,A.J., Lucibello,F.C., Patel,I. and Rider,S.H. (1987) Chromosome 5 allele loss in human colorectal carcinomas. *Nature*, **328**, 616.

510. Stanbridge,E.J., Der,C.J., Doersen,C., Nishimi,R.Y., Peehl,D.M., Weissman,B.E. and Wilkinson,J.E. (1982) Human cell hybrids: analysis of transformation and tumorigenicity. *Science*, **215**, 252.

511. Srivatsan,E.S., Benedict,W.F. and Stanbridge,E.J. (1986) Implication of chromosome 11 in the suppression of neoplastic expression in human cell hybrids. *Cancer Res.*, **46**, 6174.

512. Weissman,B.E., Saxon,P.J., Pasquale,S.R., Jones,G.R., Geiser,A.G. and Stanbridge, E.J. (1987) Introduction of a normal human chromosome 11 into a Wilms' tumor cell line controls its tumorigenic expression. *Science*, **236**, 175.

513. Dyson,P.J., Cook,P.R., Searle,S. and Wyke,J.A. (1985) The chromatin structure of Rous sarcoma proviruses is changed by factors that act in *trans* in cell hybrids. *EMBO J.*, **4**, 413.

514. Noda,M., Selinger,Z., Scolnick,E.M. and Bassin,R.H. (1983) Flat revertants isolated from Kirsten sarcoma virus-transformed cells are resistant to the action of specific oncogenes. *Proc. Natl. Acad. Sci. USA*, **80**, 5602.

515. Weissman,B.E. and Stanbridge,E.J. (1983) Complementation of the tumorigenic phenotype in human cell hybrids. *J. Natl. Cancer Inst.*, **70**, 667.

516. Sporn,M.B. and Roberts,A.B. (1988) Peptide growth factors are multifunctional. *Nature*, **332**, 217.

517. Sporn,M.B., Roberts,A.B., Wakefield,L.M. and de Crombrugghe,B. (1987) Some recent advances in the chemistry and biology of transforming growth factor-beta. *J. Cell. Biol.*, **105**, 1039.

518. Roberts,A.B., Anzano,M.A., Wakefield,L.M., Roche,N.S., Stern,D.F. and Sporn,M.B. (1985) Type β transforming growth factor: a bifunctional regulator of cellular growth. *Proc. Natl. Acad. Sci. USA*, **82**, 119.

519. Takehara,K., LeRoy,E.C. and Grotendorst,G.R. (1987) TGF-β inhibition of endothelial cell proliferation: alteration of EGF binding and EGF-induced growth-regulatory (competence) gene expression. *Cell*, **49**, 415.

520. Shipley,G.D., Pittelkow,M.R., Wille,J.J., Jr, Scott,R.E. and Moses,H.L. (1986) Reversible inhibition of normal human prokeratinocyte proliferation by type β transforming growth factor-growth inhibitor in serum-free medium. *Cancer Res.*, **46**, 2068.

521. Ohta,M., Greenberger,J.S., Anklesaria,P., Bassols,A. and Massague,J. (1987) Two forms of transforming growth factor-β distinguished by multipotential haematopoietic progenitor cells. *Nature*, **329**, 539.

522. Kehrl,J.H., Roberts,A.B., Wakefield,L.M., Jakowlew,S., Sporn,M.B. and Fauci,A.S. (1986) Transforming growth factor β is an important immunomodulatory protein for human B lymphocytes. *J. Immunol.*, **137**, 3855.

523. Kehrl,J.H., Wakefield,L.M., Roberts,A.B., Jakowlew,S., Alvarez-Mon,M., Derynck, R., Sporn,M.B. and Fauci,A. (1986) Production of transforming growth factor β by human T lymphocytes and its potential role in the regulation of T cell growth. *J. Exp. Med.*, **163**, 1037.

524. Anzano,M.A., Roberts,A.B., Meyers,C.A., Komoriya,A., Lamb,L.C., Smith,J.M. and Sporn,M.B. (1982) Synergistic interaction of two classes of transforming growth factors from murine sarcoma cells. *Cancer Res.*, **42**, 4776.

525. Fanger,B.O., Wakefield,L.M. and Sporn,M.B. (1985) Structure and properties of the cellular receptor for transforming growth factor type β. *Biochemistry,* **25**, 3083.
526. Like,B. and Massague,J. (1986) The antiproliferative effect of type β transforming growth factor occurs at a level distal from receptors for growth-activating factors. *J. Biol. Chem.,* **261**, 13426.
527. Kimchi,A., Wang,X.-F., Weinberg,R.A., Cheifetz,S. and Massague,J. (1988) Absence of TGF-β receptors and growth inhibitory responses in retinoblastoma cells. *Science,* **240**, 196.
528. Lee,W.-H., Shew,J.-Y., Hong,F.D., Sery,T.W., Donoso,L.A., Young,L.-J., Bookstein, R. and Lee,E.Y.-H. (1987) The retinoblastoma susceptibility gene encodes a nuclear phosphoprotein associated with DNA binding activity. *Nature,* **329**, 642.
529. Foulkes,J.G. (1983) Phosphotyrosyl-protein phosphatases. *Curr. Top. Microbiol. Immunol.,* **107**, 163.
530. Geiser,A.G., Der,C.J., Marshall,C.J. and Stanbridge,E.J. (1986) Suppression of tumorigenicity with continued expression of the c-Ha-*ras* oncogene in EJ bladder carcinoma-human fibroblast hybrid cells. *Proc. Natl. Acad. Sci. USA,* **83**, 5209.
531. Meijlink,F., Curran,T., Miller,A.D. and Verma,I.M. (1985) Removal of a 67-base-pair sequence in the noncoding region of protooncogene fos converts it to a transforming gene. *Proc. Natl. Acad. Sci. USA,* **82**, 4987.
532. Goodman,D.S. (1987) Retinoids and retinoid-binding proteins. *The Harvey Lectures,* Series 81.

5

Oncogenes of the DNA tumor viruses: their interaction with host proteins

David P.Lane

---

1. Introduction

The DNA tumor viruses are a large and varied superfamily whose genomes range in size from about 5000 bp for the oncogenic papova viruses, such as simian virus 40 (SV 40) or polyoma virus through to about 150 000 bp in the herpes virus family. This diversity of size is reflected in the diversity of the oncogenes they encode and in their mode of action. Some common themes are emerging that serve to unify the group reflecting the common need of all members of the group to replicate a DNA genome free from the restrictions that control host DNA synthesis. First, quite unlike those of the oncogenic retroviruses, the oncogenes of the DNA viruses do not have direct cellular counterparts and their evolutionary origins are ancient and obscure. The second theme reflects the common need of these viruses to subvert the host cell. Consequently, the oncogene products of the DNA viruses have evolved to play an essential or beneficial role in the virus life cycle. This is in contrast to the oncogenes of the acutely transforming retroviruses which are essentially cellular 'passengers' that carry no benefit to the virus. Since the DNA tumor viruses are very successful parasites, mechanisms must exist to limit the pathological impact of viral oncogene expression. The major benefit to the virus of the oncogenic action of these gene products is to create a cellular environment for the propagation of the virus by activating quiescent cells into a pseudo S-phase. Finally, the oncogene products of the DNA tumor viruses frequently interact directly with specific host cell proteins. This chapter will focus on this last theme since recent results make it clear that the oncogenes of the papova adeno- and papillomaviruses act in major part by complexing to, and probably functionally inactivating, the products of a small group of 'recessive' or 'anti'-oncogenes of the host cell. The fact that these same genes are in turn frequent targets for mutational inactivation in human neoplasia lends

191

Table 1. DNA tumor virus oncogene-host protein complexes

Viral oncoprotein	Host protein	Function
SV40 large T	p53	Oncogene (Anti-?)
	p105 RB	Anti-oncogene
	AP-2	Transcription factor
	DNA polymerase α	Replicative DNA polymerase
	HSP70	Heat shock protein
SV40 small t	56K, 32K	Unknown
BK small t	56K, 32K	Unknown
Py small t	56K, 32K	Unknown
Py middle t	p61	Unknown
	c-*src*, c-*yes*, c-*fyn*	Tyrosine kinases
	p81	P.I. kinase
Adenovirus E1A	p105 RB	Anti-oncogene
	28K,40K,50K,60K	Unknown
	80K,90K,107K,130K,	Unknown
	300K	Unknown
	HSP 70	Heat shock protein
Adenovirus E1B	p53	Oncogene (Anti-?)

a special significance to these new data. To emphasize the unifying effect of these concepts and the central importance of these viral oncogene – host protein complexes, this chapter has been organized from the viewpoint of the cellular targets of these complexes. As such it represents a personal appraisal of the field and is not intended as a broad review. The chapter concentrates on the papova adeno- and papillomavirus groups and does not address the action of the herpes viruses. *Table 1* lists the host protein – viral oncogene complexes so far identified in cells infected or transformed by viruses from these groups. Separate sections of this chapter are devoted to the best studied of these host proteins; the properties of the less well-characterized host cell proteins are summarized at the end of the chapter. In the final section of the chapter, I discuss the implications of our knowledge of these host proteins as anti-oncogenes for the diagnosis and treatment of cancer.

2. The transforming proteins of the papova and adenoviruses

The properties of these proteins are discussed very briefly below to serve as an introduction to the study of their interactions with host cell proteins. For more comprehensive descriptions of these proteins and the viruses from which they derive the reader is strongly advised to consult the relevant reviews cited below.

2.1 SV40 large T-antigen

SV40 large T is a complex multidomain protein of 708 amino acids (1, 2 for reviews). The protein is predominantly located in the cell nucleus by virtue of its having a specific nuclear transport sequence. In addition, a small fraction of the protein is present in the cell membrane. The protein is found in multiple oligomeric forms and undergoes an extensive set of post-translational modifications including multiple phosphorylations, ADP ribosylation, adenylation, glycosylation, and the covalent attachment of RNA. SV40 virus contains a small (5200 bp) closed circular DNA molecule that can replicate in primate cells and fully transform a wide range of vertebrate cell types. The large T protein plays a direct role in viral DNA replication in the permissive host cell, and it is the only viral protein required for viral DNA synthesis. The protein binds dsDNA, and specifically interacts with DNA sequences at the viral origin of replication. It then, through its ATP-dependent DNA helicase activity, unwinds the DNA double helix at the origin and, through its interaction with DNA polymerase α (see section 4), triggers the initiation of DNA replication. T-antigen may also function as a helicase at the advancing bidirectional replication fork since the protein can be detected at that site, and anti-T antibodies can block the elongation reaction. Point mutants of T-antigen that are defective in all of these specific replicative functions are still efficient at inducing cell transformation so this process must be mediated by other activities of the protein. A growing consensus is emerging (see sections 3 and 5) that the transforming activity of T-antigen resides in its ability to complex to, and by implication, alter the function of, two host proteins p53 and p105 RB (the product of the retinoblastoma gene). Studies with temperature-sensitive mutants of the protein have demonstrated that large T-antigen function is required not only to induce but also to maintain cellular transformation.

2.2 Polyoma virus large T-antigen

The polyoma virus large T-antigen, a protein of 785 amino acids, is related to that of SV40 in its amino acid sequence and in known biochemical properties and activities. It appears to play a similar role in the replication process. The protein is, however, a far less potent oncogene than that of SV40 and while it can bind the p105 RB protein (E. Harlow, personal communication) it cannot bind p53. Polyoma large T-antigen can immortalize but not fully transform primary rodent cells and this anti-senescence activity requires the continued expression of the protein. Polyoma large T-antigen can effectively complement other oncogenes such as activated *ras* to fully transform primary cells (see reference 3 for a review).

2.3 Polyoma virus middle T-antigen

This 432 amino acid protein is the strong transforming oncogene of polyoma virus. It is able to efficiently transform established cell lines in culture and can complement polyoma large T-antigen in the transformation of primary cells. The protein is associated with cell membranes, and this association is essential for its transforming activity. The protein undergoes extensive post-translational modification; it is phosphorylated and mystrilated. As discussed in detail in section 8, its transforming activity is closely linked with its capacity to bind to host proteins in particular members of the tyrosine kinase family and probably also to a phosphoinositol kinase.

2.4 SV40 small t-antigen

The small t-antigen is a 174 amino acid protein produced by alternative splicing of the early region transcripts of SV40 that also encode large T-antigen. Thus the two proteins share a common N terminus of 82 amino acids. The small t-protein is found in both the nucleus and cytoplasm of infected and transformed cells. Studies on viral genomes that cannot express small t-protein as a consequence of deletion mutagenesis, have established that the protein is not required for cellular transformation or for lytic viral growth, but it can be shown to enhance both processes substantially. As discussed in detail in section 10, the small t-protein of SV40 and the related proteins of BK and polyoma virus form complexes with a specific set of host proteins of as yet unknown function.

2.5 The adenovirus E1A proteins

The E1A region of adenoviruses encode a series of proteins which are related as a consequence of a complex pattern of alternate splicing of the primary transcript. In adenovirus 5 the two most abundant proteins are the 289 amino acid product of the 13S RNA and the 243 amino acid product of the 12S RNA. These proteins share a common N terminus and C terminus, but the larger protein contains an insert of 46 amino acids. Both proteins are produced early in infection and the larger protein acts as a positive transcriptional regulator to promote the expression of the late viral genes. Both proteins can act to reduce the enhancer-dependent transcription of some other genes, and both have been unequivocally demonstrated to have transforming activity. This activity, like that of polyoma large T-antigen, is relatively weak when compared to SV40 large T-antigen. The E1A proteins will act as immortalizing genes and will efficiently complement other oncogenes such as activated *ras* (see Chapters 1 and 4) in the transformation of primary cells. As described in sections 5 and 9, an extensive analysis of mutant E1A proteins has tightly linked the transforming activity of these proteins with their ability to complex to certain host proteins in particular the product of the retinoblastoma gene, p105 RB.

2.6   The adenovirus E1B proteins

The E1B region of adenovirus encodes at least two proteins, the 58 kd protein and the 19 kd protein. This region can complement the E1A region for efficient transformation of cells but does not by itself have transforming activity. As described below the 58 kd protein complexes the host protein p53 in cells transformed by adenovirus 2 and adenovirus 5 (see section 3.1), but the significance of this interaction for the transforming activity of this region has not been established.

2.7   The papillomavirus E7 protein

The human papillomaviruses HPV16 and HPV18 are implicated as factors in the genesis of human cervical cancer. These viruses have complex transcriptional patterns, and only recently has progress been made in identifying the active oncogenic proteins. Current work is focused on the product of the E7 open reading frame, as this 97 amino acid protein shares sequence homology with SV40 T-antigen and adenovirus E1A in the region of these proteins that is implicated in p105 RB binding (4).

3.   The p53 host protein

The first host protein shown to bind specifically to the product of a DNA tumor virus oncogene was p53 (5,6). Since the properties of the p53 protein have been the subject of several recent reviews (7,8,9,10), this chapter will emphasize new information together with any interpretations that might be at variance with those previously published. First detected in a complex with SV40 large T-antigen (5,6,11) the protein was subsequently found to complex to the E1B 58 kd product of adenovirus 5 and 2 (12). Cells transformed by these viruses contained very elevated levels of p53 of up to 100 000 molecules per nucleus. This is in contrast with normal tissues, primary cell cultures, or established but not transformed cells where p53 is present at low levels, perhaps only a few thousand molecules per nucleus. The principle mechanism accounting for the high levels of p53 in the SV40- and adenovirus-transformed cells appears to operate post-translationally since mRNA levels are similar in non-transformed and SV40-transformed 3T3 cells (13). In non-transformed cells and in normal tissues (14) p53 characteristically has a very short half-life of approximately $10-20$ min, whereas its half-life in cells transformed by SV40, adenovirus 2, or adenovirus 5 is at least 20 h (13). In non-virally induced tumors the half-life of p53 is also often extended (15). This is indicative of mutation of the p53 amino acid sequence since many naturally arising (16,17,18) and artificially engineered (19,20) mutations of the p53 sequence have the effect of extending the half-life of the protein, presumably by rendering the protein a poor substrate for the pathway of ATP-dependent degradation that is responsible for the rapid turnover of wild-type p53 (21). Tumor-bearing animals (5,6,22) and human patients (23) frequently produce auto-antibodies to p53, perhaps because the overproduction of this protein is sufficient to overcome immune tolerance to it. Alternatively, the response may be due to the formation of p53 complexes with other proteins as a result of mutations in the protein coding sequence.

The expression of p53 has been examined in transgenic mice into which the SV40 gene for large T-antigen has been introduced under the control of an insulin promoter. High levels of p53 are detected in all cells that express T-antigen so that elevation of p53 concentration by T-antigen is not restricted to tissue culture systems (24). Further attention has been focused on p53 by the finding that tumors induced by a wide range of different non-viral agents also contain elevated levels of the protein. It has recently been realized that mutation of the p53 gene is probably responsible for the elevated level of its gene product. This is beginning to suggest that alterations of the p53 gene may accompany a very wide range of neoplasias.

3.1 Complexes of viral oncogene products with p53

SV40 large T-antigen was the first viral oncogene product demonstrated to bind to p53 and this complex remains one of the most closely studied in the field. The association between these two proteins is very tight and specific so that p53 can be readily isolated from all other host proteins in a cell extract by affinity chromatography on T-antigen columns (25), and complex formation can take place in the absence of any other protein (J.V.Gannon and D.P.Lane, unpublished observation). The T-antigen binding activity of p53 has been highly conserved in vertebrate evolution since the p53 homolog from *Xenopus* can bind tightly to T-antigen (T. Soussi, personal communication). The binding site for p53 on T-antigen has been mapped to lie between amino acid residues 271 and 517 using a combination of T-antigen mutants, T-antigen fusion proteins, and T-antigen proteolytic fragments (26,27). The binding site for p53 on the E1b protein has not yet been mapped. Interestingly, in adenovirus-infected cells p53 does not seem to be complexed to the E1B protein but is found in association with a 25 kd adenovirus E4 protein. It is probable that this protein competes with p53 for binding to the E1B protein but this has not been formally established (28).

3.2 p53 as a recessive oncogene

Cloning of cDNAs for p53 allowed the design of recombinant-DNA constructs that lead to high-level expression of the p53 protein in eukaryotic cells. Transfection of cultures of primary chondrocytes with such constructs allowed the outgrowth of colonies from cultures that would otherwise become senescent (29). Thus p53 acts as an 'immortalizing' gene. Such immortalized cells, but not their mortal parents, can be transformed by an activated *ras* oncogene. Direct cotransfection of the p53 and activated *ras* genes results in the transformation of primary cells in culture (29 – 31). The p53 gene thus appeared to be a dominant transforming oncogene of the same class as *myc*, the polyoma gene for large T-antigen and the adenovirus E1A gene. However, it has become clear very recently that both the p53 genomic and cDNA constructs used in these experiments contained point mutations (16). Wild-type p53 does not complement *ras* in transformation assays (16). Thus the p53 gene may in fact be a recessive oncogene. The point mutants in turn may be activated to transform by their acquisition of a dominant negative phenotype (as discussed by Hershowitz, 32). In this model the mutant protein is rendered oncogenic by virtue of its ability to inactivate the function of the wild-type p53 protein, a concept which is discussed in more detail below.

3.3 p53 as a target in non-virally induced tumors

Expression of the p53 protein may be very frequently altered in a large number of tumors (7 – 10 for reviews). Gross rearrangements of the p53 gene have only been noted in one human neoplasia, osteosarcoma (33), but circumstantial evidence suggests that point mutations of the p53 gene may be very frequent. In the best-characterized animal model system, Benchimol and coworkers have demonstrated that alteration of the p53 gene is a very frequent occurrence in the development of erthryoleukemia induced by Friend virus. Mutation of p53 accompanies the progression of the disease from a polyclonal premalignant hyperproliferative phase to a mono- or oligo-clonal malignant state (17). The alterations in p53 gene expression range from overt absence of the polypeptide to the expression of truncated forms (18, 34). A minority of the malignant clones express the full-length polypeptide, but two lines of evidence suggest that even in these cases p53 may be mutant. Firstly, the p53 product of these clones is found to be complexed to heat-shock protein, HSP70, and secondly the p53 lacks the epitope recognized by the monoclonal antibody, PAb 246 (35). Point mutants of p53 that are activated for transformation show at least one or both of these properties (16, 20), strongly implying that the malignant clones are not expressing a normal wild-type p53. Similar circumstantial evidence suggests that point mutations of the p53 gene may occur very frequently in tumors. Indeed of the seven mouse cDNA and genomic clones that have been sequenced, four contain point mutations. Interestingly, all of these mutations lie within one of the five blocks of amino acid sequence that are conserved in *Xenopus*, mouse and human p53 sequences (36). All of these mutations activate the p53 gene as an oncogene, thereby allowing it to complement *ras* in transformation assays. These blocks of conserved sequences are also altered in the truncations and deletions which activate the p53 gene in the malignant clones of cells in erythroleukemia induced by Friend virus (18,34). These results strongly imply that the p53 protein is a hot spot for mutational alteration in neoplasia. The fact that the activating mutations are diverse and scattered throughout the molecule, but with a particular concentration in the conserved domains, implies that activation for transformation is synonymous with inactivation of the normal function of the protein. If p53 function is inactivated rather than enhanced by the mutations, then why do these mutant p53 proteins appear to act as dominant transforming oncogenes? Furthermore, can this be reconciled with the dominant transforming action of the p53 binding proteins, SV40 large T-antigen and the adenovirus E1B protein? An attractive model is that the mutant p53 proteins, SV40 large T-antigen, and the E1B protein act similarly by binding to and inactivating the wild-type p53 in the cell (*Figure 1*). As discussed in the next section, the acute removal of p53 is toxic in the absence of E1B protein or T-antigen. It is consistent with these models that transfection of mutant p53 alone has also been reported to be toxic (30).

3.4 Normal function of p53

Several tumor cell lines have been isolated that express no p53 protein. These include the human HL60 line (38), an Abelson virus-transformed mouse lymphoid cell line, L12 (39), and some of the Friend virus erthyroleukemia lines (17). It is therefore clear that all essential catabolic and anabolic processes can proceed in the absence of p53. However, no normal cell line has been described that does not express p53. In cells that do express the protein it appears to be essential for cell growth and division, since the acute loss of p53 by the introduction of anti-sense mRNA (40) or the microinjection of antibodies (41) prevent the cell DNA synthesis response to serum growth factors. The implication is that p53 plays an essential role in normal cell growth and division but that, in the course of the development of a tumor, cells can adapt to, and profit by, its absence. Since abortive infection with SV40 is not acutely toxic, the implication is either that p53 is not inactivated from its normal function when complexed to T-antigen or, more attractively that T-antigen replaces and supplants p53 function. Some support for such an idea comes from the provocative observation that primary cell cultures established as permanent lines as a consequence of the expression of a temperature sensitive mutant T-antigen abruptly cease growth when shifted to the non-permissive temperature (42,43). There are striking parallels here to the properties of the product of the retinoblastoma gene, the protein p105 RB. This protein is also found in all normal cells but lost from certain tumors (see section 5) although at the time of writing no account of the effect of its acute removal on the growth of normal cells has been published.

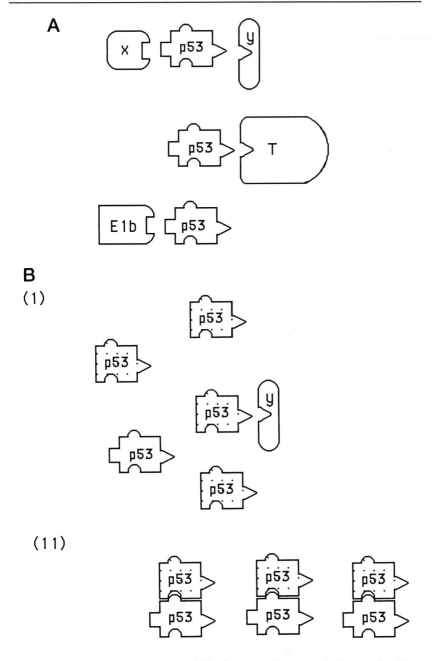

Figure 1. A model for p53 function. (**A**). The normal function of p53 is depicted in a complex to two host proteins designated x and y. Both large T-antigen and E1B can block this process by directly binding to p53. (**B**). The dominant suppression of normal p53 function by mutant p53. In (1) the mutant protein (dotted) competes with the wild-type protein for binding to y but is unable to bind x thus inactivating wild-type function. In (11), the mutant p53 oligomerizes with the wild-type protein to inactivate its normal function.

4. DNA polymerase α

4.1 The complex between SV40 T-antigen and DNA polymerase α

SV40 large T-antigen can bind to DNA polymerase α both *in vivo* (26,44,45) and *in vitro* (45 – 47), but to date this host enzyme has not been reported to interact with any other viral oncogene. The interaction with SV40 T-antigen occurs in the absence of DNA and appears specific since the antigenic structure of T-antigen detected by a panel of monoclonal antibodies is specifically modified on binding to DNA polymerase α (45,47). Certain antibodies can specifically block the binding of T-antigen to DNA polymerase α, and these antibodies can block the replication of SV40 DNA *in vitro*. Other antibodies cannot recognize T-antigen that is complexed to DNA polymerase α, and these antibodies do not block the *in vitro* replication reaction (47). The nature of the interaction between T-antigen and DNA polymerase α is not clear. The DNA polymerase α enzyme consists of a minimum of four discrete polypeptide chains. A 180 kd subunit that is the catalytic polymerase core component is very tightly associated with a 70 kd subunit of unknown function (48). It is tantalizing that in the case of the purified *Drosophila* enzyme, removal of this subunit reveals a cryptic 3′ – 5′ exonuclease activity in the 180 kd component and renders the free enzyme more accurate and precessive than the 180 kd/70 kd complex (49). The remaining two subunits of approximately 55 kd and 50 kd are less tightly associated and contain a primase activity (50). Under certain conditions of isolation, much larger and more complex holoenzyme forms have been isolated that contain additional activities (51). It is not yet established which polypeptide of the enzyme or holoenzyme is bound by T-antigen nor whether the binding alters the subunit composition of the enzyme. It appears from both immunochemical studies and analysis by affinity chromatography that only a small fraction of DNA polymerase α appears able to bind to T-antigen. This may reflect the relatively low affinity of the enzyme for T-antigen. An intriguing alternative could be that this is due to only a subset of the enzyme being available for binding, either because of the presence of a masking (T-antigen like) protein subunit or due to a limited supply of a 'bridging' protein required to join T-antigen to DNA polymerase α. Some evidence exists to support the idea that only a fraction of the cellular DNA polymerase α is active (52). The interaction of T-antigen with DNA polymerase α is essential for replication and may control virus host range since extracts of non-permissive murine cells are able to replicate DNA containing the SV40 origin *in vitro* if supplemented with the DNA polymerase α of a permissive species (human) and pure T-antigen (53). The importance of the interaction for the oncogenic action of T-antigen has not yet been investigated. T-antigen can interact with the DNA polymerase α of non-permissive cells (47), and such complexes are formed in SV40 transformed mouse cells. Clearly the T-antigen–DNA polymerase α interaction could be mutagenic if the complex acts to replicate the host cell DNA. Such an effect is unlikely to play a key role in the transformation process, however, since experiments with temperature sensitive (tsA) mutant T-antigens have established the reversibility of SV40 transformation (see 1 for review).

4.2 Competititon between DNA polymerase α and p53 for binding T-antigen

The multiple interactions of T-antigen with different host proteins raise the questions:

(i)   Are any of the interactions exclusive or competitive?

(ii) Do the mixed complexes have any specific function? That is to say, does T-antigen act as a molecular glue to stick p53 to the retinoblastoma gene product, for instance?

These same questions also arise in the case of the adenovirus E1A proteins.

The first clear example of competition is the binding of p53 and DNA polymerase α to SV40 large T-antigen. When T-antigen binds to p53, certain epitopes on T-antigen are lost (54) in so far as a number of monoclonal antibodies fail to bind the T-antigen–p53 complex. These same antibodies are also unable to bind the T-antigen–DNA polymerase α complex. Furthermore, a subset of these antibodies are able to inhibit the formation of both the complexes *in vitro*. Direct competition experiments have demonstrated that pure p53 can inhibit the binding of DNA polymerase α to T-antigen (45,47). An *in vivo* correlate of this result is the recent finding that p53 can prevent SV40 T-antigen and replication-origin-dependent replication of episomes in transfected cos cells (55). Cos cells were derived from CV-I monkey cells by transformation with a replication-origin defective SV40 genome. They constitutively express wild-type large T-antigen and thus permit the replication of any DNA that contains an intact SV40 origin (56). Strikingly, when a plasmid containing such an origin linked to the gene encoding murine p53 was transfected into cos cells it failed to replicate. The replication block required expression of p53 since it could act in *trans* on any cotransfected plasmid containing an SV40 replication-origin. In a further extension of these experiments, a range of mutant p53 proteins were analyzed. It was found that only plasmids encoding p53 that could bind to T-antigen showed the inhibitory effect. However, binding alone was not sufficient since plasmids encoding human p53 were able to replicate (55). While it is reasonable to equate the results of the polymerase competition experiments with these *in vivo* data, it is by no means established that polymerase competition is the basis of the inhibition in the *in vivo* assay. Indeed at face value the failure of the human p53 to inhibit would argue against this most simple model. It is interesting to note in the light of these results that the complex between the p53 of permissive cells and T-antigen appears less stable than the complex formed with p53 from non-permissive species (57).

5. The retinoblastoma gene product, p105 RB

5.1 The normal role of p105 RB

The p105 RB protein is not essential for growth since certain tumors and cell lines derived from them that do not express p105 RB can grow efficiently *in vivo* and *in vitro* (58,59). The genetics of hereditary retinoblastoma (RB) strongly imply that functional heterozygosity at the RB locus on chromosome 13 predisposes to the tumor because it increases the chance that cells will arise that show complete absence of the wild-type function through mutation of the remaining allele (60,61). In this sense then p105 RB is acting as a tumor suppressor since in the normal functionally diploid individual the incidence of retinoblastoma is extremely low, yet the retinal cells of such individuals are subject to the same mutagenic burden as those of the heterozygous individual. This presupposes that the retinoblastoma gene product, p105 RB, does not play a direct role in the protection or suppression of mtuations at other independent loci. The biochemical activities of p105 RB are undefined. The protein is phosphorylated and binds to DNA cellulose, properties shared by all the nuclear proto-oncogenes (59). Recent work has shown that both SV40 T-antigen and the adenovirus E1A protein will form complexes with p105 RB (see below). If T-antigen and E1A transform cells by inactivating normal RB function then it should be possible to titrate out the effect by introducing excess free p105 RB. Such a result would truly justify the title of anti-oncogene.

5.2 Complexes formed with p105 RB

Both the adenovirus E1A proteins (58) and the large T-antigen of SV40 (62) form specific protein complexes with the product of the RB gene p105 RB (59). A very careful analysis of an extensive set of adenovirus E1A mutants has localized the sequence requirements for p105 RB binding to two small regions of the molecule (63,64). The consensus sequence of the stronger of these motifs is present not only in a range of adenovirus serotypes, but also in the T-antigens of SV40, BK, JC, and polyoma virus, and in the E7 protein of human papilloma virus (HPV16) (4,65,66). Indeed it is probable that all these proteins bind to p105 RB. The crucial importance of this motif has been underlined by the finding that, as with the E1A protein, disruptions of the motif in SV40 T-antigen severely compromise the transforming activity of the protein (62). In a striking illustration of the functional conservation of the motif, it proved possible to restore the activity of an E1A mutant protein lacking the motif by substituting the SV40 motif (66). The location of the motif on SV40 also is consistent with reports that N-terminal fragments of T-antigen that contain the RB binding site are competent to transform and/or immortalize certain cell types (67,68). In the most recent of these reports, transformation was achieved with a fragment of T-antigen extending from amino acids 1 through 147 (69). Such a fragment has none of the other known biochemical activities of T-antigen intact but contains the RB binding site and the nuclear transport sequence (66). Importantly this fragment is also very stable and reaches high levels in the cell. This is consistent with at least part of the transforming activity of these oncoproteins lying in their ability to complex p105 RB. If that is the case then it represents a tremendous unification of the field and focuses attention on the role of p105 RB in normal cell growth and tumorigenesis.

6. Relationships between the p105 RB, p53, and DNA polymerase α complexes

The finding that adenovirus and SV40 both produce early proteins that can bind p53 and p105 RB suggests that chelation of these host proteins may be very beneficial for the viral life cycle. Two sets of observations had earlier indicated a potential synergy between these viruses. Firstly, SV40 large T-antigen can promote the replication of adenovirus in monkey cells (the helper effect) (70), and secondly, adenovirus early proteins can promote the replication of SV40 in human cells. The biochemical basis of these effects remains unclear, but it is now easier to conceive of some interesting and testable models. The effect of adenovirus early proteins on SV40 DNA replication in human cells was detected when plasmids containing the SV40 replication-origin and T-antigen gene were transfected into human embryonic kidney (HEK) cells or their adenovirus-transformed derivatives 293 and JW2 (71,72). In the transformed cells, SV40 replication was extraordinarily efficient and proceeded with a very short lag phase. No replication was seen in the HEK cells. Studies with mutant plasmids demonstrated that replication in this system was dependent on replication-origin and T-antigen. Surprisingly however, immunofluorescence studies revealed an antagonism between replication and levels of expression of T-antigen. Thus T-antigen could not be detected in the 293 cells by immunofluorescence when replication was taking place. Yet clearly the amount of T-antigen being synthesized was sufficient to promote very efficient replication in this cellular environment. It is provocative to suggest that only low levels of T-antigen were required because the E1A and E1B proteins present in the transformed cells had already chelated p105 RB and p53 respectively. This would have the effect of reducing the T-antigen requirement since here T-antigen is needed only for the replication reaction and not to chelate the two cellular regulator proteins. Such a model implies that both p53 and p105 RB must be removed to permit optimal viral replication.

7. The complex formed between T-antigen and the transcription factor, AP-2

Recent experiments have identified an interaction between SV40 large T-antigen and the cellular transcription factor AP-2 (73). AP-2 was first identified as one of a group of enhancer-binding proteins that acted to stimulate transcription in *in vitro* systems. The consensus binding site for AP-2 has been identified in the 5' region of many genes and the AP-2 protein has been purified by DNA affinity chromatography and identified as a 52 kd protein. It is clear that AP-2 is not p53 since it is present in cells that do not express p53; it does not bind to anti-p53 antibodies; and from recent cDNA cloning experiments its deduced amino acid sequence is distinct from that of p53. The binding of T-antigen to AP-2 blocks the binding of the transcription factor to its target sequence as indicated by DNA footprinting assays. Furthermore, a direct physical association between the two proteins has been deduced by sedimentation analysis on sucrose gradients. T-antigen will inhibit the transcriptional activation by AP-2 in *in vitro* assays, but there is not any clear evidence yet for the effect of T-antigen on AP-2 activity *in vivo*, nor for the existence of the complex *in vivo*. The finding is striking, however, particularly in the light of the recent identification of the cellular transcription factor AP-1 as a proto-oncogene c-*jun* (74) and the evidence that it, in turn, forms a physical complex with the product of another cellular proto-oncogene c-*fos* (75). The binding site for AP-2 on T-antigen has not been identified. It will be of great interest to determine where AP-2 binds, and if it shows any interference with the binding of p53, p105 RB or DNA polymerase α by large T-antigen.

8. Complexes with polyoma middle T-antigen

8.1 The complex with c-*src*

In contrast to SV40, the major transforming activity of the polyoma virus is associated not with the large T-antigen (which appears able to immortalize but not transform primary cells in culture) but with the middle T-antigen. Middle T-antigen, a 60 kd membrane-associated protein, is able to act alone to transform established rodent cells in culture (3,76, for reviews). The protein is associated with host proteins in a complex that sediments with an approximate molecular weight of 200 kd in sucrose gradients (77,78). When immunoprecipitated with specific antibodies from infected or transformed cells, polyoma middle T-antigen has an associated tyrosine kinase activity. Initially it was uncertain whether this activity was intrinsic to the viral protein or associated with it. Remarkably, studies with antibodies to the c-*src* protein demonstrated that this cellular proto-oncogene product was specifically associated with polyoma middle T-antigen (78). This was the first report of a physical complex between a DNA tumor virus oncogene and a known proto-oncogene. Subsequent analysis of a wide range of transformation defective mutants of the polyoma middle T-antigen protein have established that the ability to complex to c-*src* is essential to the transforming activity of middle T-antigen. Recently two other closely related members of the c-*src* family have also been shown to complex to middle T-antigen, namely c-*yes* (79) and c-*fyn* (80,81). It is, however, unclear at present how important these complexes are for transformation. When c-*src* is complexed to middle T-antigen its tyrosine kinase activity is stimulated (82, 83). An attractive model for this stimulation has emerged from studies of the phosphorylation of the c-*src* protein both in its free and complexed forms together with an investigation of the mutations that are found in the activated v-*src* protein (84). In contrast to the free form, the c-*src* protein associated with polyoma middle T-antigen is not phosphorylated on tyrosine 527 (85). This may be the result of a steric blockade of the site by the associated polyoma middle T-antigen, as residues essential for complex formation have been mapped to a region bounded by amino acids 518 and 525 at the C terminus of the c-*src* protein (86). The dephosphorylation or mutation of c-*src* at residue 527 results in the activation of its tyrosine kinase and transforming activity (83,85,87 – 91).

Superficially it seems most likely that polyoma middle T-antigen acts as a dominant transforming oncogene by activating rather than inhibiting c-*src* activity. However, in the light of the new data described above for the p53 and p105 RB interactions with DNA tumor virus oncogenes, it is important to reserve judgement. One must at least consider the possibility that the enzymic activation may be misleading and that the critical effect is inactivation of c-*src*'s normal function. As discussed above, the critical experiment is to look for suppression of transformation by over-expression of the normal allele. In other words would cotransfection of c-*src* reduce the transforming activity of polyoma middle T-antigen? Clearly in this case where the oncogene is an active enzyme, the requirements for restoration of normal function may prove particularly elusive.

8.2 The complex with the p81 phosphoinositol kinase

When extracts of polyma infected or transformed cells are immuno-precipitated with antibodies to either polyoma middle T-antigen or to c-*src*, the c-*src* middle T-antigen complex contains a third protein of 81 kd. This host protein is phosphorylated on tyrosine residues and can be immunoprecipitated from extracts of fibroblasts stimulated with platelet-derived growth factor (PDGF) using anti-phosphoryltyrosine antibodies. The p81 protein either has an intrinsic phosphoinositol kinase activity or is closely bound to such an enzyme. As in the case of the c-*src* complex, genetic studies with transformation defective mutants of the polyoma middle T-protein suggest that complexing to p81 is essential for the transforming activity of middle T-antigen (92,93).

8.3 The complex with heat-shock proteins

Polyoma middle T-antigen shares a common property with adenovirus E1A and SV40 large T-antigen in its ability to form complexes with members of the HSP70 group of host heat shock proteins (94,95,96). In the case of polyoma middle T-antigen and SV40 large T-antigens, the complexes are more prominent when the viral protein is mutant (95). The significance of these complexes for viral transformation is hard to evaluate in the light of the known proclivity of the HSP70 proteins to bind to proteins of distorted conformation. Judgement upon the functional significance of this interaction should be reserved, however, in view of the size of the HSP70 gene family and the realization of the essential role played by the closely homologous dnaK protein in *E.coli* and bacteriophage λ DNA-replication (96).

8.4 Other host proteins that bind polyoma middle T-antigen

In addition to the tyrosine kinases c-*src*, c-*yes*, and c-*fyn* and the p81 phosphoinositol kinase, another protein has also been consistently seen in immune precipitates of polyoma middle T-antigen. This protein, p61, so far identified only by its molecular weight of 61 kd, is of considerable interest because it is only detected in complexes with transformation competent middle T-antigen and not in complexes with certain defective mutants (97,98). Intriguingly, two-dimensional tryptic peptide maps suggest that p61, while clearly of host origin, may be structurally related to middle T-antigen itself. Clearly the isolation of cDNAs encoding p61 is a priority since they may provide clues to the host origin of a DNA tumor virus oncogene.

9. Other host proteins that bind adenovirus E1A protein

In addition to the product of the retinoblastoma gene, p105 RB, several other proteins have also been identified that form specific complexes with the adenovirus E1A protein. These are at present only identified by their molecular weights as polypeptides of 28 kd, 40 kd, 50 kd, 60 kd, 80 kd, 90 kd, 107 kd, 130 kd, and 300 kd (99, 100). An extensive analysis of these complexes on high-resolution two-dimensional gels suggests that none of these host proteins is highly abundant. This implies that the complexes are specific. It is not yet known if all the proteins bind E1A protein directly or are present in the complex by virtue of their association with other host proteins in the complex. The binding sites for the host proteins is being mapped using a large battery of mutant E1A proteins. In addition, the separation of different forms of the complex is being attempted by isolating antibodies to the individual host proteins. Clearly in the light of the results obtained with p53 and p105 RB it will be a crucial strategy to look for these other proteins in complex with papova and papilloma virus oncogenes (101).

10. Host proteins that bind to SV40 small t-antigen

The SV40 small t protein enhances the transforming and viral propagation activities of SV40 large T-antigen though it is not essential for viral growth or transformation (102, 103, 104). It can directly act to disrupt the actin filament component of the cytoskeleton (105, 106) and overcome the inhibitory action of theophylline upon DNA synthesis (107). It is found in both the nucleus and cytoplasm of SV40-infected and -transformed cells (108,109). At least three host proteins appear to bind specifically to small T-antigen; one of these has been identified as tubulin (110) but the other two (111,112) of 56 kd and 32 kd remain unidentified. The complexes can be assembled *in vitro* using pure or impure small t-antigen and the 56 kd and 32 kd proteins. It is as yet unclear how important these complexes are for the function of small *t*-antigen.

11. Summary, implications for therapy, and conclusions

The study of host proteins that bind to oncogenes of the DNA tumor viruses has led to the discovery of one new cellular proto-oncogene p53 and provided vital insight into the mode of action of two others, c-*src* and the retinoblastoma gene. The concordance in the host proteins targeted by the different groups of DNA tumor viruses strongly suggests that these target proteins must be attacked to permit some process essential for viral proliferation to take place. Many new host proteins that complex to the oncogenes of the DNA tumor viruses have been identified and others may be as yet undetected for a variety of technical reasons. Further investigation of these host proteins will provide a fertile area of research as will the identification of cellular proteins equivalent to the viral oncogenes in that they might normally bind to these proteins and be displaced by the viral oncogene. These networks of protein interactions certainly act to regulate normal cell growth and the disturbance of any component is potentially oncogenic. The complexities of protein regulation and interaction mean that considerable care needs to be exercised in the interpretation of the results. Thus alterations in p53 levels would not be readily detected by analysis at the DNA or RNA level and a cell producing abnormally high levels of (mutant) p53 may be in the same functional state as a cell expressing none of the (wild-type) protein at all. Direct sequencing using the polymerase chain reaction and detailed immunochemical studies are required. With this caution in mind the therapeutic possibilities of introducing extra copies of the wild-type p53 and RB genes into the DNA of tumor cells are readily apparent since the loss of normal function of one or both of these gene products is likely to prove a very frequent event in the development of human cancer. In the future it may prove possible to provide extra copies of these protective genes prophylacticaly to all somatic cells thus greatly reducing the risk of developing cancer.

12. Acknowledgements

I thank the ICRF for financial support and my lab, family, co-authors and editors for their patience.

13. Note added in proof

Since completing this review support for several of the items of conjecture have been published.

(i)   p61 protein that binds to polyoma middle t-protein has been shown to be identical to the 56 kd protein bound by the papova virus small t-antigens. Middle T also binds the 32 kd protein that binds to small t (114).

(ii)   Direct evidence for the binding of the papillomavirus E7 protein to p105 RB *in vitro* has been presented (115).

(iii)   Mutation of the p53 gene has been directly demonstrated in human colon cancers. The tumors have undergone selective allele loss of the region of chromosome 17p encoding p53 and the remaining allele contains point mutations consistant with those known to activate p53 in the mouse systems. Allele loss of 17p is a frequent occurence in colon, lung and breast tumors (116).

14. References

1. Rigby,P.W.J. and Lane,D.P. (1983) Structure and function of simian virus 40 large T-antigen. In *Advances in Viral oncology, Volume 3*, Klein,G. (ed.), Raven Press, New York, pp. 31–57.
2. Livingstone,D.M. and Bradley,M.K. (1987) Review: the simian virus 40 large T antigen—a lot packed into a little. *Mol. Biol. Med.*, **4**, 63–80.
3. Fried,M. and Prives,C. (1986) The biology of simian virus 40 and polyoma virus. *Cancer Cells*, **4**, 1–16.
4. Phelps,W.C., Yee,C.E., Munger,K. and Howley,P.M. (1988) The human papillomavirus type 16 E7 gene encodes transactivation and transformation functions similar to those of adenovirus E1A. *Cell*, **53**, 539–547.
5. Lane,D.P. and Crawford,L.V. (1979) T-antigen is bound to a host protein in SV40-transformed cells. *Nature*, **278**, 261–263.

6. Linzer,D.I.H. and Levine,A.J. (1979) Characterization of a 54K dalton cellular SV40 tumor antigen present in SV40 transformed cells and uninfected embryonal carcinoma cells. *Cell*, **17**, 43–52.
7. Jenkins,J.R. and Stürzbecher,H.-W. (1988) The p53 oncogene. In *The Oncogene Handbook*, Reddy,E.P. (ed.), Elsevier, Amsterdam.
8. Crawford,L. (1983) The 53,000-dalton cellular protein and its role in transformation. *Int. Rev. Exp. Path.*, **25**, 1–50.
9. Rotter,V. and Wolf,D. (1985) Biological and molecular analysis of p53 cellular encoded-tumor antigen. *Adv. Cancer Res.*, **43**, 113–141.
10. Oren,M. (1985) The p53 cellular tumor antigen; gene structure, expression and protein properties. *Bichim. Biophys. Acta.*, **823**, 67–78.
11. McCormick,F. and Harlow,E. (1980) Association of a murine 53,000 dalton phosphoprotein with simian virus 40 large T antigen in transformed cells. *J. Virol.*, **34**, 213–224.
12. Sarnow,P., Ho,Y.S., Williams,J. and Levine,A.J. (1982) Adenovirus E1b-58Kd tumor antigen and SV40 large tumor antigen are physically associated with the same 54Kd cellular protein in transformed cells. *Cell*, **28**, 387–394.
13. Oren,M., Maltzman,W. and Levine,A.J. (1981) Post translational regulation of the 54K cellular tumor antigen in normal and transformed cells. *Mol. Cell. Biol.*, **1**, 101–110.
14. Rogel,A., Popliker,M., Webb,C.G. and Oren,M. (1985) p53 cellular tumor antigen: analysis of mRNA levels in normal adult tissues, embryos, and tumors. *Mol. Cell. Biol.*, **5**, 2851–2855.
15. Reich,N.C., Oren,M. and Levine,A.J. (1983) Two distinct mechanisms regulate the level of a cellular tumor antigen, p53. *Mol. Cell. Biol.*, **3**, 2143–2150.
16. Finlay,C.A., Hinds,P.W., Tan,T.-H., Eliyahu,D., Oren,M. and Levine,A.J. (1988) Activating mutations for transformation by p53 produce a gene product that forms an hsc70-p53 complex with an altered half-life. *Mol. Cell. Biol.*, **8**, 531–539.
17. Mowat,M., Cheng,A., Kicumca,N., Bernstein,A. and Benchimol,S. (1985) The arrangements of the cellular p53 gene in erythroleukaemic cells transformed by Friend virus. *Nature*, **314**, 633–636.
18. Rovinski,B., Munroe,D., Peacock,J., Mowat,M., Bernstein,A. and Benchimol,S. (1987) Deletion of 5'-coding sequences of the cellular p53 gene in mouse erythroleukemia: a novel mechanism of oncogene regulation. *Mol. Cell. Biol.*, **7**, 847–853.
19. Jenkins,J.R., Rudge,K., Chumakov,P. and Currie,G.A. (1985) The cellular oncogene p53 can be activated by mutagenesis. *Nature*, **317**, 816–818.
20. Stürzbecher,H.-W., Chumakov,P., Welch,W.J. and Jenkins,J.R. (1987) Mutant p53 proteins bind hsp 72/73 cellular heat-shock-related proteins in SV40-transformed monkey cells. *Oncogene*, **1**, 201–211.
21. Gronostajski,R.M., Goldberg,A.L. and Pardee,A.B. (1984) Energy requirement for degradation of tumor-associated protein p53. *Mol. Cell. Biol.*, **3**, 442–448.
22. DeLeo,A.B., Jay,G., Appella,E., Dubois,G.C., Law,L.W. and Old,L.J. (1979) Detection of a transformation-related antigen in chemically induced sarcomas and other transformed cells of the mouse. *Proc. Natl. Acad. Sci. USA*, **76**, 2420–2424.
23. Crawford,L.V., Pim,D.C. and Bulbrook,R.D. (1982) Detection of antibodies against the cellular protein p53 in sera from patients with breast cancer. *Int. J. Cancer*, **30**, 403–408.
24. Efrat,S., Baekkeskov,S., Lane,D. and Hanahan,D. (1987) Coordinate expression of the endogenous p53 gene in β cells of transgenic mice expressing hybrid insulin -SV40 T genes. *EMBO J.*, **6**, 2699–2704.
25. McCormick,F., Clark,R., Harlow,E. and Tjian,R. (1981) SV40 T antigen binds specifically to a cellular 53K protein *in vitro*. *Nature*, **292**, 63–65.
26. Mole,S., Gannon,J., Ford,M. and Lane,D. (1987) Structure and function of SV40 large T antigen. *Phil. Trans. R. Soc. Lond.*, **B317**, 455–469.
27. Schmieg,F.I. and Simmons,D.T. (1988) Characterization of the in vitro interaction between SV40 T antigen and p53: Mapping the p53 binding site. *Virology*, **164**, 132–140.
28. Sarnow,P., Hearing,P., Anderson,C.W., Halbert,D.N., Shenk,T. and Levine,A.J. (1984) Adenovirus early region lb 58,000-dalton tumor antigen is physically associated with an early region 4 25,000-dalton protein in productively infected cells. *J. Virol.*, **49**, 692–700.

29. Jenkins,J.R., Rudge,K. and Currie,G.A. (1984) Cellular immortalization by a cDNA clone encoding the transformation-associated phosphoprotein p53. *Nature*, **312**, 651–654.
30. Eliyahu,D., Raz,A., Gruss,P., Givol,D. and Oren,M. (1984) Participation of p53 cellular tumour antigen in transformation of normal embryonic cells. *Nature*, **312**, 646–649.
31. Parada,L.F., Land,H., Weinberg,R.A., Wolf,D. and Rotter,W. (1984) Cooperation between gene encoding p53 tumour antigen and ras in cellular transformation. *Nature*, **312**, 649–651.
32. Hershowitz,I. (1987) Functional inactivation of genes by dominant negative mutations. *Nature*, **329**, 219–222.
33. Masuda,H., Miller,C., Koeffler,H.P., Battifora,H. and Kline,M.J. (1987) Rearrangement of the p53 gene in human osteogenic sarcomas. *Proc. Natl. Acad. Sci. USA*, **84**, 7716–7719.
34. Munroe,D.G., Rovinski,B., Bernstein,A. and Benchimol,S. (1988) Loss of a highly conserved domain on p53 as a result of gene deletion during Friend virus induced erythroleukemia. *Oncogene*, **2**, 621–624.
35. Yewdell,J.W., Gannon,J.V. and Lane,D.P. (1986) Monoclonal antibody analysis of p53 expression in normal and transformed cells. *J. Virol.*, **59**, 444–452.
36. Soussi,T., Caron de Fromentel,C., Mechali,M., Hay,P. and Kress,M. (1987) Cloning and characterization of a cDNA from Xenopus laevis coding for a protein homologous to human and murine p53. *Oncogene*, **1**, 71–78.
37. Eliyahu,D., Goldinger,N., Pinhasi-Kimhi,O., Shaulsky,G., Skurnic,Y., Arai,N., Rotter,V. and Oren,M. (1988) Meth A fibrosarcoma cells express two transforming mutant p53 species. *Oncogene*, **3**, 313.
38. Wolf,D. and Rotter,V. (1985) Major deletions in the gene encoding the p53 tumor antigen cause lack of p53 expression in HL-60 cells. *Proc. Natl. Acad. Sci. USA*, **82**, 790–794.
39. Wolf,D. Admon,S. Oren,M. and Rotter,V. (1984) Abelson murine leukemia virus-transformed cells that lack p53 protein synthesis express aberrant p53 mRNA species. *Mol. Cell. Biol.*, **4**, 552–558.
40. Shohat,O., Greenberg,M., Reisman,D., Oren,M. and Rotter,V. (1987) Inhibition of cell growth mediated by plasmids encoding p53 anti-sense. *Oncogene*, **1**, 277–283.
41. Mercer,W.E., Avignolo,C. and Baserga,R. (1984) Role of the p53 protein in cell proliferation as studied by microinjection of monoclonal antibodies. *Mol. Cell. Biol.*, **4**, 276–281.
42. Hirakawa,T. and Ruley,H.E. (1987) Rescue of cells from ras oncogene-induced growth arrest by a second complementing oncogene. *Proc. Natl. Acad. Sci. USA*, **85**, 1519–1523.
43. Ridley,A.J., Paterson,H.F., Noble,M. and Land,H. (1987) *Ras*-mediated cell cycle arrest is altered by nuclear oncogenes to induce Schwann cell transformation. *EMBO J.*, **7**, 1635–1645.
44. Jones,C. and Su,R.T. (1982) DNA polymerase α from the nuclear matrix of cells infected with simian virus 40. *Nucleic Acids Res.*, **10**, 5517–5532.
45. Gannon,J.V. and Lane,D.P. (1987) p53 and DNA polymerase α compete for binding to SV40 T antigen. *Nature*, **329**, 456–458.
46. Smale,S.T. and Tjian,R. (1986) T-antigen-DNA polymerase α complex implicated in simian virus 40 replication. *Mol. Cell. Biol.*, **6**, 4077.
47. Gough,G., Gannon,J.V. and Lane,D.P. (1988) Competition between DNA polymerase α and p53 for binding to SV40 T antigen. *Cancer Cells*, **6**, 153–158.
48. Wong,S.F., Paborsky,L.R., Fisher,P.A., Wang,T. and Korn,D. (1986) Structural and enzymolological characterisation of immunoaffinity purified DNA polymerase α. DNA primase from KB cells. *J. Biol. Chem.*, **261**, 7958–7968.
49. Cotterill,S.M., Reland,M.E., Loeb,L.A. and Lehman,I.R. (1987) A cryptic proofreading 3′-5′ exonuclease associated with the polymerase subunit of the DNA polymerase-primase from *Drosophila melanogaster*. *Proc. Natl. Acad. Sci. USA*, **84**, 5635–5639.
50. Nasheuer,H.P. and Grosse,F. (1988) DNA polymerase α-primase from calf thymus. Determination of the polypeptide responsible for primase activity. *J. Biol. Chem.*, **263**, 8981–8988.
51. Ottiger,H., Frei,P., Hassig,M. and Hubscher,U. (1987) Mammalian DNA polymerase α: A replication competant holoenzyme from calf thymus. *Nucleic Acids Res.*, **15**, 4789–4807.

52. Sylvia,V., Curtin,G., Norman,J., Stec,J. and Busbee,D. (1988) Activation of a low specific activity form of DNA polymerase by inositol-1,4, bisphosphate. *Cell,* **54,** 651–658.
53. Murakami,Y., Wobbe,C., Weissbach,L., Dean,F. and Hurwitz,J. (1986) Role of DNA polymerase α and DNA primase in simian virus 40 replication in *vitro. Proc. Natl. Acad. Sci. USA,* **83,** 2869–2873.
54. Lane,D.P. and Gannon,J. (1986) Monoclonal antibody analysis of the SV40 large T antigen-p53 complex. *Cancer Cells,* **4,** 387–393.
55. Braithwaite,A.W., Stürzbecher,H.-W., Addison,C., Palmer,C., Rudge,K. and Jenkins,J.R. (1987) Mouse p53 inhibits SV40 origin-dependent DNA replication. *Nature,* **329,** 458–460.
56. Gluzman,Y. (1981) SV40 transformed cells support the replication of early SV40 mutants. *Cell,* **23,** 175–182.
57. Harlow,E., Pim,D.C. and Crawford,L.V. (1981) Complex of simian virus 40 large-T antigen and host 53,000-molecular-weight protein in monkey cells. *J. Virol.,* **37,** 564–573.
58. Whyte,P., Buchovich,K., Horowitz,J., Friend,S., Raybuck,M., Weinberg,R. and Harlow,E. (1988) Association between an oncogene and an anti-oncogene; the adenovirus E1A proteins bind to the retinoblastoma gene product. *Nature,* **334,** 124–129.
59. Lee,W.-H., Shew,J.-Y., Hong,F.D., Sery,T.W., Donoso,L.A., Young,L.-J., Bookstein,R. and Lee,E.Y.-H.P. (1987) The retinoblastoma susceptibility gene encodes a nuclear phosphoprotein associated with DNA binding activity. *Nature,* **329,** 642–645.
60. Knudson,A.G., Jr. (1971) Mutation and cancer: statistical study of retinoblastoma. *Proc. Natl. Acad. Sci. USA,* **68,** 820–823.
61. Caveneem,W.K., Dryja,T.P., Phillips,R.A., Benedict,W.F., Godbout,R., Gillie,B.L., Murphree,A.L., Strong,L.C. and White,R.L. (1983) Expression of recessive alleles by chromosomal mechanism in retinoblastoma. *Nature,* **305,** 779–784.
62. DeCaprio,J.A., Ludlow,J.W., Figge,J., Shew,J.-H., Huang,C.-M., Lee,W.-H., Marsilio,E., Paucha,E. and Livingston,D.M. (1988) SV40 large tumour antigen forms a specific complex with the product of the retinoblastoma susceptibility gene. *Cell,* **54,** 275–283.
63. Whyte,P., Williamson,N.M. and Harlow,E.(1989) Cellular targets for transformation by the Adenovirus E1A proteins. *Cell,* **56,** 67–75.
64. Whyte,P., Ruley,H.E. and Harlow,E. (1988) Two regions of the adenovirus early region E1A proteins are required for transformation. *J. Virol.,* **62,** 257–265.
65. Stable,S., Argos,P. and Philipson,L. (1985) The release of growth arrest by microinjection of adenovirus E1A DNA. *EMBO J.,* **4,** 2329–2336.
66. Moran,E. (1988) A region of SV40 large T antigen can substitute for a transforming domain of the adenovirus E1A products. *Nature,* **334,** 168–170.
67. Clayton,C.E., Murphy,D., Lovett,M. and Rigby,P.W.J. (1982) A fragment of the SV40 T-antigen gene transforms. *Nature,* **229,** 59–61.
68. Colby,W.W. and Shenk,T. (1982) Fragments of the simian virus 40 transformation of rat embryo cells. *Proc. Natl. Acad. Sci. USA,* **79,** 5189–5193.
69. Sompayrac,L. and Danna,K.J. (1988) A new SV40 mutant that encodes a small fragment of T antigen transforms established rat and mouse cells. *Virology,* **163,** 391–396.
70. Rabson,A.S., O'Conor,G.T., Berezesky,I.K. and Paul,F.J. (1964) Enhancement of adenovirus growth in African green monkey kidney cells by SV40. *Proc. Soc. Exp. Biol. Med.,* **116,** 187.
71. Lewis,E.D. and Manley,J.L. (1985) Repression of simian virus 40 early transcription by viral DNAS replication in human 293 cells. *Nature,* **317,** 172–175.
72. Lebkowski,J.S., Lancy,S. and Calos,M. (1985) Simian virus 40 replication in adenovirus transformed human cells antagonises gene expression. *Nature,* **317,** 169–171.
73. Mitchell,P.J., Wang,C. and Tijan,R. (1987) Positive and negative regulation of transcription in *vitro:* enhancer-binding protein AP-2 is inhibited by SV40 T antigen. *Cell,* **50,** 847–861.
74. Bohmann,D., Bos,T.J., Admon,A., Nishimura,T., Vogt,P.K. and Tjian,R. (1988) Human proto-oncogene c-*jun* encodes a DNA binding protein with structural and functional properties of transcription factor AP-1. *Science,* **238,** 1386–1392.

75. Rauscher,F.J., Cohen,D.R., Curran,T., Bos,T.J., Vogt,P.K., Bohmann,D., Tjian,R. and Franza,R. (1988) Fos-associated protein p39 is the product of the *jun* proto-oncogene. *Science,* **240,** 1010–1016.
76. Smith,A.E. and Ely,B.K. (1983) The biochemical basis of transformation by polyoma virus. *Adv. Viral. Oncol.,* **3,** 3–30.
77. Walter,G., Huthinson,M.A., Hunter,T. and Eckhart,W. (1982) Purification of polyoma middle T tumor antigen by immunoaffinity chromatography. *Proc. Natl. Acad. Sci. USA,* **79,** 4025–4029.
78. Courtneidge,S.A. and Smith,A.E. (1983) Polyoma virus transforming protein associates with the product of the cellular *src* gene. *Nature,* **303,** 435–439.
79. Kornbluth,S., Sudol,M. and Hanafusa,H. (1987) Association of polyomavirus middle-T antigen with c-yes protein. *Nature,* **325,** 171–173.
80. Kypta,R.M., Hemming,A. and Courtneidge,S.A. (1988) Identification and characterisation of p59^{fyn} (a *src*-like protein tyrosine kinase) in normal and polyoma transformed cells. *EMBO J.,* **7,** 3837–3844.
81. Cheng,S.H., Harvey,R., Espino,P.C., Semba,K., Yamamoto,T., Toyoshima,K. and Smith,A.E. (1988) Peptide antibodies to the human c-fyn gene product demonstrate pp59^{c-fyn} is capable of complex formation with the middle-T antigen of polyomavirus. *EMBO J.,* **7,** 3845–3856.
82. Bolen,J.B., Theile,C.J., Israel,M.A., Yonemoto,W., Lipsich,L.A. and Brugge,J.S. (1984) Enhancement of cellular src gene product associated tyrosyl kinase activity following polyoma virus infection and transformation. *Cell,* **38,** 767–777.
83. Courtneidge,S.A. (1985) Activation of the pp60 c-src kinase by middle T antigen binding or by dephosphorylation. *EMBO J.,* **4,** 1471–1477.
84. Hunter,T. (1987) A tale of two src's: *mutatis mutandis. Cell,* **49,** 1–4.
85. Cartwright,C.A., Kaplan,P.L., Cooper,J.A., Hunter,T. and Eckhart,W. (1986) Altered sites of tyrosine phosphorylation in pp60^{c-src} associated with polyoma virus middle tumor antigen. *Mol. Cell Biol.,* **6,** 1562–1570.
86. Cheng,S.H., Piwinica-Worms,H., Harvey,R.W., Roberts,T.M. and Smith,A.E. (1988) The carboxy terminus of pp60^{c-src} is a regulatory domain and is involved in complex formation with the middle-T antigen of polyomavirus. *Mol. Cell Biol.,* **8,** 1736–1747.
87. Cooper,J.A. and King,C.S. (1986) Dephosphorylation or antibody binding to the carboxy terminus stimulates pp60^{c-src}. *Mol. Cell Biol.,* **6,** 4467–4477.
88. Piwnica-Worms,H., Saunders,K.B., Roberts,T.M., Smith,A.E. and Cheng,S.H. (1987) Tyrosine phosphorylation regulates the biochemical and biological properties of pp60^{c-src}. *Cell,* **49,** 75–82.
89. Kmiecik,T.E. and Shalloway,D. (1987) Activation and suppression of pp60c-src transforming ability by mutation of its primary sites of tyrosine phosphorylation. *Cell,* **49,** 65–73.
90. Cartwright,C.A., Eckhart,W., Simon,S. and Kaplan,P.L. (1987) Cell transformation by pp60^{c-src}W mutated in the carboxy-terminal regulatory domain. *Cell,* **49,** 83–91.
91. Reynolds,A.B., Vila,J., Lansing,T.J., Potts,W.M., Weber,M.J. and Parsons,J.T. (1987) Activation of the oncogenic potential of the avian cellular *src* protein by specific structural alteration of the carboxy terminus. *EMBO J.,* **6,** 2359–2364.
92. Kaplan,D.R., Whitman,M., Schaffhausen,B., Pallas,D.C., White,M., Cantley,L. and Roberts,T.M. (1987) Common elements in growth factor stimulation and oncogenic transformation: 85 kd phosphoprotein and phosphatidylinositol kinase activity. *Cell,* **50,** 1021–1029.
93. Courtneidge,S.A. and Heber,A. (1987) An 81 kd protein complexed with middle T antigen and pp60^{c-src}: A possible phosphatidylinositol kinase. *Cell,* **50,** 1031–1037.
94. White,E., Spector,D. and Welch,W. (1988) Differential distribution of the adenovirus E1A proteins and colocalization of E1A with the 70-kilodalton cellular heat shock protein in infected cells. *J. Virol.,* **62,** 4153–4166.
95. Walter,G., Carbone,A. and Welch,W.J. (1987) Medium tumor antigen of polyoma virus transformation-defective mutant NG59 is associated with 73-kilodalton heat shock protein. *J. Virol.,* **61,** 405–410.
96. Yamamoto,T., McIntyre,J., Sell,S.M., Georgopoulos,C., Skowyra,D. and Zylicz,M. (1987) Enzymology of the pre-priming steps in λdv DNA replication in vitro. *Proc. Natl. Acad. Sci. USA,* **84,** 7996–7999.
97. Grussenmeyer,T., Scheidtmann,K.H., Hutchinson,M.A., Eckhart,W. and Walter,G.

(1985) Complexes of polyoma virus medium T antigen and cellular proteins. *Proc. Natl. Acad. Sci. USA*, **82**, 7952–7954.

98. Dilworth,S.M. and Griffin,B.E. (1982) Monoclonal antibodies against polyoma virus tumor antigens. *Proc. Natl. Acad. Sci. USA*, **79**, 1059–1063.

99. Yee,S.-P. and Branton,P.E. (1985) Detection of cellular proteins associated with human adenovirus type 5 early region E1A polypeptides. *Virology*, **147**, 142–153.

100. Harlow,E., Whyte,P., Franza,B.R. and Schley,C. (1986) Association of adenovirus early region 1A proteins with cellular polypeptides. *Mol. Cell Biol.*, **6**, 1579–1589.

101. Whyte,P. and Harlow,E. (1987) Regions of the Adenovirus E1A proteins required for transformation are recognize sites for cellular proteins. In *Nuclear Oncogenes*, Alt,F.W., Harlow,E. and Ziff,E.B. (eds), Cold Spring Harbor Press, pp. 106–111.

102. Topp,W.C. (1980) Variable defectiveness for lytic growth of the dl 54/59 mutants of simian virus 40. *J. Virol.*, **33**, 1208–1210.

103. Bikel,I., Montano,X., Agha,M.E., Brown,M., McCormac,M., Boltax,J. and Livingston,D.M. (1987) SV40 small t antigen enhances the transformation activity of limiting concentrations of SV40 large T antigen. *Cell*, **48**, 698–702.

104. Gauchat,J.F. and Weil,R. (1986) On the functional roles of simian virus 40 large and small T-antigen in the induction of a mitotic lost response. *Nucl. Acid. Res.*, **14**, 9339–9351.

105. Graessmann,A., GraessmanM., Tjian,R. and Topp,W.C. (1980) Simian virus 40 small-t protein is required for loss of actin cable networks in rat cells. *J. Virol.*, **33**, 1182–1191.

106. Bikel,I., Roberts,T.M., Blandon,M.T., Green,R., Amann,E. and Livingston,D.M. (1983) Purification of biologically active simian virus 40 small tumor antigen. *Proc. Natl. Acad. Sci. USA*, **80**, 906–910.

107. Rundell,K. and Cox,J. (1979) Simian virus 40 t antigen affects the sensitivity of cellular DNA synthesis to theophylline. *J. Virol.*, **30**, 394–396.

108. Montano,X. and Lane,D.P. (1984) Monoclonal antibody to simian virus 40 small t. *J. Virol.*, **51**, 760–767.

109. Ellman,M., Bikel,I., Figge,J., Roberts,T., Schlossman,R. and Livingston,D.M. (1984) Localization of the simian virus 40 small t antigen in the nuleus and cytoplasm of monkey and mouse cells. *J. Virol.*, **50**, 623–628.

110. Murphy,C.I., Bikel,I. and Livingston,D.M. (1986) Cellular proteins which can specifically associate with simian virus 40 small t antigen. *J. Virol.*, **59**, 692–702.

111. Yang,Y.-C., Hearing,P. and Rundell,K. (1979) Cellular proteins associated with simian virus 40 early gene products in newly infected cells. *J. Virol.*, **32**, 147–154.

112. Rundell,K. (1987) Complete interaction of cellular 56,000- and 32,000-M_r proteins with simian virus 40 small-t antigen in productively infected cells. *J. Virol.*, **61**, 1240–1243.

113. Rundell,K., Major,E.O. and Lampert,M. (1981) Association of cellular 56,000- and 32,000-molecular-weight proteins with BK virus and polyomavirus t-antigens. *J. Virol.*, **37**, 1090–1093.

114. Walter,G., Carbone-Wiley,A., Joshi,B. and Rundell,K. (1988) Homologous cellular proteins associated with simian virus 40 small T antigen and polyomavirus medium T antigen. *J. Virol.*, **62**, 4760–4762.

115. Dyson,N., Howley,P.M., Munger,K. and Harlow,E. (1989) The human papilloma virus-16 E7 oncoprotein is able to bind to the retinoblastoma gene product. *Science*, **243**, 934–936.

116. Baker,S.J., Fearon,E.R. Nigro,J.M. Hamilton,S.R. Preisinger,A.C., Jessup,J.M. vanTuinen,P., Ledbetter,D.H., Barker,D.F., Nakamura,Y., White,R. and Vogelstein,B. (1989) Chromosome 17 deletions and p53 gene mutations in colorectal carcinomas. *Science*, **244**, 217–221.

Index

in Burkitt's lymphoma, 2
induction of expression, 157
loss of exon 1, 80
transcriptional deregulation, 80
translocation in T cells, 88
c-*neu*, amplification of EGF receptor
    and, 143
colony stimulating factor receptor
    (CSF-1^R), 10, 124
colony stimulating factor-1, *fms* and,
    116
colorectal carcinoma, 98
c-*raf*1, 4
c-*raf*2, 4
c-*rel*, 5
c-*ros*, 4
c-*sea*, 4
CSF-1 receptor, *see* colony stimulating
    factor receptor
c-*sis*, 4
c-*ski*, 5
c-*src*, 1, 2, 143, 192
    activation of, 143
c-*yes*, 4, 192
cytoplasmic tyrosine kinases, 15

Dash protein, 124
dbl, 5
diacylglycerol, 10, 118, 121, 122, 140,
    154
DNA polymerase, 192
DNase I hypersensitive sites, 39, 80, 81
    within c-*myc*, 81
double minute, 95
Drosophila, *abl* and *sevenless*
    protein – tyrosine kinases in, 146
dsi-1, 35
D*scr*28C, 124
D*scr*64B, 124

EGF receptor, *see* epidermal growth
    factor receptor
EIA protein of adenovirus, association of
    with retinoblastoma protein, 101
elongation factor-2 (EF-2), 121
env, 26, 27
eph, 123
epidermal growth factor receptor
    (EGF^R), 10, 124, 159
    and *erb*B, 116
Epstein – Barr virus (EBV), 74
*erb*A, 153
*erb*B, 35
    gene, EGF receptor and, 41
ets-1, 72
evi-2, 35

familial polyposis coli (FAP), 99
fes, 72
FGF, *see* fibroblast growth factor
fgr, 124

fibroblast growth factor (FGF), 128
fim-1, 5, 35
fim-2, 51
fim-3, 35, 51
fis-1, 5, 35
fms, 72, 116
    and CSF-1, 116
follicular lymphomas, 70, 83
fos, 72, 153, 154
fps, 124
Friend murine leukemia virus (F-MuLV),
    44, 51, 52
Friend virus-induced erythroleukemias,
    51
fyn, 123, 124

gag, 26
    gag/erb-A, thyroid hormone (T3)
    receptor, 119
gastrin-releasing peptides, 128
germinal vesicle breakdown in *Xenopus
    laevis*, 139
gin-1, 35
gli, 5
glucocorticoid receptor, 159
G-proteins, *see* guanine nucleotide
    binding proteins
granulocyte-macrophage CSF-1 receptor,
    10
GTPase activity of the p21*ras* proteins,
    7, 117
guanine nucleotide binding proteins
    (G-proteins), 10, 117, 118, 164
    as anti-oncogenes, 163

Ha-*ras*-1, 7
H(Harvey)-*ras*, 72, 131
    mutations in human bladder
    carcinomas, 132
hck, 123, 124
heat-shock protein, 192
homogenously staining regions, 95
HSP 70, *see* heat-shock protein
hst, 4, 128
    homology of to basic FGF, 116
    human stomach cancer and, 129
    Kaposi's sarcoma and, 129
HTLV, 30, 90
human adenocarcinomas, 97

IGF-1 receptor, *see* insulin-like growth
    factor receptor
Igh, 81, 83, 87, 88, 94
inositol trisphosphate (IP3), 10, 14, 118,
    140
Ins 1,3,4-P, 127
Ins 1,4,5-P, 127
Ins 1,3,4,5-P4, 118
insertional mutagenesis by hepatitis-B
    virus, 52